THE COMMONWEALTH AND INTERNATIONAL LIBRARY

Plant Metabolism

SECOND EDITION

Plant Metabolism

SECOND EDITION

BY

H. E. STREET

Professor of Botany and Chairman of the School of Biological Sciences, University of Leicester

AND

W. COCKBURN

Lecturer in Plant Physiology, School of Biological Sciences University of Leicester

PERGAMON PRESS

OXFORD · NEW YORK · TORONTO

SYDNEY · BRAUNSCHWEIG

Pergamon Press Ltd., Headington Hill Hall, Oxford
Pergamon Press Inc., Maxwell House, Fairview Park, Elmsford, New York 10523
Pergamon of Canada Ltd., 207 Queen's Quay West, Toronto 1
Pergamon Press (Aust.) Pty. Ltd., 19a Boundary Street, Rushcutters Bay, N.S.W. 2011, Australia
Vieweg & Sohn GmbH, Burgplatz 1, Braunschweig

First published 1963
Reprinted with corrections 1966
Reprinted 1967
First revised edition 1970
Second edition 1972
Library of Congress Catalog Card No. 76–174629

Printed in Great Britain by A. Wheaton & Co., Exeter

08 016753 5 (flexicover)
08 016752 7 (hard cover)

Contents

PREFACE TO THE FIRST EDITION vii

PREFACE TO THE SECOND EDITION viii

UNITS AND ABBREVIATIONS ix

1 *Introduction* 1

2 *Cell Structure and Function* 19

3 *Enzymes—The Catalysts of Metabolism* 59

4 *Catabolism* 84

5 *Anabolism* 131

6 *Secondary Plant Products* 186

7 *Absorption, Secretion and Translocation* 212

8 *The Regulation of Metabolism* 247

9 *Growth and Differentiation* 263

INDEX 303

Preface to the First Edition

THERE is first the book and then its readers. For whom then is this book intended? Primarily I have had in mind those studying biological sciences during their first year at a university or equivalent institution. I have therefore felt able to assume that my readers have an elementary knowledge of organic chemistry and of biology such as is acquired in the United Kingdom by preparation for the Advanced Level Examinations of the General Certificate of Education.

The biochemistry of metabolic processes has purposely been presented only in outline. This means, of course, that the university student will need to develop a parallel knowledge of biochemistry. It also means, I venture to hope, that almost all of the chapters will be within the grasp of students in our grammar schools, particularly those in their second and third years in the sixth form.

Newer aspects of metabolic physiology are now being introduced and emphasised in the Advanced Level syllabuses of Biology and Botany. The book, therefore, may be of service to those whose task it is to develop the teaching of these revised syllabuses.

Might I also hope that trained biologists working in other fields may find here a not too forbidding account of the present status of our knowledge of the metabolism of plants and that some advanced university students may be stimulated to further reading by this new presentation of familiar material.

For the presentation and for all errors I take full responsibility. I am, however, very grateful to my colleagues, Drs. Helgi Opik and E. G. Brown who have criticised the manuscript during its preparation. May I also thank Patricia Phillips and Adéle Fishman who have typed the manuscript and Marlene Jones who has helped to prepare the text figures.

I am also grateful to the many publishers and authors who granted permission to use copyright material.

Swansea, 1963 H. E. STREET

Preface to the Second Edition

RAPID progress in the field of the metabolism of higher plants has called for major revision of some sections, particularly those relating to photosynthesis, protein synthesis, ion and sugar transport, regulation and cell differentiation. The explosive increase, in recent years, of studies on the secondary products of plant metabolism has led to the addition of a new chapter. In making these major revisions we have endeavoured to retain the style and approach distinctive of the first edition.

For all errors we accept responsibility, but gratefully acknowledge helpful criticism of parts of our manuscript by our colleagues Dr. M. Fowler and Dr. M. C. Elliott. We also gratefully acknowledge the secretarial help of Mrs. Sandra Lewis and assistance in preparing new text-figures from Miss S. Pearcey.

Leicester

H. E. STREET
W. COCKBURN

Units and Abbreviations

THE following list is included for easy reference by the reader. Abbreviations are defined when first introduced in the text.

UNITS

Units in common usage	*Corresponding Système International d'Unités (SI)*
μ = micron = 10^{-6} m	μm
μl = microlitre = 10^{-6} litre	10^{-9} m^3
mμ = millimicron = 10 Å	nm
Å = Ångström unit	0·1 nm
g = gramme	10^{-3} kg
g-mol = gramme-molecule	mol = mole
hv = quantum	
kcal = kilocalorie = 10^3 calorie	4·18 J $\times$ 10^3
s = second	s
atm = atmospheres	

ABBREVIATIONS

A = adenine
A′ = anion
ADP = adenosine diphosphate
AMP = adenosine monophosphate, adenylic acid
ATP = adenosine triphosphate
C = cytosine
Chl = chlorophyll

CoA = CoA.SH = co-enzyme A
CoQ = co-enzyme Q
Cyt = cytochrome
DNA = deoxyribonucleic acid
e^- = electron
E^N = Nernst potential
EMP = Embden–Meyerhof–Parnas pathway of respiration
ER = endoplasmic reticulum
ES = enzyme substrate complex
F = Faraday constant
F-6-P = fructose-6-phosphate
FAD = flavin adenine dinucleotide
FMN = flavin adenine mononucleotide
G = guanine
G-1-P = glucose-1-phosphate
G-6-P = glucose-6-phosphate
ΔG° = change in Gibbs free energy of a reaction: if negative, energy is released; if positive, energy must be supplied
$\Delta G'$ = ΔG° for a reaction in solution under standard conditions but at a specified pH (here pH 7·0)
IAA = indol-3yl-acetic acid
IPP = isopentenyl pyrophosphate
J = fluxes of ions
K_m = Michaelis constant
K_s = dissociation constant of enzyme–substrate complex
m-RNA = messenger RNA
MVA = mevalonic acid
NAD = nicotinamide adenine dinucleotide (diphosphopyridine nucleotide, co-enzyme I)
NADP = nicotinamide adenine dinucleotide phosphate (triphosphopyridine nucleotide, co-enzyme II)
~P = high energy phosphate bond
PAL = phenylalanine ammonia lyase
PAPS = 3′-adenosine-5′-phosphosulphate
P_i = inorganic orthophosphate
PP_i = inorganic pyrophosphate
PEP = phosphoenolpyruvate

PGA = 3-phosphoglyceric acid
Q_{10} = temperature coefficient
RNA = ribonucleic acid
R.Q. = respiratory quotient
RuDP = ribulose diphosphate
T = thymine
t-RNA = transfer RNA
TCA = tricarboxylic acid (Krebs) cycle
U = uracil
UDP = uridine diphosphate
UDPG = uridine diphosphate glucose
UTP = uridine triphosphate
V_{max} = maximum velocity of reaction
Ψ = electrical or diffusion potential
Ψ_p = protoplast water potential
Ψ_s = osmotic potential of cell sap
Ψ_t = turgor potential
Ψ^i = electrical potential of an ion within the cell
$\bar{\mu}$ = electrochemical potential
π = osmotic potential

CHAPTER 1

Introduction

"It is sure that if he can add to what the eye itself reveals, an adequate mental picture of the invisible molecular events which underlie the visible, the biologist will gain increased understanding of the behaviour of every living thing."

F. Gowland Hopkins, lecture on *The Influence of Chemical Thought in Biology* delivered at Harvard, 1936.

LIVING organisms are built up of molecules and while organisms remain alive they are centres of intense and complex chemical activity. Their growth, development, movements and reproductive activities are the outcome of these highly complex and organised chemical changes. Visible patterns of development arise out of the invisible patterns of chemical activity. The sum total of these chemical reactions of living organisms comprises their *metabolism*. This book describes and discusses some of the more important and most actively investigated aspects of the metabolism of plants, having in mind particularly the green flowering plant.

The study of the functioning of living organisms is usually referred to as their physiology and hence the study of the functioning of plants is known as plant physiology. If you look at some textbooks of plant physiology, you will find that their contents are, in nearly every case, grouped under the two main headings: Metabolism and Growth. This separation of the study of growth from the study of metabolism reflects the different status of our knowledge of these two aspects of plant physiology. The phenomena of growth and development are, however, certainly the expressions in time and in plant structure of the changing metabolism of the organism and our, at present, limited ability to interpret growth and development in this way is discussed in Chapter 9 of the present work.

The emphasis placed on chemical activities in the opening paragraph of this Introduction draws attention to a difficult problem of demarcation; the question of the distinction between metabolism and biochemistry. When biochemistry is concerned with the structure and chemical properties of isolated compounds of biological origin and when metabolism is interpreted as embracing the integrated chemical activities of the whole organism, these two aspects of biology are clearly working at different levels of complexity. However, our understanding of metabolism comes not only from the study of whole organisms and of the structure and physiology of their individual cells but from advances in biochemistry. The student of metabolism assumes as a guiding principle that the reactions observed in isolated enzyme systems reflect physiological events and are not artefacts of isolation. Nevertheless, he uses such biochemical knowledge with discretion when attempting to interpret the physiology of living cells. From this, it follows that as the biochemist goes on to study the behaviour of complex systems containing, sometimes, many enzymes and other biological molecules and as the student of metabolism becomes concerned with the sequences of individual chemical reactions which underlie such processes as respiration and photosynthesis, the two approaches come very close together. It is sometimes argued that the distinction between the physiologist and the biochemist can be drawn by saying that the physiologist is concerned with metabolism at the levels of the cell, tissue and organism, whereas the biochemist studies the metabolic spectra of subcellular systems ranging from the organelles of the cell (nucleus, chloroplast, mitochondrion, microsome) down to single enzymes and their specific substrates. It is, however, impossible to observe strictly such a boundary when developing a discussion of plant metabolism. Nevertheless, insofar as the distinction drawn above emphasises that the metabolic physiologist is concerned with the interpretation of the activities of living cells, it will guide the emphasis developed in the present volume. This will impose a welcome restriction on the scope of our subject matter and give a useful but not excessive overlap with present and future introductory texts of plant biochemistry.

THE METABOLIC PROCESSES INVOLVED IN GERMINATION

"*. . . enzymes played a most important part in all metabolic changes . . . for each reserve the protoplasm was able to call into existence an appropriate enzyme.*"

R. J. Harvey-Gibson, *Outlines of the History of Botany*, 1919.

Starting our story with a living mature seed and by tracing very briefly the development from this of a new and independent plant, we can review the main aspect of plant metabolism. Such a review will provide a general background for the more detailed later chapters.

The essential structures of selected angiosperm seeds are illustrated in Fig. 1.1. The multicellular embryo is differentiated into embryonic shoot (plumule) and root (radicle) and is associated with storage tissue from which it will receive the organic food material essential for the early growth of these essential organs. The mature seed has a low water content (10–20%) and associated with this the dormant tissues have a very low rate of metabolic activity (for instance, the rate of respiration of the dry barley grain as measured by its rate of oxygen absorption at 22°C is of the order of $0 \cdot 06\,\mu l\ O_2/g/hr$). When such a seed is placed at a suitable temperature and in the presence of oxygen and water it germinates; its embryo awakens to active life and begins to grow into the young plant or seedling. The process of germination is initiated by a rapid uptake of water leading to a swelling of the seed tissues and a stretching of the enclosing seed coat. The progress of this uptake of water with time for the seeds of barley, is shown in Fig. 1.2. The initial attractive force (*imbibition* or matrix potential) exerted by the "dry" seed for water may be very great indeed (500–1000 atm) and involves a binding of water to organic molecules which can be compared to that exhibited by inorganic molecules for their water of crystallisation. This initial,

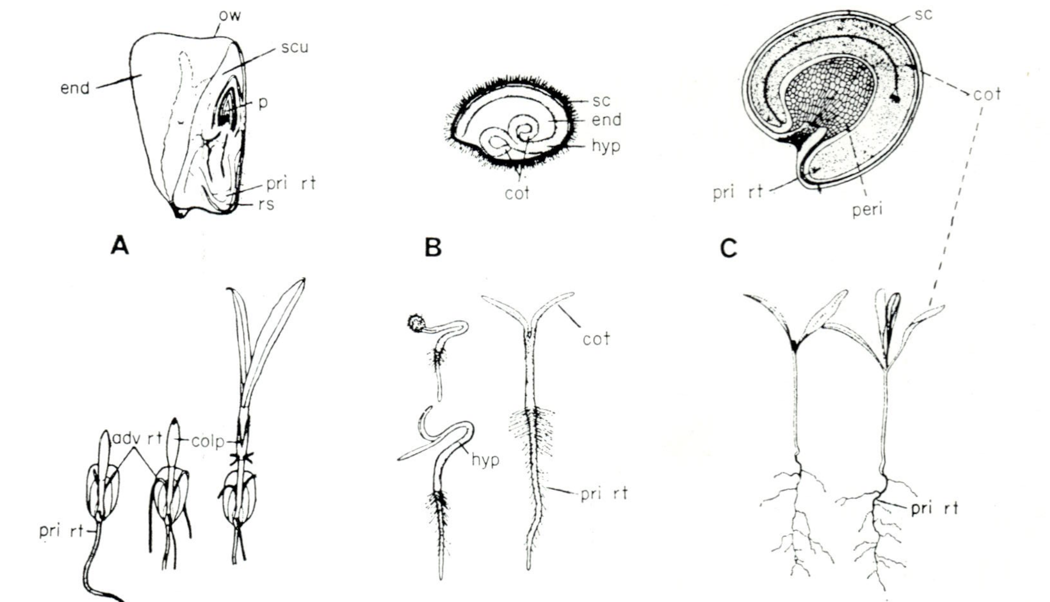

FIG. 1.1. Seeds and seedlings. A. *Zea mays* (maize) (germination after Avery). B. *Lycopersicon esculentum* (tomato). C. *Beta vulgaris* (beet) (seed structure after Bennett and Esau).

Key: end, endosperm; cot, cotyledon; scu, scutellum; peri, perisperm; hyp, hypocotyl; p, plumule; pri rt, primary root; rs, root sheath; colp, coleoptile; adv rt, adventitious root; ow, ovary wall; sc, seed coat. (From H. E. Hayward, *The Structure of Economic Plants*, Macmillan, New York, 1938.)

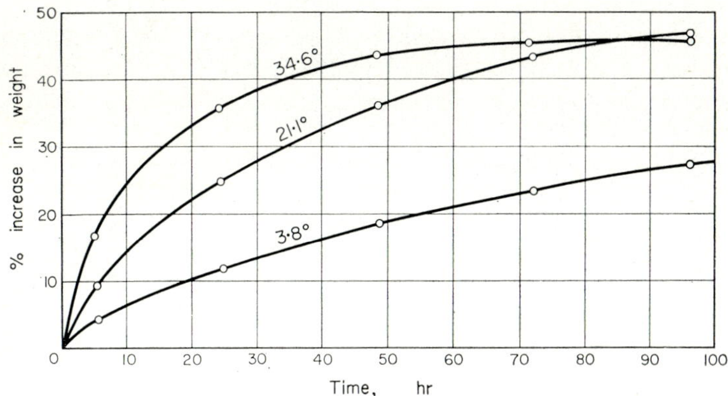

FIG. 1.2. The uptake of water at various temperatures by grains of *Hordeum vulgare* (barley). (After A. J. Brown and F. P. Worley, *Proc. Roy. Soc.* B, **85:** 546, 1912.)

imbibitional uptake of water leads to the liberation of a small amount of heat indicative of the loss of kinetic energy by the absorbed water molecules. As hydration of the cells proceeds, *osmotic forces* come into play and the forces motivating water uptake are of a lower order of magnitude (of the order of 10–30 atm). This hydration of the tissues is associated with a rise in their metabolic activity first occurring in the radicle region of the embryo. The enhanced metabolic activity is indicated by an increased respiration rate (in contrast to the figure quoted above for "dry" barley grain, the oxygen uptake of the germinating barley grain is of the order of 100 μl O_2/g/hr). The relationship of respiration rate to moisture content in the oat grain is shown in Fig. 1.3.

In some grains, such as maize, the soluble sugar, sucrose, has been shown to be uniformly distributed in the dry embryo and reducing sugars like glucose cannot be detected in appreciable amounts until the embryo begins to elongate. Sucrose may, therefore, be the initial respiratory substrate involved in the rise in respiration rate. The activation of the metabolic process of respiration early in germination implies not only the availability of a respiratory *substrate* like sucrose in the embryo but the activation of the essential biological catalysts termed *enzymes*. These enzymes are protein molecules and hydration of these molecules is essential for their activity. Thus, respiratory enzymes are present preformed in the mature "dry" embryo and become active during the imbibitional phase of water uptake.

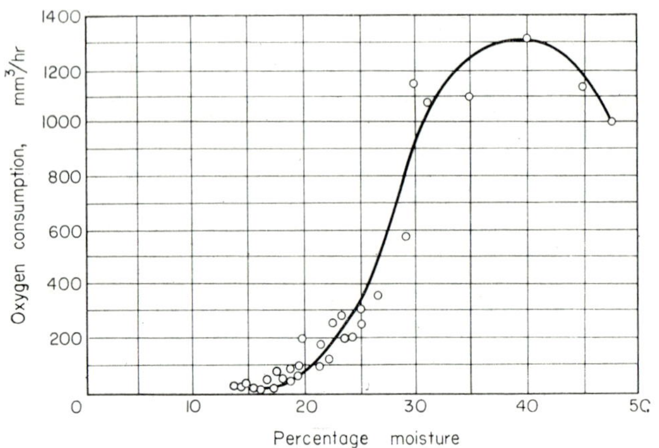

FIG. 1.3. The relationship between water content and the rate of respiration of germinating grains of *Avena sativa* (oat). Oxygen uptake determined at 25°C expressed as mm^3 per hr per 10 g dry weight. (After A. C. Bakke and N. L. Noecker, *Iowa Agr. Expt. Sta. Res. Bull.*, **165**: 319, 1933.)

Now, following the uptake of water, the cells of the embryo become active, expand and even begin to elongate before the seed coat is ruptured or the main food reserves of the grain are mobilised. The food reserves stored in cotyledons or endosperm or less frequently in perisperm tissue are mainly in an insoluble form as polysaccharides (particularly as starch), as fats in the form of oil globules or as granules of protein. Their ultimate utilisation to promote the growth of the tissues of the embryonic root and shoot depends upon their conversion to soluble compounds and the transport of these compounds to the regions of cell expansion and cell division in the embryo. The conversion of insoluble food substances to simpler soluble compounds is another expression of enzymic activity. Some of the enzymes involved are present in the dry storage cells and are, like the respiratory enzymes, activated during the phase of rapid water uptake by imbibition. Other enzymes involved in the mobilisation of food reserves are, however, synthesised during the germination process. This can be illustrated by reference to the digestion of starch grains (which contain two polysaccharides; amylopectin and amylose, the latter usually constituting 15–35% of the grain). Probably in all, at least four enzymes are involved in the degradation of starch to the soluble monosaccharide, glucose. The two involved in the initial attack are *α-amylase* and *β-amylase* and their actions are, to an extent, complementary. α-amylase attacks both amylose and amylopectin to give rise to a complex mixture of molecules of lower molecular weight, called dextrins. β-amylase releases disaccharide units (maltose) from these compounds. Studies of α-amylase and β-amylase activity in germinating grains of *Avena sativa* show that the β-amylase is preformed in the grain and is activated during the initial rapid uptake of water. α-amylase development is, however, dependent upon synthesis of enzyme protein and this synthesis takes place in the endosperm cells in response to a hormonal stimulus from the germinating embryo. The embryo synthesises and releases a gibberellin (see Chapter 9) which moves into the endosperm and initiates α-amylase synthesis in the aleurone layers. The gibberellin also seems to be implicated in promoting the synthesis in the endosperm of other enzymes involved in the degradation of its food reserves.

This discussion of the activities of the amylases enables us to draw attention to an important property of all enzymes. Proteins lose irreversibly their essential biological properties when heated. This denaturation of proteins, which involves a change in molecular architecture rather than in chemical composition, can also be brought about by irradiation in the ultra-violet, by short exposures to extremes of acidity and alkalinity or even by mechanical agitation in solution. Enzymes are, therefore, inactivated by heat; they are thermolabile. However, the exact temperature conditions which cause thermal denaturation are characteristic of each individual enzyme. Such a difference in the heat instability of the amylases was exploited in the researches mentioned above on the changing activity of these enzymes during germination. It was possible, by heating the seedling extract to 70°C for 20 min, to completely inactivate β-amylase without significantly reducing the α-amylase activity.

The germination of the embryo initiated by the utilisation of its own food constituents is now maintained by the larger supply of soluble compounds generated by enzyme activity in the storage tissue of the seed. Not only does the embryo stimulate the development of enzymes in the storage tissue but it directs the flow of soluble compounds from the storage cells to its own growing regions. The storage cells of the endosperm or cotyledons have a relatively low rate of respiration but by the action of their hydrolysing enzymes they become enriched with soluble food substances. The movement or translocation of these substances to the embryo is, therefore, movement from a region where they occur at high concentration to a region where they are rapidly consumed in embryo metabolism. The storage tissue is a "source", the embryo a "sink" for these essential compounds. This, however, neither specifies the nature of the actual compounds which flow to the embryo nor the mechanism of the translocation process. The storage fats are probably attacked by "lipase" enzymes and the fatty acids so produced are converted to sugar. This raises the possibility that both storage polysaccharides like starch *and* fats are ultimately converted to the disaccharide, sucrose (cane-sugar), since there is very strong evidence that it is as sucrose that the carbon of carbohydrates is transported in the phloem of the conducting strands. In studies on barley germination there is the corroborative evidence that the embryo preferentially utilises sucrose for respiration and to build its cell walls.

The proteins of the storage cells are acted upon by a group of proteolytic enzymes to yield a mixture of free amino acids, together with the amides of glutamic acid and aspartic acid (glutamine and asparagine) and, probably, simple peptides. Whether all these simple organic nitrogen compounds are equally translocated is not known but the evidence does not point to a single transported compound as is the case for carbohydrates. Further, we know that the embryo cells rather than the storage cells contain the enzymes which promote the interconversions of amino acids required to produce the particular mixtures of amino acids involved in the synthesis of the proteins of the embryo cells. In quantitative terms the transport of carbohydrates and of soluble organic nitrogen from the storage cells is dominant. The storage tissues are, however, sources of other substances essential to the embryo, including phosphorus compounds and certain essential vitamins. Further, there is strong evidence that the growth hormones equally essential for the growth of the embryo are formed from precursors which are released from the storage cells early in the germination process. These growth hormones in their active form occur at the points of most active growth within the embryo and may, in some way not yet understood, direct the flow of soluble food substances to these growth centres.

The exact nature of the translocation of organic solutes has yet to be satisfactorily worked out but current views on possible mechanisms involved will be discussed in a later chapter (Chapter 7, p. 212). Clearly, in the germinating seed food materials must pass from cell to cell within the storage tissues, be conducted within the vascular strands of the developing seedling and pass out from these strands to the dividing, expanding and differentiating cells of the growing regions. Longitudinal transport of organic solutes within the vascular strands takes place predominantly in the sieve tubes of the phloem. One very strong possibility is that within these special conducting units transport of materials is effected either by a mass flow involving the bulk of their fluid contents or by "channel" or "strand" flow of fluid and cell particles (these latter could be special "carrier" particles) within a mainly stationary fluid mass. There may well be, in the near future, a "break-through" in our understanding of the mechanism of sieve tube transport following improved techniques for the microscopic observation of living phloem tissue and the application of the electron microscope to the study of sieve tube structure. Entry of food materials into the phloem at points adjacent to the storage tissue and their removal in the regions of active growth of the seedling probably involves processes of secretion and absorption which demand energy derived from the respiration of living cells. Such movements of substances across cell boundaries, often in directions opposed to concentration gradients and at speeds altogether more rapid than would occur by diffusion, are often referred to as "*active*" movements. Work is done by the cell in promoting such "active" movements and in performing such work a redistribution of the energy released by respiration occurs. There is now an impressive body of evidence that the absorption and secretion by plant cells of both inorganic ions and organic substances are "active" processes in this sense. The movement of organic solutes *within* living tissues such as the growing regions of the young shoot and root of the seedling and through the storage tissues of the seed probably, at some points, involves diffusion, at others, "active" transport across cell membranes, and at others, mass movement over longer distances by protoplasmic streaming within the cells. It remains a controversial question as to whether any flow of fluid occurs in the fine cytoplasmic strands (*plasmodesmata*) which interconnect the protoplasts of adjacent cells in all living plants.

The cells of the embryo are capable of increase in size (expansion growth) and differentiation; many of them are capable of cell division (mitosis). With the increase in their water content which occurs during the initial phase of germination this expansion growth commences and is soon reinforced by the formation of new cells, first at the apical growing point of the root and then at the corresponding growing region of the embryo shoot. These processes of cell division and cell expansion are dependent upon a release of energy and of essential small reactive molecules from the complex food substances (first of the embryo cells and later of the storage tissues). In *the process of growth* this energy and these reactive molecules, both of which arise from a degradation of complex food substances, participate in the synthesis of new cell material, particularly of the protein and lipid molecules essential to protoplast structure and of the complex polysaccharide and polyuronic acid molecules of the cell wall. The metabolism of the embryo cells, set in motion by the absorption of water, involves a multitude of interconnected chemical reactions. Some of the reaction sequences result in the degradation of complex food molecules (this aspect of metabolism is called *catabolism*), other sequences result in the synthesis of the unique molecules from which living structures are built (*anabolism*).

To illustrate the activation of metabolism which follows the uptake of water by the air-dry seed we quoted rates of respiration and expressed these in terms of oxygen absorption per unit of time. For most seeds oxygen is as essential as water for germination because of its requirement in the process of *respiration*. This catabolic process, whereby food materials are degraded and their energy released, is an oxidative process. The storage carbohydrates, fats and proteins serve as substrates for respiration and are oxidised to yield as end products carbon dioxide and water. Oxygen absorption and carbon dioxide evolution proceed continuously at the respiratory centres within the living cells. During respiration, energy locked up within the structure of the food substance molecules is released. Some of this energy is dissipated as heat and this evolution of heat is most easily demonstrated by confining a mass of germinating seeds in a Dewar flask. Some of the released energy is, however, conserved because the degradation of the food substance molecules is linked at the respiratory centres to a simultaneous synthesis of certain energy-rich phosphorus compounds which act as mobile "power-houses" able, at other points in the cells, to "drive" the synthetic reactions of anabolism. Further, during the degradation of respiratory substrates many reactive intermediate compounds arise and some of these are withdrawn from the oxidative pathways of degradation to be used as "raw materials" for the synthesis of new essential organic molecules like the enzyme and structural proteins of the cell.

The swelling of all the living cells within the seed combined with the early extension growth of the embryo root results in the rupture of the seed coat. The young root emerges and quickly penetrates downwards into the soil. The rapid downward growth of the root anchors the seed and immediately creates an absorbing surface for the uptake of water and of essential inorganic ions from the soil. This is soon followed by the upward growth into the air of the young shoot and the opening of its first green leaves. The establishment of the root in the soil and the emergence into the light of the young shoot enables the seedling to draw upon its external environment for all its *essential nutrients*. The organic food materials of the seed, essential for the establishment of the young plant, are then no longer required for its further growth and development.

The carbon essential for the formation of all organic compounds is available to the green plant as the gas, carbon dioxide. This gas is present in the air in low concentration (3 parts in 10,000 by volume) and its absorption by land plants takes place via their green leaves and young stems. Submerged water plants obtain their carbon dioxide already in solution from the water in which they live. Water is essential to all living organisms as the medium in which essential cell constituents dissolve and move and because the biological properties of many molecules like those of the proteins depend upon their association with water molecules and with hydrogen and hydroxyl ions. Water is also the source of hydrogen and oxygen atoms and these atoms are second only to carbon in importance in the architecture of organic molecules. In the land plants this water is absorbed mainly from the soil by the root system and transported to the shoot via the conducting elements of the xylem. The molecules of protein contain, in addition to carbon, hydrogen and oxygen, the elements nitrogen and sulphur. Many other essential organic molecules contain nitrogen and/or sulphur. The lipo-proteins involved in the formation of cell membranes and the nucleoproteins involved in the transmission of hereditary characters and in enzyme synthesis also contain phosphorus. Other phosphorus compounds, as already mentioned, play an essential part in the energy relationships of cells. Calcium is essential for the proper formation of the middle lamella. Magnesium is contained in the molecules of the chlorophylls, the essential photocatalysts of photosynthesis and also is implicated as part of certain enzyme systems. Potassium is an essential and mobile element within the plant. Active leaves and growth centres are usually rich in potassium. Although there is some evidence that it is involved in protein synthesis we do not know of essential potassium-containing organic molecules or of enzyme systems whose activity depends upon the presence of potassium. The essential elements discussed above are all required in differing but moderately large amounts for the growth and development of the green plant. There are, however, a number of other essential elements which are required in relatively very small amounts, as *micro*-rather than as *macro*-nutrients. The recognised micro-nutrient elements are: iron, manganese, zinc, copper, boron, molybdenum and chlorine. It seems that in all cases these micro-nutrients are essential to plant growth because they function as essential constituents of particular enzyme systems. These essential macro-nutrient and micro-nutrient elements from nitrogen to molybdenum are absorbed as anions and cations from the soil or water in which the plant is growing.

Reference has already been made to the osmotic nature of the water uptake by the fully imbibed seed and osmotic forces are also involved in the absorption of water from the soil by the seedling root. Further discussion of the mechanism of water uptake by plant cells and of the movement of water through plant tissues will, however, be best deferred to the next chapter (p. 27) where the central theme will be the relationship between cell structure and cell function. Reference has also already been made to salt absorption, citing it as an "active" process, a process dependent upon the energy released in respiration. The "active" nature of salt absorption is emphasised by two important aspects of this process. Firstly, the absorption of anions and cations is *selective*, they are absorbed in amounts not determined by their relative concentrations in the soil or other nutrient solution. Secondly, the uptake of ions leads to their *accumulation*, to the development of internal concentrations of free ions within the cell often vastly in excess of their concentrations in the external solution. This selective accumulation of inorganic ions by the absorbing cells of the seedling root raises the whole question of how the energy released in respiration is harnessed to perform cellular work, in this particular case to power a number of selective "salt pumps". Later, when we explore further the inter-relationships between metabolic processes it will be appropriate to examine critically how far we really understand the mechanism of salt uptake by plant cells (Chapter 7, p. 212 *et seq.*).

The essential elements absorbed by the root are not only involved in its own metabolism, but are, in part, translocated either as free ions or as soluble organic compounds to the growing regions of the shoot, being metabolised most actively in its apical meristems and developing leaves. The root therefore plays its special role in the nutrition of the whole plant. In turn, and to an increasing extent, as the organic food reserves passed on from the parent plant are depleted, the development of the young plant becomes dependent upon the synthetic activity of its leaves and green stems, upon the primary synthesis of organic molecules from carbon dioxide and water by *photosynthesis*. Respiration has been described as an oxidative release of the energy locked up in the structure of organic molecules, such as in the molecules of sugars. Photosynthesis is, by contrast, a reductive process in which the energy-rich molecules of sugar are built up from carbon dioxide and water. The *overall* equation for both these processes can therefore be represented

$$\underset{\text{sugar}}{C_6H_{12}O_6} + 6O_2 \underset{\text{photosynthesis}}{\overset{\text{respiration}}{\rightleftharpoons}} 6CO_2 + 6H_2O \quad (1)$$

Further, as clearly indicated by its name, the unique phenomenon in photosynthesis is not the assimilation of carbon dioxide, nor the synthesis of sugars but the utilisation for these processes of the radiant energy of the sun. This means that photosynthesis involves, in contradistinction to the other metabolic processes so far mentioned, certain *photochemical* reactions, insensitive to temperature and mediated through a photocatalytic system. The chlorophyll pigments organised in special cellular structures called chloroplasts represent this system and give to photosynthetic tissues their green colour.

Study of the reactions involved in respiration has revealed that the oxidation of organic food substances is linked to the reduction of a special pyridinedinucleotide molecule (a co-enzyme) and that much of the available energy becomes concentrated in the reduced molecules of this compound. The subsequent regeneration of the co-enzyme by oxygen not only results in the formation of water (eqn. 1) but is coupled to the simultaneous synthesis of an energy-rich phosphorus compound (a polyphosphate of the nucleotide, adenylic acid, the substance, *adenosine triphosphate*—abbreviated to **ATP**). This compound is one of those phosphorus compounds which can act as a mobile "power house" and drive the synthetic reactions of anabolism. It is, therefore, of great interest that it has recently been shown, particularly in experiments with isolated chloroplasts, that the photochemical reactions of photosynthesis result in the synthesis of ATP and in the reduction of a pyridinenucleotide co-enzyme. The energy trapped in ATP and the reducing power of the reduced co-enzyme then effect carbon dioxide assimilation and reduction by thermochemical (enzyme-activated) reactions which can occur in the absence of light (are "dark" reactions) and even in non-photosynthetic plant cells. There is a common energy currency in all plant cells; photosynthetic cells have a unique way of minting this currency.

Once its leaf area is sufficiently developed the young plant acquires the ability to grow and develop when supplied with its essential nutrients entirely in inorganic form. It becomes independent of the organic food reserves of the seed. The sugars synthesised in the chloroplasts become the primary sources of energy and of reactive intermediates for the synthesis of all the multitude of other organic molecules which go to make a living cell. From the leaves the sugars which are in excess of those required for the nutrition of the leaf cells pass via the sieve tubes of the phloem and in the form of the disaccharide sucrose, to nurture the growing points of the shoot and to the root.

With the emergence of the shoot of the seedling into the air not only does photosynthesis become possible but there occurs an inevitable loss of water from the plant by evaporation, a process termed *transpiration.* Transpiration is not only a consequence of the exposure of the external surface of the shoot system to the air but is enhanced in magnitude by the development in the shoot surface of breathing pores (stomata) which facilitate the exchange of oxygen and carbon dioxide between the shoot tissues and the external air. The stomata are bordered by special cells (guard cells) whose size and form can change in a way which increases or decreases the pore area or closes the pore completely. This regulation of stomatal aperture is of importance both in the retention of respiratory carbon dioxide for subsequent photosynthesis and for water conservation at times of drought.

If the water content of the actively metabolising leaf cells is not to fall significantly as a result of transpiration a continuous flow of water to the leaves must be set in motion; a "transpiration stream" must take place in the xylem strands stretching from the veins of the leaf down to the absorbing regions of the root. Such a transpiration stream will inevitably not only demand an enhanced water uptake by the root but will lead to transport of inorganic nutrient ions in the mass flow of liquid in the conducting cells of the xylem. A cut shoot set under conditions conducive to active transpiration absorbs water at the cut stem surface with considerable force, a suction pull for water is set up in the shoot system. The attraction for water develops in the leaf cells as water is evaporated from their surfaces into the air space system of the leaf tissue and is transmitted down through the continuous columns of liquid in the xylem vessels and tracheids. The removal of a plant from a water-saturated to a dry atmosphere leads to a marked enhancement of the rate of water absorption and to the force with which water is removed from the soil. Under conditions of active transpiration, plant roots may desiccate the soil to the point where it retains its small residual water content with a force as high as 15 atm. There is strong evidence for the view that under such conditions of active transpiration, a *tension* is developed in the xylem fluid and that then the root acts in water absorption purely as a physical system, a wick or sponge, from which water is withdrawn by the transpiration of the leaf cells. Water absorption by the root under these conditions has, therefore, been described as a "passive" process.

However, during the passage of the seedling shoot through the soil its water supply cannot derive from transpiration. The same applies during the initial spring growth of perennating organs or during the early growth of the buds of shrubs and trees in the spring. Further, many seedlings and herbaceous plants when placed in a saturated atmosphere show a release of droplets of water from their leaves (the phenomenon of guttation). Under these conditions there is evidence that the upward movement of water follows from the development of a positive pressure (a *compression*) of water in the xylem; a positive root pressure. Further, this positive root pressure is only demonstrable when the temperature and oxygen supply are favourable to root metabolism, supporting the contention that a positive root pressure is indicative of an "active" water uptake by the root.

By means of the above outline discussion of the physiology of germination and of seedling establishment, it has been possible to introduce or remind the reader of the major metabolic processes which proceed in the green plant. This enables us to pose most of the problems which will be considered in the present volume. This can perhaps best be illustrated by formulating a number of important questions which arise directly out of the present chapter and which will to some extent be answered in the subsequent chapters.

What is the significance of the fact that the reactions of living cells are catalysed by a large number of specific enzymes? What special properties of protein molecules make them able to play this unique catalytic role in plant metabolism? If enzyme proteins catalyse the synthesis and degradation of sugars and other carbohydrates, of fats and other lipids, of organic and amino acids, of vitamins and of the many other essential organic molecules of the cell, how then are the enzyme molecules themselves synthesised? If these enzymes control metabolism, is it through the controlled synthesis of enzymes that the genes of the nucleus control plant growth, form and development? Are compounds generally, always or never degraded and synthesised by the same chemical reactions? If there is a common pathway of synthesis and degradation what controls the direction of metabolism along such a pathway? As an example, and leaving aside the photochemical reactions, are the reactions involved in the photosynthesis of sugars the reverse of those in sugar respiration? How does the cell convert radiant into chemical energy in the process of photosynthesis? How exactly is energy conserved in the cell and transferred from energy-yielding to energy-absorbing reactions and with what efficiency is this cellular redistribution of energy effected? If energy can be transferred from one chemical system to another in this way does this explain how cells perform the work which must be involved in such "active" processes as cell growth, growth movements, accumulation of solutes against concentration gradients, and rapid transport of organic molecules within the plant? Where do the reactions of metabolism take place within the cell and is cell structure capable of being interpreted in terms of the maintenance of orderly patterns of degradation and synthesis within the quite extraordinarily complex chemical factory represented by the plant cell? What determines that some cells remain meristematic while others undergo enlargement and change in the several ways which give rise to the diverse types of tissue cells found in the body of the green flowering plant? What are the nutritive and other physiological interrelationships between the different tissue systems? Why is the duration of life of the multicellular plant finite or, phrased another way, can we explain in metabolic terms senescence and death?

The extent to which we can answer such questions and others which the thoughtful reader will formulate for himself, is the measure of our present understanding of plant metabolism.

FURTHER READING

H. E. Street and H. Öpik. *The Physiology of Flowering Plants*. Edward Arnold, London, 1970.

J. Bonner and A. W. Galston. *Principles of Plant Physiology*. W. H. Freeman & Co., San Francisco, 1952.

B. S. Meyer, D. B. Anderson and R. H. Bohning. *Introduction to Plant Physiology*. D. van Nostrand Co., New York, 1960.

P. J. Kramer. *Plant and Soil Water Relationships*. McGraw-Hill Book Co. Inc., New York, 1949.

D. M. Mayer and A. Poljakoff-Mayber. *The Germination of Seeds*. Pergamon Press, Oxford, 1963.

CHAPTER 2

Cell Structure and Function

"The vital processes of the individual cells form the first indispensable and fundamental basis for both vegetable physiology and comparative physiology in general." M. J. Schleiden (1838)

"We have seen that all organisms consist of essentially like parts, the cells . . . each cell . . . capable of developing independently if only there be provided the external conditions under which it exists in the organism. . . . The question as to the fundamental power of organised bodies resolves itself into that of the individual cells."

Theodore Schwann (1839)

INTRODUCTION

THESE writings of the botanist Schleiden and the zoologist Schwann, from which our above quotations are taken, formed the first clear expression of the view that the properties of complex organisms are an expression of the separate activities of their cells, each cell living in the special environment created by the association together of the total cell population. This enunciation of the "cell theory" had, as its background, the brilliant researches of earlier microscopists right back to the first use of the term "cell" by Robert Hooke in 1665.

The early microscopists recognised that living material was built up from, or divided into, minute compartments or cells but not until the 19th century were microscopes of sufficient resolving power available for the living contents of cells to be observed. Robert Brown first observed the nucleus in the epidermal cells of certain plants in

1831. The recognition of the general occurrence of a thin layer of mucilaginous material within the cell boundary and the use of the term *protoplasm* for this substance, which was only to be seen in life (and not like the cell boundary or wall also in death), followed from the gifted studies of Hugo von Mohl in the middle of the 19th century. Still further progress followed upon the manufacture and distribution of Abbé microscopes by Zeiss of Jena from around 1878. It was from studies with microscopes of this kind and by using basic analine dyes as stains that Eduard Strasburger first described the chromosomes of the nucleus and their behaviour during the division of cells. Such microscopes made possible very rapid advances in our knowledge of the range of cell structure and form in plant tissues and laid the foundations of our present knowledge of nuclear and cytoplasmic structure as they can be seen with the ordinary light microscope with its resolving power of around 0·275 μ (2750 Å).

Cells in the living state have always been difficult objects to study microscopically because they and their contents are for the most part transparent to visible light. On the other hand, examination of fixed and stained cells always raises the problem of how far the structures seen correspond to structures in the living cell. Various extensions of optical microscopy have, therefore, been developed to overcome the lack of contrast within the living cell. The use of polarised light has enabled cell structures in which the molecules are highly oriented, e.g. the layers of the cell wall, and the lamellae of plastids, to be recognised by virtue of their birefringence. Perhaps more important has been the development of the phase-contrast microscope by means of which such processes as nuclear and plastid division and the movement and change in form of bodies like mitochondria can be followed and photographically recorded in the living cell (Fig. 2.1). In the phase-contrast microscope we use a monochromatic light source, reduce the intensity of the directly transmitted component of the light and retard the deflected beam to bring it a half-wavelength out of phase with the transmitted light and, thereby, through destructive or constructive interference, yield strong contrasts. The resolving power is not diminished, the contrast is dramatically increased.

One of the factors limiting resolution in the light microscope is the wavelength of visible radiation, and, therefore, to increase resolution it is necessary to use radiation of shorter wavelengths. This led Köhler to devise in 1904 an ultra-violet microscope involving a monochromatic source of ultra-violet radiation, lenses of fused

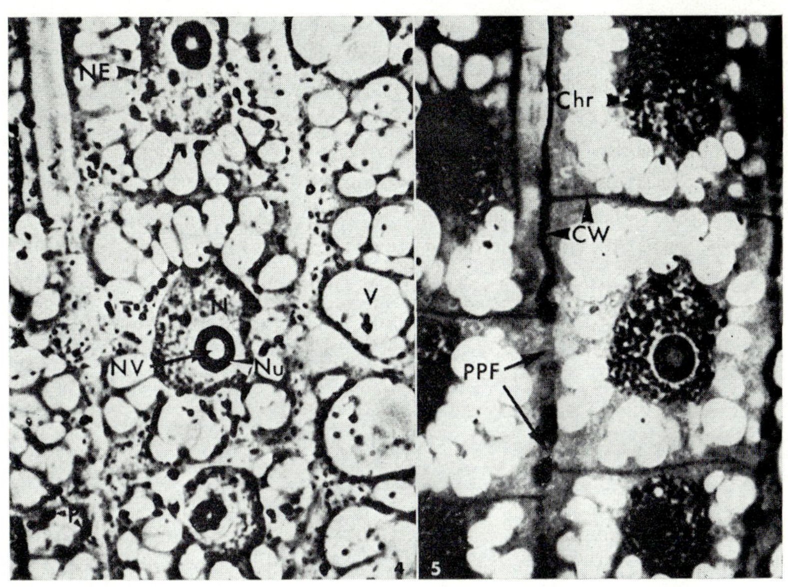

FIG. 2.1. *Left*. Interphase cells from the root tip of *Vicia faba* (broad bean), stained with iodine and photographed under phase contrast. *Right*. The same section stained with toluidine blue.
Key: N, nucleus; NE, nuclear envelope; Nu, nucleolus; NV, nucleolar vacuole; P, plastid; V, vacuole; Chr, chromatin; CW, cell wall; PPF, primary pit field.
(Both from T. P. O'Brien and M. E. McCully, *Plant Structure and Development*, Macmillan, London, 1969.)

quartz which will transmit radiation down to a wavelength of 2400 Å and photographic recording of the images. Such ultraviolet microscopes have, however, been used in cytological work not so much because of their increased resolution but because when combined with a microspectrophotometer they permit estimates of changes in the nucleic acid content (characteristically absorbing in the region of 2600 Å) in the nucleus and cytoplasm during cell division and differentiation.

The search for higher resolution has found its modern answer in the electron microscope whose resolution is more than 200 times greater than that attainable with light. The electron microscope uses a beam of high-speed electrons focused by electromagnetic lenses. To be examined in the electron microscope the specimen must withstand evacuation (living material is excluded) and be unusually thin. To prepare ultra-thin sections, special microtomes are used which employ glass or diamond knives, and the material, usually fixed in neutral osmic acid, must be embedded in a plastic which is polymerised after the monomer has penetrated the dehydrated cells. There is still a need to devise methods of dehydration and embedding which do not distort or disrupt cell structures. Nevertheless, the picture of cell structure revealed by the electron microscopy of ultrathin sections can be accepted with growing confidence as it increasingly links up with knowledge obtained from other forms of microscopy and from studies of the chemistry and metabolic activity of fractions isolated from cell extract by high-speed centrifuging. A new technique of specimen preparation for electron microscopy is now adding to our knowledge of fine structure, particularly of the surface fine structure of cellular membranes (Fig. 2.11). This technique, termed freeze-etching, involves very rapid freezing of the plant tissue, cutting (splintering) on the microtome, etching of the exposed surface by freeze-drying (to a depth of tenths of a mμ) and the taking of a replica (depositing on the surface a film of platinum and carbon) which is subsequently released from the specimen surface, mounted and examined in the electron microscope.

In this chapter we shall consider the structure of plant cells as revealed by light (resolving power down to $\simeq$ 3000 Å) and electron (resolving down to $\simeq$ 15 Å) microscopy and consider what we can learn of the relationship of structure to function by observations on the intact cell and on cell "fractions" isolated by centrifuging.

GENERAL STRUCTURE OF PLANT CELLS

Many different kinds of cells make up the body of a flowering plant. All the various specialised cells, are however, derived by cell division and subsequent enlargement and differentiation from groups of meristematic cells situated at the apices of the root and shoot or making up the vascular and cork cambia of the larger root and shoot axes. The meristematic cells of the apical growing points are roughly isodiametric and frequently not more than 10 μ in diameter, those of the cambia are small rectangles in transverse section but are elongated, often 100 μ or more in length. Such cells have thin *cell walls*, rich in hemicelluloses and pectin. Within the cell wall the living material (the *protoplast*) consists of *cytoplasm* and a large *nucleus*, the latter occupying two-thirds to three-quarters of the protoplast volume. Minute *vacuoles* (rich in fatty or proteinaceous material), spherical or cylindrical *mitochondria* and *proplastids* can be distinguished by appropriate techniques in the ground mass of the cytoplasm (the *hyaloplasm*). Mature tissue cells, such as parenchymatous cells, are derived from such cells. The marked increase in cell size (this may be several hundredfold) and the changes in protoplast and wall structure which lead to the development of a parenchymatous cell exemplify the processes of cell growth and differentiation. Further, since cells involved in water and salt absorption and in photosynthesis and food storage are of this kind their structure is of great interest to the physiologist.

Figure 2.2 shows some drawings of parenchymatous cells, in which the structures described below are labelled. The *cell wall* is composed of a framework of the polysaccharide cellulose. The water content of the wall accounts for 92–94% of its fresh weight. Associated with the cellulose (constituting perhaps 25% of the dry weight) are variable quantities of hemicelluloses, pectins, proteins and fats. During the marked increase in size that accompanies the differentiation of a parenchymatous cell, the cell wall grows. The increase in cell wall area is accompanied by an increase in the amount of cell wall substance and there is usually a further increase in cell wall material after cell enlargement is complete.

The process of cell enlargement is associated with a massive uptake

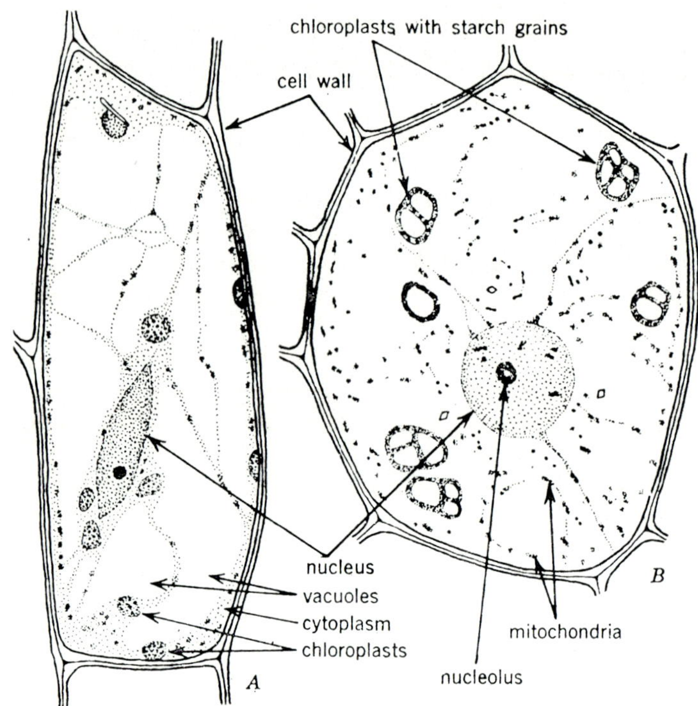

FIG. 2.2. Structure of parenchymatous plant cells. A. Cell from the petiole of a sugar-beet leaf. It has vacuolated cytoplasm with mitochondria, chloroplasts and nucleus. B. Starch sheath cells from young stem of tobacco, showing prominent starch grains in the chloroplasts. (Both ×1190.) (From K. Esau, *Plant Anatomy*, John Wiley & Sons, Inc., New York, 1953.)

of water and as this takes place the vacuoles fill with fluid, become more prominent and ultimately fuse to give a large central vacuole which may occupy at least 90% of the total cell volume. The vacuolar fluid, often referred to as the cell sap, is a solution containing inorganic ions and soluble organic substances like sugars, organic acids and amino acids. The cytoplasm also increases in amount during the growth of the parenchymatous cell. This is indicated by the increase in protein content of the cell during the expansion phase of its differentiation. Nevertheless, the cytoplasm rarely occupies more than 5% of the total cell volume and typically takes the form of a thin layer (often not more than 5 μ thick) immediately within the cell wall and surrounding the central vacuole. Some 95% of the fresh weight of cytoplasm is water, and some 3% is protein. On the assumption that one million can be taken as an average figure for the molecular weight of protein molecules, it has been calculated that a mature cell of the root cortex may contain some 10^8 protein molecules and at this concentration some 20 successive mono-molecular layers of protein could be present in the thickness of the cytoplasmic layer. The nucleus (8–10 μ diameter) is contained in the cytoplasmic layer or suspended in the centre of the cell by trans-vacuolar strands of cytoplasm.

The lining layer of cytoplasm has a granular appearance due to the presence of various inclusions or cellular "particles". Some of these, the *mitochondria*, are spherical or cylindrical bodies with a maximum dimension up to 2 μ and characteristically stained in the living cell by the "vital" stain, Janus Green B. Others of similar size but not usually stained by this dye are regarded as *proplastids*. These are the bodies from which it seems develop, during differentiation, the various forms of plastid, most important of which are the *chloroplasts* (in photosynthetic parenchyma) and the starch-forming *amyloplasts* (of storage parenchyma). As will be discussed later, mitochondria and plastids are centres of metabolic activity and can increase by division. They are constant features of the cytoplasmic complex. By contrast, other inclusions are inactive and represent reserve food substances or by-products of cell metabolism. Examples of this type of inclusion are starch grains, protein "crystals" (aleurone grains) and minute oil droplets. These various inclusions are suspended in the optically clear ground-mass of the cytoplasm, the *hyaloplasm*. This hyaloplasm, from its staining properties, is clearly rich in protein and shows physical properties (such as reversible sol $\rightleftharpoons$ gel change at constant temperature) corresponding closely with those of a hydrophilic colloidal system in which the disperse phase is composed of fibre-like units capable of weak chemical association.

The surfaces of contact between the cytoplasm and the cell wall and between the cytoplasm and the central vacuole seem to be almost free of visible inclusions and appear from staining reactions to be relatively rich in lipoidal (fatty) material. By appropriate treatment the fatty material at the vacuolar–cytoplasmic surface can, in some cases, be demonstrated by causing the formation of myelin processes extending into the vacuole—such myelin processes are strongly indicative of the presence at the interface of phospholipids like lecithin. By plasmolysis, followed by micromanipulative rupture of the cell wall, naked protoplasts can be obtained. When such naked protoplasts are immersed in potassium chloride solution the cytoplasmic layer can be dispersed. The result is to isolate the aqueous vacuole. This, however, does not immediately disperse and the spherical mass can be induced to swell or shrink by changing the osmotic potential of the bathing solution. Dispersion, however, immediately follows the introduction of a fat solvent and minute oil droplets appear. The vacuole appears to be enclosed by a membrane (the *tonoplast*) which has semi-permeable properties, elasticity and an essential content of lipids. Experiments with dyes like acid fuchsin and aniline blue also suggest the occurrence of a membrane (the *plasmalemma*) at the cell wall–cytoplasm surface. These dyes do not readily penetrate cells and if injected into the vacuole or the lining of cytoplasm spread only within the limits of each phase. There seem to be barriers to their diffusion at both cytoplasmic surfaces. Micromanipulation has also demonstrated the presence of a well-defined elastic membrane (the *nuclear envelope*) at the surface of the nucleus.

By grinding up (homogenising) plant cells in an appropriate medium (such as a buffered sucrose solution of appropriate strength) it is possible to obtain a *brei* in which nuclei, plastids and mitochondria can still be distinguished. By centrifuging such a brei at successively higher speeds we can sediment in turn the cell wall fragments, the nuclei, the plastids and the mitochondria. The behaviour of these cytoplasmic inclusions strongly points to the presence at their surfaces of differentially-permeable, elastic, lipid-rich membranes. The cytoplasmic complex is rich in membranes and studies on the distribution within this complex of absorbed radioactive substances strongly suggests that these membranes restrict the movement into and within the cell of molecules and ions.

Cells in tissues are firmly cemented together by the *middle lamella*. Therefore, each protoplast is separated from its neighbouring protoplasts by the two intervening cell walls and the middle lamella between them. Nevertheless, in many tissues, the light microscope reveals the presence of many fine strands of cytoplasm (0·2–0·5 μ

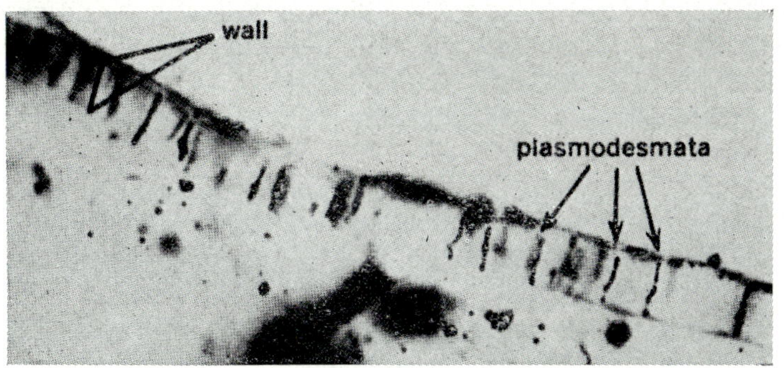

FIG. 2.3. Plasmodesmata in parenchymatous cells of *Solanum tuberosum* (potato) (×900). (From A. S. Crafts, *Plant Physiology*, **8**: 81, 1933.)

thick) crossing this wall system and maintaining cytoplasmic continuity between the tissue cells. These cytoplasmic strands or *plasmodesmata* (Fig. 2.3) are frequently concentrated together in certain areas and it is apparently at such sites that pits develop when the initially thin walls are thickened during subsequent differentiation (see primary pit fields, Fig. 2.1). It has been calculated that a meristematic cell (about 20 μ long on each face) from the root of onion is probably connected with adjacent cells by some 20,000 cytoplasmic strands.

THE OSMOTIC BEHAVIOUR OF CELLS

Work related to the phenomenon of osmosis dates back to the researches of the Abbé Nollet in 1748. Pfeffer was, however, the first to study quantitatively the movement of water into solutions of sucrose across copper ferrocyanide membranes, recording the hydrostatic pressures which were required to stop the osmotic flow of water in such a system. The data presented by Wilhelm Pfeffer in his book *Osmotische Untersuchungen: Studien zur Zellmechanik* published in 1877, showed that the hydrostatic pressures equal to and opposed to the osmotic pressures of the sucrose solutions were proportional to the concentrations of the sucrose solutions (i.e. proportional to $1/V$ where V is the volume of solution containing unit weight of sucrose) and proportional to the absolute temperature (T). The osmotic pressures of the sucrose solutions obeyed an equation analogous to the gas equation of Boyle; $PV = KT$. Later, in 1886, van't Hoff showed that if we use appropriate units (P in atmospheres, V in litres of solution containing one gramme-molecule of solute) then the K of the above equation is equal to the gas constant (R) and the equation $PV = RT$ applies to both gaseous and osmotic pressures. This enables us to define osmotic pressure thus: *the osmotic pressure* of a solution is equal to the gas pressure which the solute would exert if it were present as a gas, at that temperature, in a volume equal to the volume of the solution. Thus an "ideal" molar solution should at 0°C have an osmotic pressure of 22·4 atm. Since no actual pressure is developed unless the solution is placed in an osmometer it is preferable to use the term osmotic potential, designated π (Gr. *pi*). This is conventionally given a negative sign and is thus equal to the water potential (see below) of the solution at atmospheric pressure.

Pfeffer's system involved separation of the sucrose solution from water (the solvent) by a membrane allowing diffusion of water molecules but not of sucrose and a membrane of this type was described as *semipermeable*. Further, the movement of water into the sucrose solution was prevented by a hydrostatic pressure applied to the solution; the osmotic pressure was revealed as a deficit of hydrostatic pressure, as a diffusion pressure deficit (*DPD*) of water in the sucrose solution. The osmotic pressure of the solution was the excess hydrostatic pressure which when applied to the solution prevented osmosis (the net diffusion of water across the membrane) by raising the chemical potential (activity) of the water in the solution to that of pure water at the same temperature. This reduction of activity or "lowering of the pressure of water" in solutions is referred to as the water potential (Ψ, Gr. *psi*) and is given by the equation

$$\Psi = \mu_w - \mathring{\mu}_w = RT \ln(e/e^\circ)$$

where μ_w = chemical potential of the solution,
$\mathring{\mu}_w$ = chemical potential of pure water,
both the above under standard conditions,
e = vapour pressure of solution,
e° = vapour pressure of pure water,
both the above under the same conditions,
R = gas constant,
T = °K.
Since $\mathring{\mu}_w$ is greater than μ_w, Ψ is a negative quantity and *DPD* (as defined above) = $-\Psi$.

The water potential of solutions is then reflected in the lowering of their vapour pressures (elevation of their boiling points) and depressions of their freezing points below that of the solvent and both these criteria (and particularly the latter) can be conveniently used for the determination of the osmotic potentials of solutions. The osmotic potential (π) in atmospheres at 0°C is given by the equation

$$\pi = \frac{\Delta T}{1{\cdot}86} \times 22{\cdot}4 = -12{\cdot}04\, \Delta T$$

where ΔT is the observed freezing point depression (°C). Further, all these properties depend upon the number of solute particles per unit volume of solution (are colligative properties). Hence, solutions of electrolytes, depending upon their number of ions and percentage ionisation, have osmotic pressures in excess of solutions of non-electrolytes of equivalent molarity; an observation which led Arrhenius to develop the ionic theory of solutions.

The concept of the vacuolated parenchymatous plant cell as an osmotic system is based upon the classical researches of Pfeffer and of his contemporary, Hugo de Vries (working between 1871 and 1888) both of whom studied microscopically the response of plant cells to externally applied solutions. When such cells are placed in solutions of sufficient strength the protoplasts decrease in volume to such an extent that they shrink away from the cell walls; the cells are *plasmolysed*. Using cells of the red beet with their coloured vacuolar fluid, de Vries was able to follow this process accurately and detect the first evidence of separation of the protoplast from the cell wall (the point of *limiting plasmolysis*). When working with uniform layers of tissue, the individual cells are not all quite equally

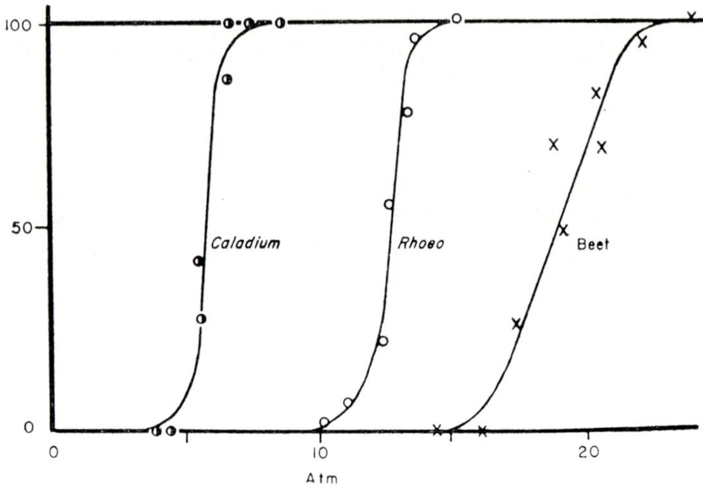

FIG. 2.4. Percentages of cells found plasmolysed (ordinate) in solutions of different osmotic potentials (atm): *Caladium* petiole, *Rhoeo* leaf and beet root. The osmotic potential bringing 50% of the cells to plasmolysis brings the tissue to "limiting plasmolysis". (From T. A. Bennet-Clark, A. D. Greenwood and J. W. Baker, *New Phytologist*, **35**: 277, 1936.)

susceptible to plasmolysis and, therefore, when dealing with such a population of cells, the solution which just plasmolyses 50% and just fails to plasmolyse the remaining 50% of the cells can be said to bring the tissue to the condition of "limiting plasmolysis" (Fig. 2.4). Such solutions are isotonic with the tissue and such isotonic solutions can be shown to be solutions of equal osmotic potential. Therefore, the ability of solutions to withdraw water from cells is a function of their osmotic potentials and when the plasmolysed protoplast ceases to decrease in volume (comes to equilibrium with the external plasmolysing solution) it can be postulated that the *DPD* of the solution is balanced by the *DPD* of the protoplast.

The phenomenon of plasmolysis indicates that the cell wall is permeable to (allows the diffusion of) both solute and solvent molecules and that the osmotic removal of water is one proceeding across membranes in the protoplast. Pfeffer referred to these as plasmatic membranes and visualised such membranes as occurring at both the outer surface of the protoplast and at the surface of the vacuole; de Vries used the terms ectoplast (we have referred to this membrane as the plasmalemma) and tonoplast. Clearly, to the extent that the protoplast behaves as an osmotic system one or both of these membranes must have semi-permeable properties.

When a cell is placed in a solution capable of plasmolysing it there occurs a decrease in cell volume prior to the onset of plasmolysis. Similarly, if a plasmolysed cell is transferred to water the protoplast first expands to fill the cell lumen and then an increase in cell volume occurs before the uptake of water ceases. The cell develops a state of turgor in which the protoplast presses against the cell wall (develops a *turgor pressure*) and the stretched elastic cell wall builds up a counter pressure (*a wall pressure*). The wall pressure builds up in the protoplast a hydrostatic pressure opposing the inward movement of water and when the cell is in equilibrium with water (is at full turgor) this wall pressure (*WP*) balances the osmotic potential of the protoplast. At any point between limiting plasmolysis and full turgor, the osmotic *suction force* (*SF*) of the cell (also referred to as the *DPD*) is given by the equation

$$SF = -\pi_p - WP$$

where π_p is the osmotic potential of the fluid phase (predominantly vacuole) of the protoplast. Expressed in terms of water potentials (Ψ) we have:

$$\Psi_p = \Psi_s + \Psi_t$$

where Ψ_p = water potential of protoplast,
Ψ_s = water potential due to protoplast solutes
(both of these are negative),
Ψ_t = potential due to turgor (a positive quantity).

Water will, therefore, leave or enter the cell according as to whether the osmotic potential of the external solution is greater or less than the protoplast water potential (Ψ_p) (suction force). In land plants, the tissues are usually not fully turgid and the suction force of each separate tissue can be determined by finding the osmotic potential to which the tissue can be subjected without gain or loss in weight (or volume) (Fig. 2.5). Usually the value so determined cannot be regarded as the true suction force for the tissue *in situ* because within

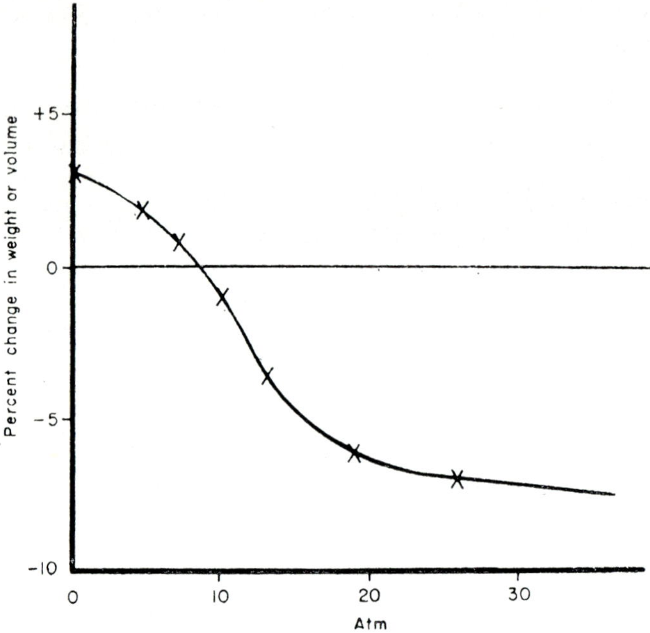

FIG. 2.5. Relationship between change in tissue or cell volume or weight from its initial (natural) value and the osmotic potential of the solution in which it is immersed. The osmotic potential in which there is no change in volume or weight gives the Ψ_p (water potential of protoplast) of the cell or tissue. Ψ_p is equal but opposite in sign to the DPD (diffusion pressure deficit) or SF (suction force). (From T. A. Bennet-Clark, in *Plant Physiology*, vol. 2, pp. 105–192, edited by F. C. Steward, Academic Press, New York, 1959.)

the plant the tissue will usually be either under compression by surrounding tissues (when the true value will be lower than that determined) or under tension (when the true value will be higher than that determined.)

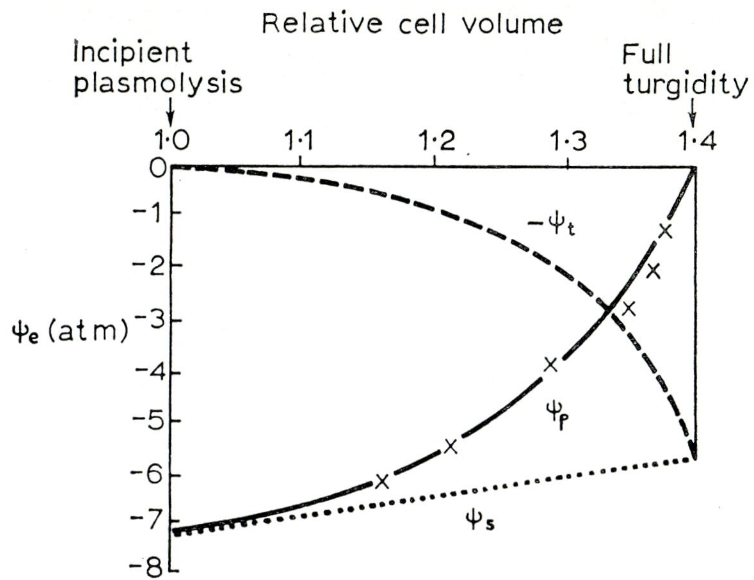

FIG. 2.6. Relationship between Ψ_p (protoplast water potential), Ψ_s (water potential due to protoplast solutes = osmotic potential of cell sap), Ψ_e (water potential of the bathing solution with which cell is in equilibrium) and Ψ_t (turgor potential). (The data refers to cells of the alga *Nitella* and is derived from H. Tamuya, *Cytologia* (*Tokyo*), **8:** 542, 1938.) Note that Ψ_s and Ψ_p become less negative as the cells become more turgid while Ψ_t becomes more positive ($-\Psi_t$ more negative).

The osmotic concept of the water relations of the vacuolated cell as outlined above can be summarised in a diagram (Fig. 2.6) which depicts the changing values for cell volume (relative cell volume), turgor potential (Ψ_t), protoplast water potential (Ψ_p), and osmotic potential of the cell sap (Ψ_s) as the cell absorbs water from the condition of incipient plasmolysis (limiting plasmolysis) to that of full turgor.

The critical test which one should apply to this osmotic hypothesis is the demonstration that equations such as $\Psi_p = \Psi_s + \Psi_t$ for the partially turgid cell and $\Psi_p = \Psi_s$ for the plasmolysed cell are quantitatively obeyed. It is, however, technically very difficult to determine the wall pressure or to obtain an undoubtedly pure sample of vacuolar fluid. A number of workers have reported that the osmotic potential of the solution causing limiting plasmolysis is considerably more negative (0·2 to 5·0 atm) than the osmotic potential of the extracted vacuolar fluid. These findings suggest that the absorbing power of the cell for water may be greater than expected from the osmotic relationship and has led to the hypothesis that the cell augments osmotic forces by a mechanism of "active" water uptake. This implies that the force motivating water uptake is $\Psi_s + a$ where a is a secretion pressure brought about by some unknown mechanism involving the expenditure of cellular energy.

Water uptake is dependent upon aerobic respiration. The stimulation of water uptake which occurs when cells expand under the influence of the plant growth, hormone, auxin (see Chapter 9) is not associated with any increase in the osmotic potential of their protoplasts. Further work has however failed to establish a direct energy relationship between respiration and water uptake; in view of the permeability of cells to water the energy demanded to maintain a level of turgor above that established by osmosis would be very high. The importance of respiration seems to be mediated via its roles in the maintenance of membrane structure and the accumulation of inorganic ions. Again the promotion of water uptake by auxin is now considered to follow from its action of increasing the plastic extensibility of cell walls (thereby reducing the wall pressure).

An active or non-osmotic force in water absorption does not therefore seem to be of general significance. Certain specialised cells may however be able to move water in opposition to the osmotic potential. For instance such an active (energy-demanding) secretion of water is probably involved in the functioning of the water-secreting glands (hydathodes) of certain leaves. Again the setting of the bladder trap of the bladderworts (*Utricularia* spp.) involves passage of water from the bladder to the exterior via the bladder wall, until a tension (negative pressure) is established. It is the relief of this tension by an inrush of water via the bladder pore that brings insects into the bladder when they motivate pore opening by touching special sensitive hairs by the bladder pore.

The picture of the cytoplasmic membranes as semipermeable, permitting diffusion of water molecules but not of solute molecules, arose from the apparent stability of plasmolysis when using as plasmolysing agents solutions of sucrose or of single salts. However, as early as 1880 de Vries recorded that the plasmolysis of epidermal cells of *Rhoeo discolor* by glycerol and urea solutions was only temporary, the cells fairly quickly regaining turgor. Klebs in 1887 recorded a similar temporary plasmolysis of the cells of filaments of the alga, *Zygnema*, by urea solutions. Clearly, here there was evidence of the entry of solute molecules into the cells enhancing their Ψ_p and enabling them to regain turgor. This subject of the permeability of cells to solutes was investigated in detail for the first time between 1890 and 1899 by Charles E. Overton at the University of Zurich and led to the realisation that many solutes can diffuse into cells and that substances with a high solubility in lipids relative to water (a high oil/water partition coefficient) penetrated most rapidly. Later work, particularly by W. Ruhland, showed that not only lipid solubility but molecular size determines the permeating power of solutes to living cells. We have already referred to the evidence for the concentration of lipoidal material in cellular membranes. Their elasticity and their sieve-like properties as exposed by Ruhland's studies on permeability strongly suggest that proteins are also involved in the structure of cell membranes. Protein mono-molecular layers would confer elasticity and also act as molecular sieves. These classical studies on permeability established that many solute molecules can move across cell membranes by a diffusion process; that cell membranes were not semipermeable (in the sense of being permeable only to solvent molecules) but structures having highly characteristic permeability properties.

From considerations of the permeability of cell membranes one is led on to problems of the uptake and retention by cells of solute molecules and ions. This, however, forms the principal subject matter of

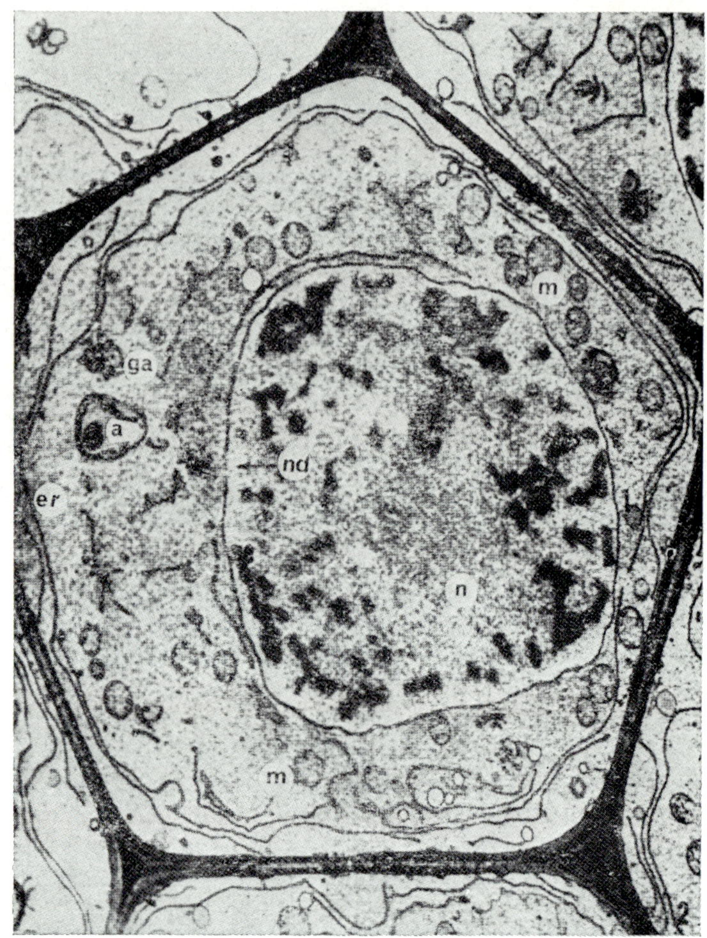

FIG. 2.7. Electron microscope (EM) picture of a transverse section of a root tip cell showing the nucleus (n), endoplasmic reticulum (er), mitochondria (m), golgi apparatus (ga), and an amyloplast (a). The point (nd) is a discontinuity in the nuclear envelope not a pore. The dark boundary region is the cell wall. (Approx. ×6000.) (From Whaley, H. G., Mollenhauer, H. H., and Leech, J. H., *Amer. J. Bot.*, **47** 401, 1960.)

Chapter 7. Here it is only necessary to emphasise that Overton clearly visualised that the interchange of solutes between the cell and its environment was regulated not only by resistances to diffusion (permeability in the narrow sense) but by an "active" transport of substances often in the opposite direction to that in which, from a purely physical standpoint, they would be expected to move. For this "active" transport mechanism, Overton used the term "adenoid" (gland-like) activity and it is with this dominant and active component that we shall be concerned when we come to discuss our knowledge of the accumulation by plant cells of essential nutrient ions.

THE FINE STRUCTURE OF CELLS

Bearing in mind the structure and osmotic behaviour of plant cells as revealed by the light microscope, consideration can now be given to the finer details of structure revealed by the electron microscope (Figs. 2.7–13, 2.16, 2.17, 2.22–24). These finer details of structure are particularly pertinent to our understanding of the localisation of metabolic events within the cell and of the nature of cell growth and differentiation.

The electron microscope resolves the optically clear hyaloplasm of the light microscope into two components. One, the matrix or ground substance, replaces the hyaloplasm of the light microscope in that, even in the electron microscope, no structural detail is revealed. The second is the *endoplasmic reticulum* (ER), a membrane-enclosed vesicular system appearing in thin sections of the cell as rounded, oblong or long slender vesicles (Figs. 2.9 and 2.10), or as branched, fenestrated sheets in freeze-etched replicas (Fig. 2.11). In meristematic cells this system is clearly a continuum extending from the *nuclear envelope*, throughout the cytoplasm to the surface membranes and even through the cell walls. In differentiated cells the ER may be less prominent and may be represented only by apparently discontinuous vesicles. The ER usually shows a characteristic pattern of structure for each cell type; the ER structure and the metabolic activities of the cell are closely linked. Thus, in cells active in protein synthesis there occur on the outer surfaces of the ER membranes (Fig. 2.9) and sometimes free in the matrix numerous particles

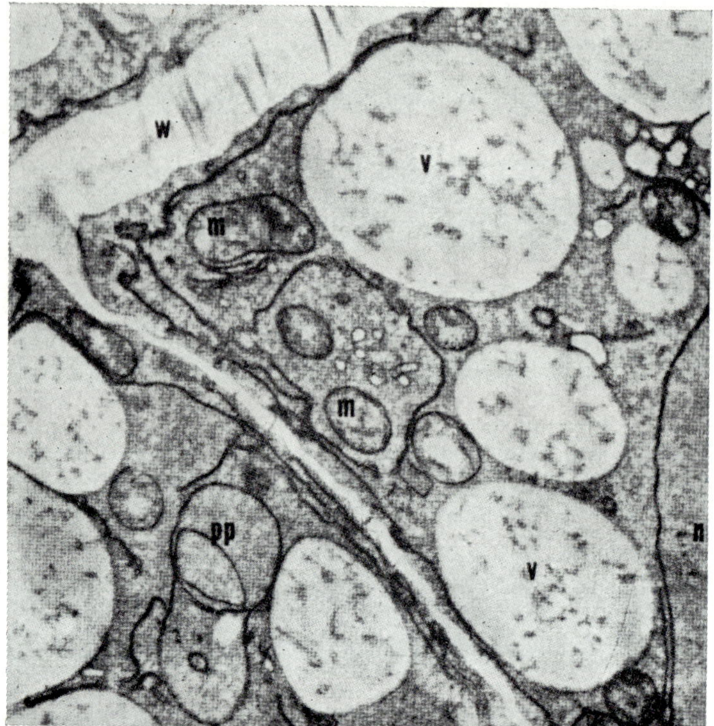

FIG. 2.8. EM picture of section of cells below the root promeristem where vacuolation is proceeding. Shows vacuoles (v), mitochondria (m), proplastid (pp), nucleus (n), and walls (w) with plasmodesmata. (Approx. ×20,000.) (From Whaley, Mollenhauer and Leech, 1960.)

(*ribosomes*), often about 150 Å in diameter and rich in ribonucleoprotein (RNA-protein). These ribosomes, together with fragments of the ER are collected as the *microsome* fraction when a suitably prepared tissue homogenate is first cleared by centrifuging at 10,000 *g* (to sediment mitochondria and larger cell inclusions) and then submitted to a much higher centrifugal force (about 100,000 *g* for 1 hr or more). We shall refer to the metabolic activity of this microsome fraction in Chapter 5. By contrast, in differentiating cells in which

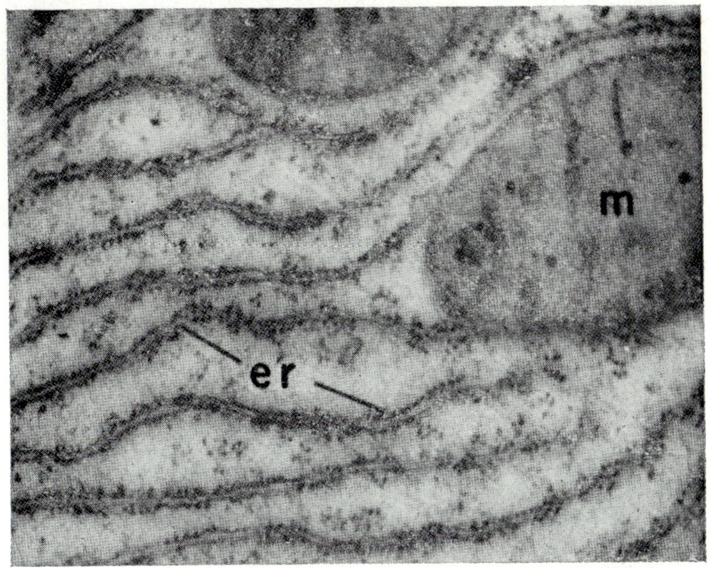

FIG. 2.9. Typical EM image of the endoplasmic reticulum (er) in which the outer surfaces are studded with microsomes. A few similar particles are seen in the cytoplasmic matrix. m, mitochondria. (Approx. ×40,000.) (From Keith R. Porter in *The Cell*, vol. 2, edited by J. Brachet and A. E. Mirsky, Academic Press, New York, 1961.)

the dominant metabolic activity is the synthesis of cell wall material the ER, particularly adjacent to the wall surface, will be relatively free of ER particles and the membranes appear quite smooth.

The ER membranes are rich in lipoprotein and about 5 mμ in thickness, each vesicle being bounded by a single membrane. The electron microscope reveals the nuclear envelope as consisting of two membranes of similar structure and the outer of these two membranes has been shown to be continuous with the membranes of the ER, so that the space (20–40 mμ in width) between the two membranes of the nuclear envelope is continuous with the space system of the ER (Fig. 2.10). The ER offers a great area of contact with the matrix of the cytoplasm which in turn has continuity into the nucleus via pores (20–40 mμ diam.) in the nuclear envelope.

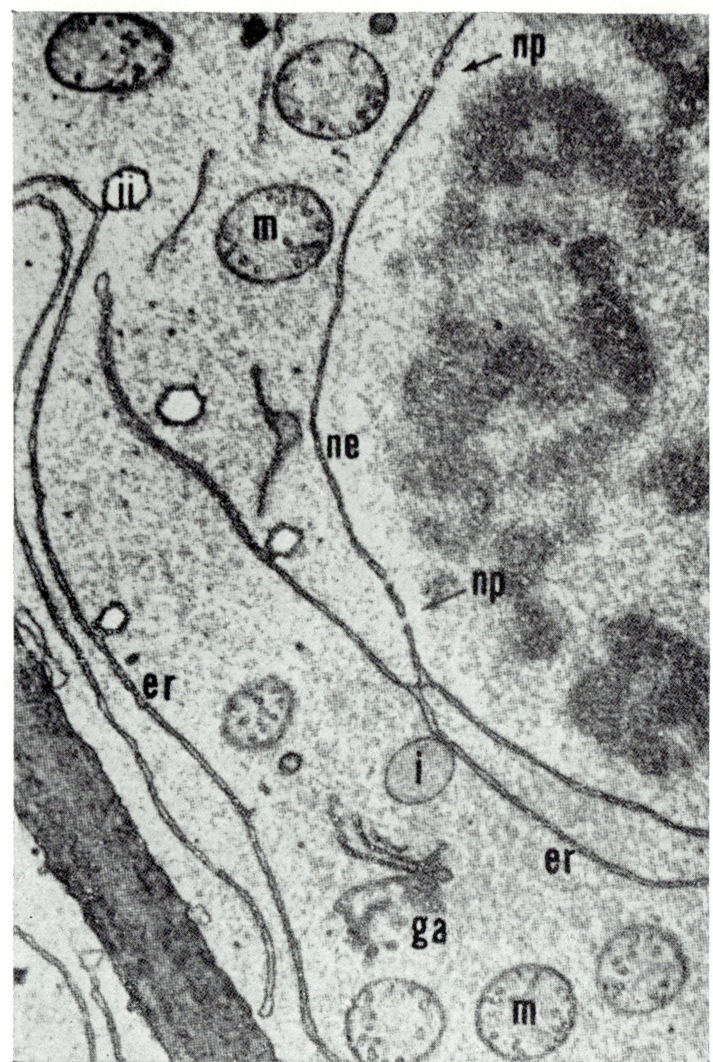

FIG. 2.10. Portion of a root meristematic cell as seen in the EM showing the double nuclear envelope (ne) with pores (np) and showing continuity of the outer nuclear membrane with the endoplasmic reticulum (er).
Key: m, mitochondria; ga, golgi apparatus; i and ii, inclusions. The cell wall runs across the left-hand bottom corner. (Approx. ×17,000.) (From Whaley, Mollenhauer and Leech, 1960.)

There is evidence of the involvement of the ER in protein synthesis (microsome particles) and in intra-cellular transport and if enzymes are built into the ER membranes then it may be a major site of cytoplasmic activity, the location of the "soluble" enzymes of the cell.

There can also be distinguished in meristematic cells, particularly in the region of the cytoplasm in which the cell plate forms, *Golgi structures* (*dictyosomes*) (Figs. 2.7 and 2.10) composed of a stack of plate-like sacks bounded by membranes like those of the ER and associated with a group of more of less spherical vesicles (Fig. 2.12). The metabolic role of the Golgi structures is uncertain, they are characteristic of meristematic cells, appear to increase in number during cell division and to disappear during the early stages of differentiation. Vesicles released from golgi bodies may be involved in the transport of polysaccharide materials to developing cell walls.

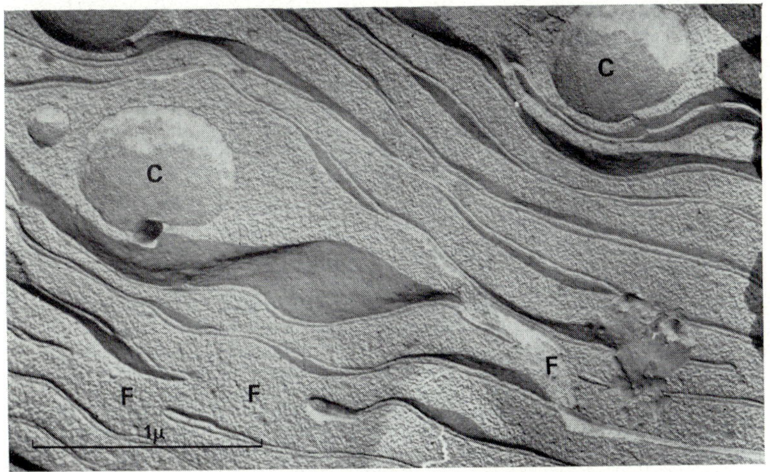

FIG. 2.11. Electron micrograph of a replica of a frozen and etched wheat (*Triticum*) root tip cell. Sheets of ER have been fractured and exposed both in transverse and surface view. Fenestrations in the ER are visible (F) and there are also casts of two unidentified cytoplasmic organelles (C). (From F. A. L. Clowes and B. E. Juniper, *Plant Cells*, Blackwell Scientific Publications, Oxford, 1968. Micrograph by Dr. D. H. Northcote.)

Sections of meristematic cells frequently reveal numerous small vacuoles, some of which appear to contain reservoirs of metabolites. As cells begin to expand these vacuoles become "empty" (presumably contain in life an aqueous solution) and are clearly enclosed in a membrane similar to the plasmalemma. The *plasmalemma* itself is seen in the electron microscope as a continuous membrane sometimes thrown into invaginations. The thickness of the plasmalemma is

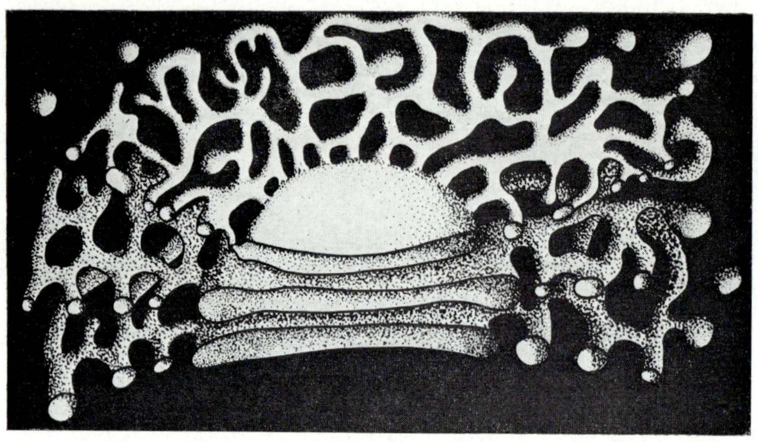

FIG. 2.12. Reconstruction of a Golgi-body cut across its diameter showing the fenestrated outer region giving off vesicles. (From F. A. L. Clowes and B. E. Juniper, *Plant Cells*, Blackwell Scientific Publications, Oxford, 1968.)

7–10 mμ and, at high resolution, is seen to consist of two electron dense layers separated by an electron transparent region. It has been suggested that the outer layer might consist of carbohydrate, the inner of protein and the central region of phospholipid. When golgi vesicles reach the plasmalemma and release their contents it seems that their membranes become incorporated into the plasmalemma. The plasmalemma may grow in area by this process; it may also arise during cell division from the membranes of the many golgi vesicles involved in cell plate formation. The vacuolar membranes (*tonoplasts*) are indistinguishable structurally from the plasmalemma, their origin is uncertain but they may be derived from the enclosing membranes of the endoplasmic reticulum.

The *mitochondria* are seen by electron microscopy (Fig. 2.13) to be enclosed by an outer membrane and within this is an inner membrane which is infolded to form across the central matrix

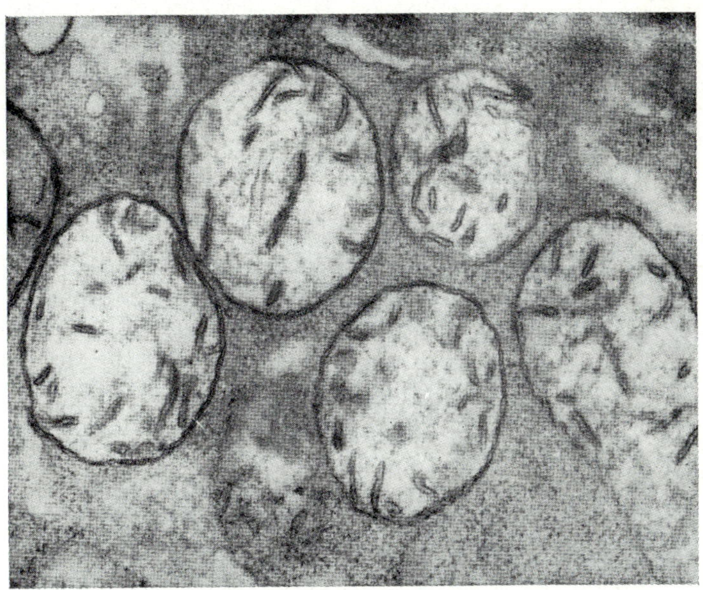

FIG. 2.13. Mitochondria of *Phaseolus aureus* hypocotyl as seen in the EM showing typical crista structure. (×40,000.) (Photo by Helgi Öpik, Swansea.)

baffles or *cristae* or to give rise to finger-like intrusions into the matrix (Fig. 2.14). The membranes are about 45 Å thick and the space between, in which the phospholipid content seems to be concentrated, is about 50 Å thick. The composition of dry mitochondria isolated by centrifuging is 40% protein, 25% phospholipid and 5% ribonucleic acid (RNA). When isolated mitochondria are burst, about 60% of the protein is released in soluble form suggesting that the central matrix contains soluble protein. The mitochondria are distributed in the matrix of the cytoplasm and are often observed in contact with the ER or nuclear envelope. The number of mitochondria present in each cell is very variable but many living plant tissues contain 500–800 per cell. When cells are starved they shrink and

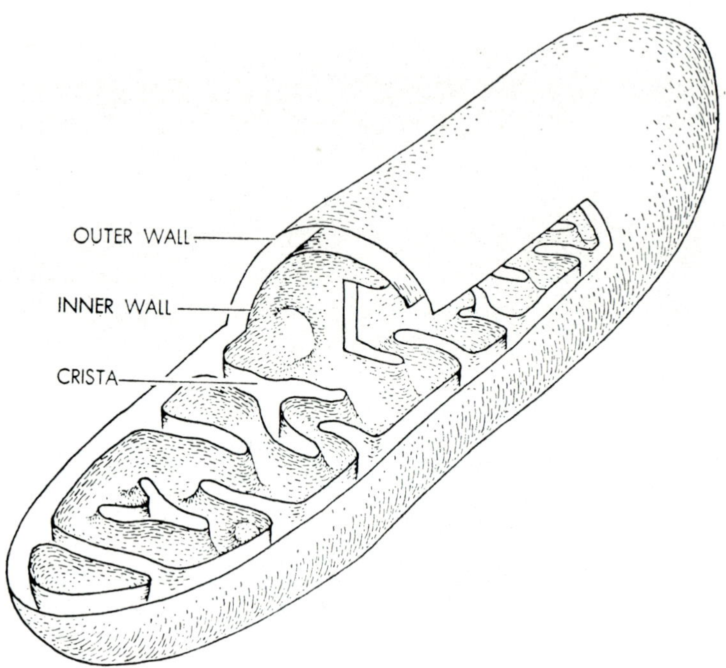

FIG. 2.14. Diagrammatic representation of the structure of a mitochondrion. (After A. L. Lehninger, *Scientific American* Reprint 91, Sept. 1961.)

when cells age they become disorganised. There is strong evidence that mitochondria can divide and the new mitochondria then grow to full size. Mitochondria contain deoxyribonucleic acid (DNA) and ribosomes and can effect protein synthesis. These mitochondria are centres of respiratory activity in the cell; when they are disrupted most of their metabolic activity is lost. The surface of the membranes of the cristae in contact with the matrix is covered with stalked granules (8·5 mμ in diameter) termed *oxysomes* which are thought to be the location of enzymes involved in electron transport (see Chapter 4, p. 115).

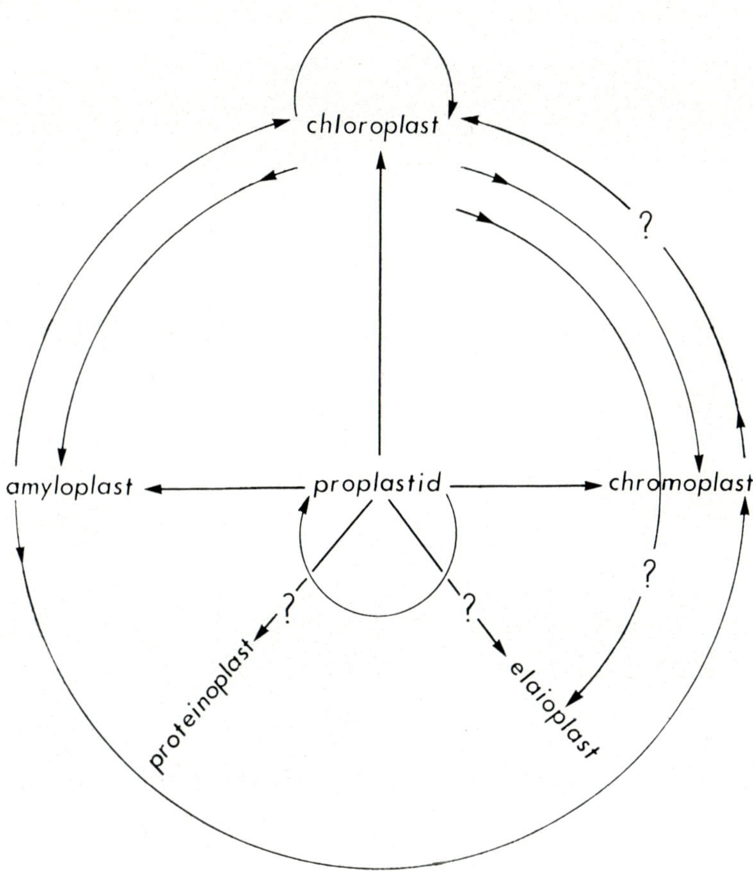

FIG. 2.15. The interrelationships of plastids. Only proplastids and mature chloroplasts are known to divide. (From F. A. L. Clowes and B. E. Juniper, *Plant Cells*, Blackwell Scientific Publications, Oxford, 1968.)

The *amyloplasts* (starch forming plastids) (Fig. 2.7), chromoplasts and chloroplasts arise from structures, *proplastids* (Fig. 2.8), similar in size to mitochondria but not readily stained with Janus Green B and containing internally an array of granules or vesicles (100–250 Å diam.) rather than defined cristae. In the cells of higher plants these proplastids are transmitted from cell to cell, increasing

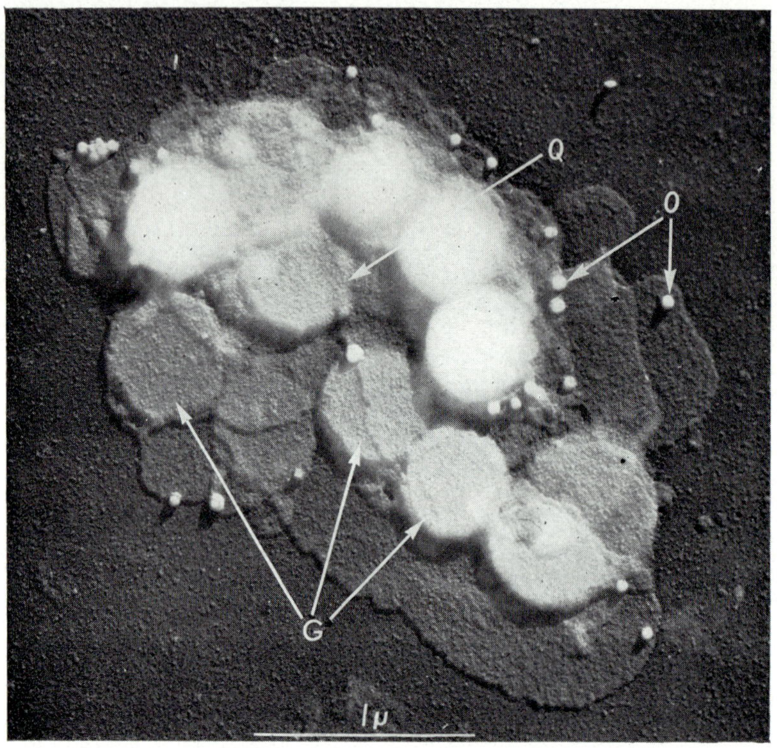

FIG. 2.16. Electron micrograph of a shadowed fragment of a *Vicia* leaf chloroplast. Note the appearance of the grana (G). On the surface of the granal lamellae can be seen small particles (Q) which may be identical to those seen in freeze-etched preparations. Osmiophilic globules (O) are also present. (From F. A. L. Clowes and B. E. Juniper, *Plant Cells*, Blackwell Scientific Publications, Oxford, 1968. Micrograph by Mr. A. D. Greenwood.)

in numbers by division and give rise to the functional plastids during cell differentiation. Chloroplasts are also capable of division. The interconversions which can take place between various plant plastids and their origin from proplastids is shown diagrammatically in Fig. 2.15. Proplastids and the plastids to which they give rise are like mitochondria rich in lipo-protein and protein and have a small content of RNA and DNA.

The *chloroplast*, the centre of photosynthetic activity, contains the two chlorophylls *a* and *b* (usually in a ratio of about 3:1) together with the accessory carotenoids and xanthophylls. A typical analysis of dry chloroplasts is protein 40–50%, phospholipid 23–25%, nucleic acids about 5%, chlorophylls *a* and *b* 5–10%, carotenoid 1–2%. The chloroplasts of higher plants are shaped like a plano-convex lens, approximately 5 $m\mu$ in diameter and 2–3 $m\mu$ thick. They lie in the cytoplasm with their broad sides parallel to the cell wall. Both the light and the electron microscope (Fig. 2.16) reveal the presence within the chloroplast membrane of a *stroma* (colourless) and disc-like *grana* (green). The electron microscope examination of thin sections has revealed details of the internal structure (Figs. 2.17, 2.18). The outer membrane is double-layered (each layer 35–50 Å thick) and within this is the proteinaceous stroma which contains minute granules (50–250 Å diam.), larger osmium-staining droplets, starch grains ("assimilation starch") and ribosomes. Embedded in this stroma is a double-membraned lamellar system (the thylakoid system) extending the width of the plastid and consisting of a system of interconnecting sacs (Figs. 2.18, 2.19). Some of the thylakoids spread throughout the whole plastid giving the intragranal or stroma lamellae. Others are smaller disc-like sacs (grana lamellae) stacked one above another to give the grana (the number of sacs per granum varies with species from 10 to 200). The whole thylakoid system is a single complex membrane bound cavity but the chlorophylls associated with the protein membranes and the intervening lipid molecules are both concentrated in the grana. The number and size of the grana per chloroplast varies with species. In the chloroplasts of spinach there are, for instance, 40–60 grana, each 6000 Å in diameter and 800 Å thick. Present evidence allows us to postulate that the photochemical reactions of photosynthesis

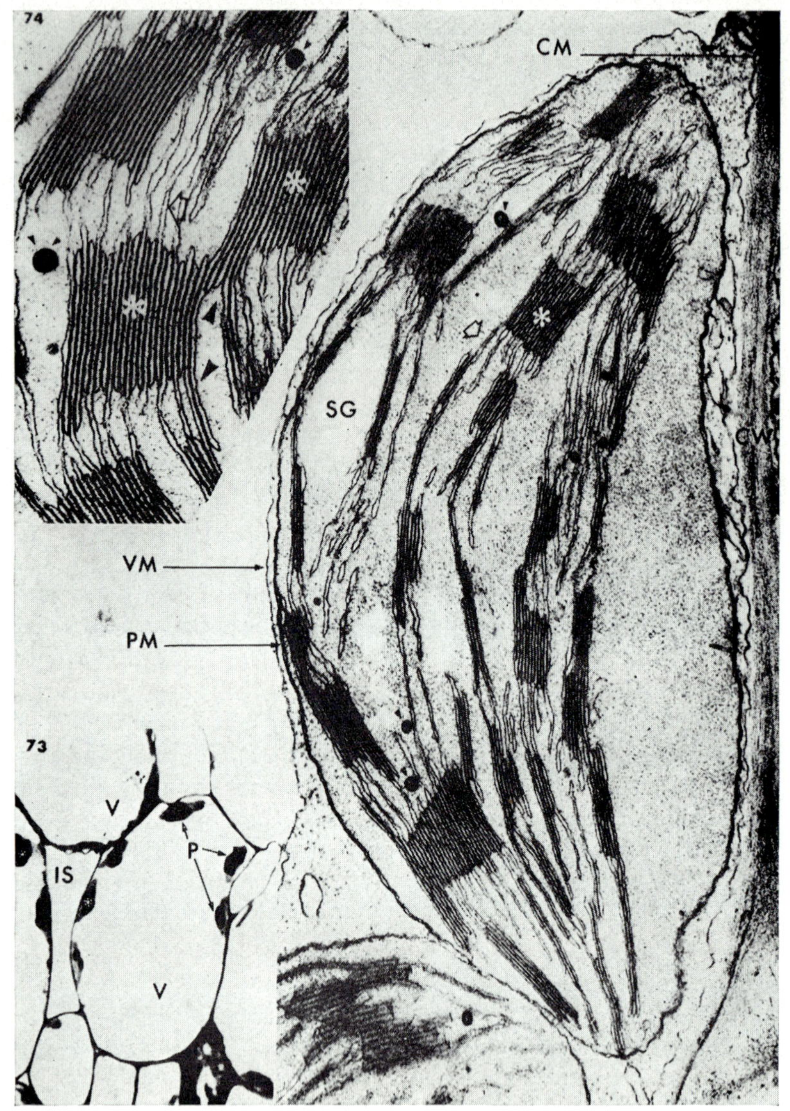
74
CM
SG
CW
VM
PM
73
V
P
IS
V

and the consequent photodecomposition of water and the production of oxygen occur in the thylakoid system, while the reactions whereby carbon dioxide is converted to starch occur in the stroma regions of the plastid. Granular structures (upper surface *ca.* 300 mμ^2) have been observed on chloroplast lamellae and termed *quantasomes*. These have been interpreted as physical expressions of a photosynthetic unit (see Chapter 5, p. 158), but their function remains uncertain.

The resting or interphase *nucleus* in electron micrographs shows the nuclear envelope, *chromatin* areas, *nucleoli* and intervening "structureless" *nucleoplasm* (Figs. 2.7 and 2.10). Chemical analyses of whole nuclei show the presence, on a dry weight basis, of 70% protein, 3–5% phospholipid, 10% deoxyribonucleic acid (DNA) and 2–3% ribonucleic acid (RNA). The chromatin material, as might be expected, is rich in DNA (30–50% dry weight) and in basic proteins (proteins rich in basic amino-acids). The nucleoli are also rich in protein and contain nucleic acid, probably entirely RNA. The RNA of the nucleus differs in base composition from that of the cytoplasm (it has, for instance, a higher adenylic acid content). A number of enzymes have been detected in nuclei although these do not include the respiratory enzymes. There are no enzymes which occur only or even mainly in the nucleus although some enzymes

---

FIG. 2.17. *Bottom left.* Chloroplasts (P) in chlorenchyma of pea leaf. Note the large starch grains within the chloroplast (asterisks). IS, intercellular space; V, vacuole. × 780. *Right.* Electron micrograph of a chloroplast of a leaf of spinach (*Spinacea oleracea*). The chloroplast is surrounded by a double membrane (PM) and the internal membrane system is differentiated into grana (asterisks) and stroma lamellae (open arrows). Osmiophilic droplets (small black arrows) occur in the plastid stroma. The structure of the grana is shown in more detail in the inset (*top left*) as are regions of continuity between the grana and stroma lamellae (large solid arrows). Key: CM, cell membrane; CW, cell wall; SG, starch grain; VM, vacuolar membrane. ×24,000; inset, ×43,600. (From T. P. O'Brien and M. E. McCully, *Plant Structure and Development,* Macmillan, London, 1969. Micrograph by Mr. A. D. Greenwood.)

(e.g. DNA-ase) may be very active in the nucleus. Although the nucleus is relatively metabolically inert by comparison with the cytoplasm, nevertheless, the maintenance of active metabolism in the cytoplasm is dependent upon the presence of the nucleus. Enucleated cytoplasm more or less quickly becomes metabolically disorganised. We shall return again to this question of the nuclear-cytoplasmic relationship when considering protein synthesis, and particularly the synthesis of enzyme proteins.

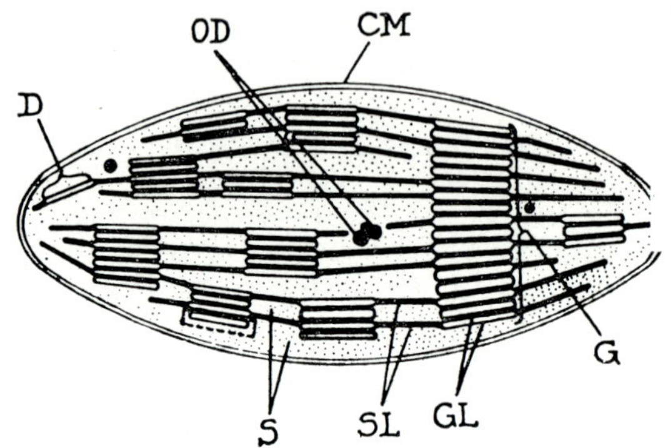

FIG. 2.18. Interpretation of the structure of a barley chloroplast (as von Wettstein).
Key: CM, chloroplast double membrane (each 35–50 Å thick); S, stroma; G, granum (made up of a cylindrical pile of discs); SL, stroma lamella; GL, grana lamella; D, disc; OD, osmiophilic droplet.
(From S. Granick, in *The Cell*, vol. 2, p. 501, as Fig. 2.9.)

Modern knowledge of *cell wall* structure comes from detailed chemical analyses, X-ray diffraction studies and work with the electron microscope. The dry primary cell wall contains hemicelluloses (up to 50%), cellulose (up to 25%) and smaller amounts of pectic substances, fats and protein. Hemicelluloses, cellulose and pectins are all polysaccharides with molecules built up from the linking together in chains of sugars or uronic acid residues. The morphology of cell walls is, however, *not* destroyed by extracting the hemicelluloses and pectins, indicating that the structural framework of the wall is

FIG. 2.19. A representation of the thykaloid system of a chloroplast of a higher plant based upon work of Heslop-Harrison (1963) and Wehrmeyer (1964). (From F. A. L. Clowes and B. E. Juniper, *Plant Cells*, Blackwell Scientific Publications, Oxford, 1968.)

built of cellulose. The molecules of cellulose consist of long chains of about 3000 glucose units and in the cell wall these molecules are associated parallel to one another to form *microfibrils*. In the microfibril the cellulose molecules are united laterally by weak chemical linkages (hydrogen bonds) (see Chapter 3, p. 63 and Fig. 3.2) and end-to-end by primary co-valent bonds. Each microfibril contains some 2000 cellulose molecules, has a diameter of 100–250 Å and a length of several microns. The cellulose molecules within the microfibril are much more closely associated in certain regions than in others. X-ray diffraction studies reveal that these regions of high association (the so-called *micelles*) are 50–60 Å wide (i.e. involve about

100 molecules parallel to one another) and about 600 Å long (Fig. 2.20).

The microfibrils are themselves associated into *macrofibrils* (these may involve up to 500 microfibrils) which can be seen with the light microscope, particularly after appropriate swelling treatment. Since primary cell walls may contain up to 95% of water, the structural

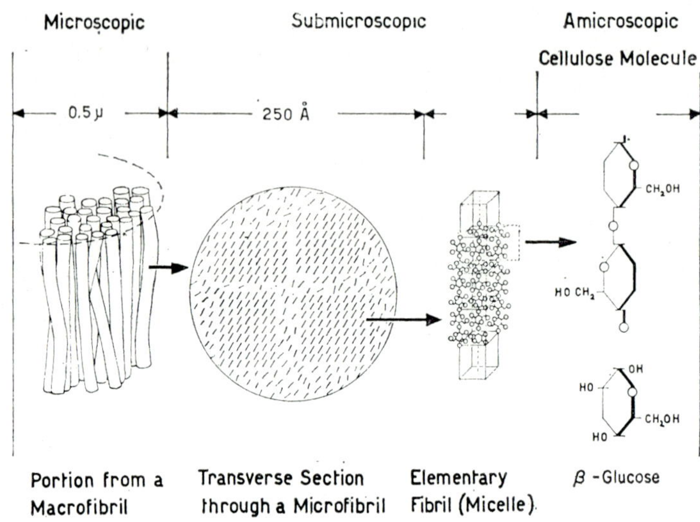

FIG. 2.20. Structural elements of cellulose involved in cell wall architecture. (From K. Muhlethaler, in *The Cell*, vol. 2, p. 85, as Fig. 2.9.)

framework of microfibrils is a very "open" one probably occupying only 2·5% of the wall volume.

The primary cell walls of the meristematic cells have their microfibrils orientated predominantly in a direction transverse to the axis of the root or shoot. When cell expansion occurs there is an increase in cell wall material, a synthesis of cellulose and other cell wall constituents. Expansion at the outer surface of the cell wall causes stretching in the direction of most active cell extension and as this happens the microfibrils become re-orientated more and more along the longitudinal axis of growth. At the same time new layers of wall

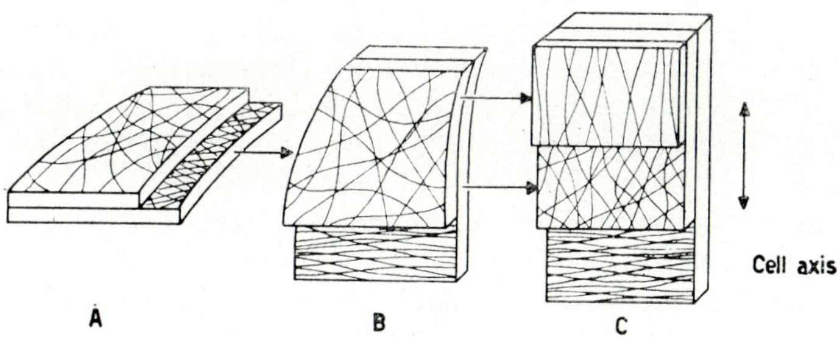

FIG. 2.21. Multinet structure of the cell wall of the cotton hair showing the orientation of the cellulose microfibrils in the successive wall layers laid down during elongation. (After A. L. Houwink and P. A. Roelofsen, *Acta. Botan. Neerl.*, **3:** 385, 1954.)

are laid down on the inner face of the wall and with their microfibrils in the original transverse direction. These, in turn, undergo re-orientation as growth proceeds so that at the end of the cell expansion phase there is a transition outwards in the wall from transversely to longitudinally scattered microfibrils (Figs. 2.21 and 2.22). This type of cell wall growth is usually termed *multinet* growth and clearly

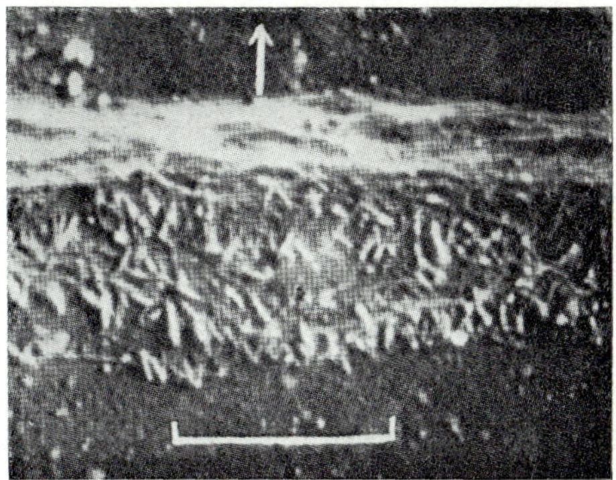

FIG. 2.22. Transverse section (EM) through the wall of epidermal cell of an onion root; the cell was situated about 1 mm from the tip. The arrow points towards the centre of the cell. The scale corresponds to 1 μ. (By courtesy of G. Setterfield and S. T. Bayley, *Canad. J. Bot.*, **35:** 435, 1957.)

involves the laying down of new wall material over the inner surface of the existing wall (growth by *apposition*). However, there is evidence, particularly from feeding expanding cells with sugars labelled with tritium (a radioactive isotope of hydrogen of mass 3), that although growth by apposition may sometimes account for the major increase in cell wall substance there is, nevertheless, some incorporation of new wall material throughout the whole thickness of the wall. Such incorporation of new wall substance into the existing framework is referred to as growth by *intussusception*.

Cell walls can continue to grow for a time after separation from contact with cytoplasm; cellulose synthesis must then occur *within* the wall. Certain electron micrographs have shown the presence within the cell wall of granular aggregations (cytoplasm ?) at which microfibrils appeared to terminate and these have been described as "islands of synthesis", possibly associated in some way with plasmodesmata and other extensions of the ER into the cell wall. Growth by intussusception could occur by growth of cellulose microfibrils from such centres into the intervening wall. It is, however, perhaps even more likely that cellulose and other wall polysaccharides can arise *in situ* and remote from such cytoplasmic islands by the action of soluble enzymes. Thereby, intussusception growth could occur without disturbing the orderly layered arrangements of microfibrils developed by apposition.

It should be emphasised that we do not yet fully understand the mechanism whereby cellulose, or hemicelluloses or pectins are synthesised nor do we yet know whether the plant hormones (auxins) promote cell expansion by some direct action on the association together or primary synthesis of cell wall constituents. Furthermore, we do not understand fully what factors determine the regular orientation of microfibrils as laid down in each wall layer. Particularly, during wall thickening in fully expanded cells, the orientation may differ from layer to layer (Fig. 2.23). The pattern for each layer seems to be determined by some pattern established at the cytoplasm-wall interface and which can be revealed prior to the actual development of the new layer of cellulose (Fig. 2.24).

Recently structures termed *microtubules* have been observed in many plant cells both within the mitotic spindle and in the peripheral

FIG. 2.23. Outermost lamellae of the wall of the alga *Valonia ventricosa*. EM picture at $\times 25{,}000$. (From R. D. Preston, *The Molecular Architecture of Plant Cell Walls*, Chapman & Hall, London, 1952.)

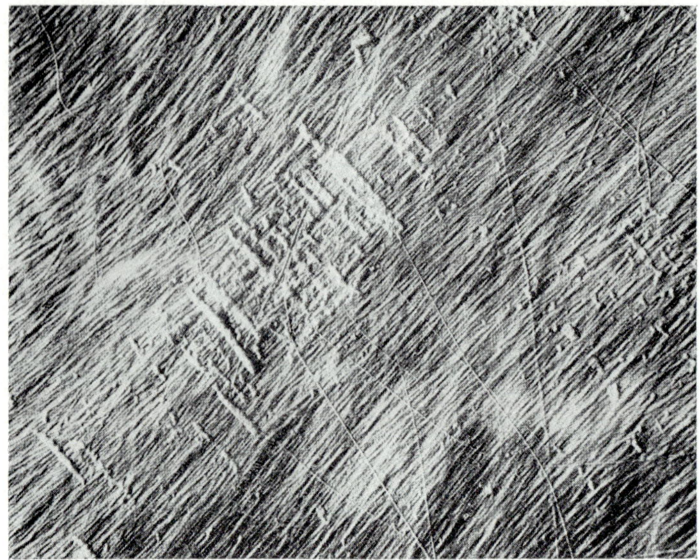

FIG. 2.24. Electron micrograph ($\times$24,000) of the innermost surface of the cell wall of the alga, *Chaetomorpha melagonium*, showing cytoplasmic remnants arranged in files at approximately right-angles to the existing microfibrils of the inner face of the wall, i.e. in the direction which the microfibrils will take up in the wall layer next to be deposited. Note the few microfibrils which would form part of this new layer. (From Eva Frei and R. D. Preston, *Proc. Roy. Soc.* B, **154:** 70, 1961.)

cytoplasm immediately within the plasmalemma (they approach within 17–20 mμ of this membrane). These peripheral cytoplasmic microtubules are 23–27 mμ in diameter and of undetermined length (individual tubules have been traced for several microns). In longitudinal section they appear as two parallel lines (each line 7 mμ thick and separated by a central space of 10 mμ diam.) (Fig. 2.25, A). In transverse section they are seen to consist of a number (13 ?) of subunits arranged in a circle (Fig. 2.25, B). It has been suggested that these microtubules exert an influence on the orientation of the newly synthesised cellulose microfibrils and a number of electron micrographs have been published showing a parallel orientation of microfibrils and peripheral microtubules in regions of rapid wall synthesis. However it must be borne in mind that the organisation of the microfibrils takes place external to the plasmalemma; there is therefore a spatial separation between the microtubules and the centres of cellulose synthesis. An alternative role for microtubules in cell wall synthesis may be that they direct the movement of Golgi vesicles and bring them into contact with the plasmalemma; if this is so they would be involved in the feeding of the matrix materials into the space between the structural microfibrils.

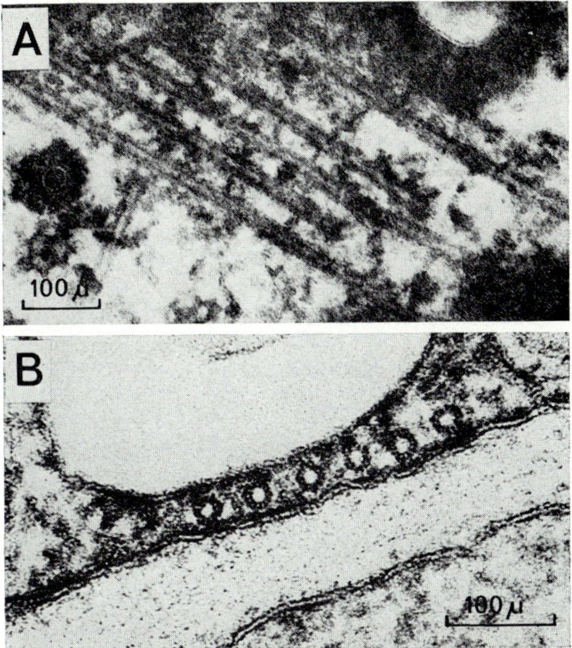

FIG. 2.25. Electron micrographs of microtubules. (A) In longitudinal section, from a meristematic cell of a root tip of *Juniperus*. (B) As seen in transverse section, from a cell of the nectary of *Euphorbia*. Each microtubule is made up of a number of more or less spherical sub-units which can be clearly seen. By courtesy of Myron C. Ledbetter, Brookhaven National Laboratory, and Blackwell Scientific Publications.

PROLOGUE TO CHAPTER 3

The living cell is the territory of metabolism; the reactions of metabolism occur each at their allotted place in this territory. Though the cell is "small" the territory of the cell in terms of molecules is vast and its pathways of metabolism (surfaces or membranes) are long and many in number. These are protein surfaces, and the catalysts of metabolism are the special proteins termed *enzymes*. Now if enzymes are built into membranes and these are distributed in patterns then we can begin to visualise the physical basis responsible for the organisation and regulation of chemical reactions in the cell. The rapid and efficient operation of enzyme systems with many enzymes participating in a regular sequence implies that the enzymes must be built into membranes in an orderly sequence, in the form of production lines. Studies of the fine structure of cells, therefore, seek to reveal the factory organisation and separate workshops in which metabolism proceeds. What of the production units—the individual enzyme proteins?

FURTHER READING

C. P. Swanson. *The Cell.* Prentice-Hall Inc., New Jersey, 1960.

J. A. V. Butler. *Inside the Living Cell.* Allen & Unwin, London, 1959.

J. Brachet. *The Living Cell.* Scientific American Reprint No. 90 (Sept. 1961). W. H. Freeman & Co., San Francisco.

T. P. O'Brien and M. E. McCully. *Plant Structure and Development.* Macmillan, London, 1969.

R. Buvat. *Plant Cells.* Weidenfeld & Nicolson, London, 1969.

MORE ADVANCED READING

H. Gordon Whaley, H. H. Mollenhauer and J. H. Leech. The ultrastructure of the meristematic cell. *Am. J. Bot.*, **47**: 401–449, 1960.

R. Brown. *The Plant Cell and its Inclusions* in *Plant Physiology*, vol. IA, pp. 3–129, edited by F. C. Steward. Academic Press, New York, 1960.

J. Brachet and A. E. Mirsky (Editors). *The Cell*, vol. 2, *Cells and their Component Parts.* Academic Press, New York, 1961.

P. A. Roelofsen. *The Plant Cell-wall.* Gebrüder Borntragen, Berlin-Nickolasse, 1959.

F. A. L. Clowes and B. E. Juniper. *Plant Cells.* Blackwell Scientific Publ., Oxford, 1968.

T. A. Bennet-Clark. *Water Relations of Cells* in *Plant Physiology*, vol. II, pp. 105–192, edited by F. C. Steward. Academic Press, New York, 1959.

CHAPTER 3

Enzymes—The Catalysts of Metabolism

"*It should be emphasised that the results obtained have, for the most part, been concordant with those furnished by isotopic and genetic studies on intact cells. There is, therefore, good reason to believe that, in general, the reactions observed in isolated enzyme systems do reflect physiological events and are not artefacts of isolation.*"

Bernard D. Davis, in *Enzymes, Units of Biological Structure and Function*. Academic Press, New York, 1966.

"*Enzymes are wonderful substances—they consist of wonderful molecules. We do not know much about these molecules as yet, . . . more must be done in the field of molecular structure . . . leading to the complete structural determination—the location of every atom—of the molecules of many proteins.*"

Linus Pauling, in *Enzymes, Units of Biological Structure and Function*. Academic Press, New York, 1956.

INTRODUCTION

THE CHEMICAL reactions which constitute cellular metabolism proceed at a temperature only fractionally above that of the natural environment and the pH of cytoplasm is close to neutrality, neither markedly acid nor alkaline. Rapid reactions occur in cells between compounds which are quite stable in mixed aqueous solution. This intense chemical activity in living cells results from the activity of numerous specific catalysts or enzymes. Such substances were first extracted, in an active form, from living cells by the Buchners in

1897. They made the discovery that when yeast is ground with sand, pressed and the expressed juice filtered, one obtains a clear liquid capable of rapidly fermenting sugar, of degrading sugar to carbon dioxide and ethanol. Fermentation, shown in Pasteur's elegant experiments to be the result of the activity of living cells is here

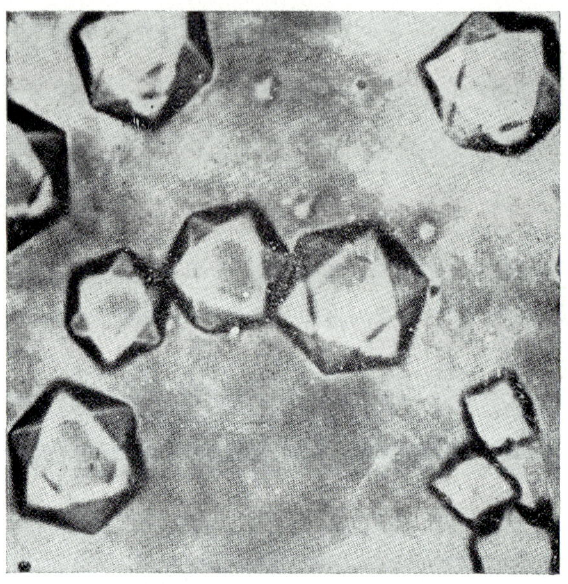

FIG. 3.1. Crystals of the enzyme urease. (From J. B. Sumner and G. F. Somers, *Chemistry and Methods of Enzymes*, Academic Press, New York, 1953.)

being effected by a clear solution of cellular origin. The ability of the yeast-press juice to promote fermentation is destroyed by heat or by rendering the liquid markedly acid or alkaline. Its catalytic activity is unstable.

During the present century such catalytic solutions derived from living cells have been submitted to intensive chemical study. From them have been obtained, in increasing purity and in increasing number, the catalytic agents or enzymes. A milestone in this progress was the obtaining of an enzyme in crystalline form, the enzyme *urease* isolated from jack beans by Sumner in 1926 (Fig. 3.1). This crystalline urease of Sumner was the first convincing demonstration of the protein nature of enzymes. This enzyme also illustrates the often high specificity of these biocatalysts. Urease promotes the hydrolysis of urea (hence its name) to ammonia and carbon dioxide. It is inactive towards substituted ureas and to all compounds except urea. Another example of enzyme specificity is the finding that enzymes which have one optical isomer as substrate are inactive to the other isomer. Thus, the enzyme which oxidises glutamic acid, the enzyme *glutamate dehydrogenase*, is active only towards the L-isomer of glutamic acid. Here we are considering the oxidation of glutamic acid to the optically inactive 2-oxoglutarate and ammonia. However, the reaction, in common with many biological reactions, is reversible and this enzyme, using 2-oxoglutaric acid as the carbon skeleton, can effect an asymmetric synthesis giving only L(+) glutamic acid free from any of the D(−) glutamic acid.

THE PROTEIN NATURE OF ENZYMES

The consideration just mentioned leads us to consider the nature of proteins which are themselves composed of units (residues) of the L-amino acids. Proteins are molecules of great size (with molecular weights ranging from 10,000 to several million.) When completely degraded by acid or enzyme hydrolysis they yield a mixture of amino acids. The number of different amino acids released and their proportions are characteristic of each protein. The questions therefore arise of how these amino acid units are linked together to form the protein molecules and of how far their molecular architecture explains their specific properties, including for the enzyme proteins their substrate and reaction specificity. Further, since enzyme proteins are giant molecules and the substrates of most enzymes are small organic molecules, it was anticipated and subsequently demonstrated that substrate specificity involves small specific active centres at which substrates are bound to their enzymes.

The α-amino acids have the general formula:

$$\begin{array}{c} H \\ | \\ R—C—COOH \\ | \\ NH_2 \end{array}$$

The α-carbon atom is bound to four different groups or atoms (hence the optical activity of all the α-amino acids except glycine where R = H). The R group is characteristic of each amino acid and may be aliphatic or aromatic and may contain only C and H atoms or may be basic (containing $—NH_2$ or =NH group(s)) or acidic (containing the —COOH group) or sulphur-containing. The α-amino group is basic, ionising thus:

$$—NH_2 + H_2O \rightarrow —NH_3^+ + OH' \quad (2a)$$

the α-carboxyl group is acidic, ionising thus:

$$—COOH \rightarrow —COO' + H^+ \quad (2b)$$

This means that amino acids are amphoteric compounds and can exist as zwitterions (ions carrying both positive and negative charges). At acid pH values the amino acid behaves as a base and is positively charged, at alkaline pH values it behaves as an acid and is negatively charged. At some intermediate pH (the iso-electric point) it has no net charge.

The amphoteric nature of α-amino acids is involved in their union together to build up the giant molecules of protein. The primary linkage is one formed by condensation of the α-amino group of one amino acid with the α-carboxyl of a second amino acid. The linkage is called a peptide bond and its formation can be represented thus:

$$\begin{array}{ccc} —COOH & & HNH— \\ \downarrow & & \\ —CO—NH— & \rightarrow & + H_2O \end{array}$$

peptide bond

In protein molecules, amino acids are arranged in chains (polypeptide chains) by such peptide bonds.

The carbon atom has four valencies and these make fixed angles with one another. The amino acids involved in polypeptide structure are all of the L series (have the same configuration as the reference compound, L-glyceraldehyde). This means that the polypeptide chain when represented in the single dimension of this page has the form:

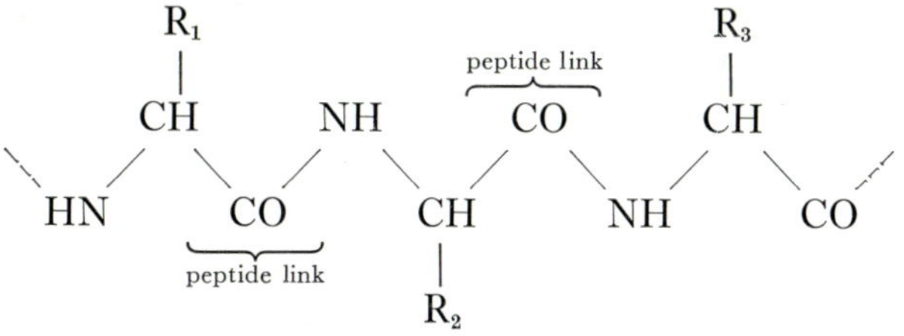

```
              R1                                        R3
              |                 peptide link            |
              CH        NH          CO                  CH
    \       /    \    /    \      /    \      /    \      /
      HN          CO          CH          NH          CO
               peptide link   |
                              R2
```

There is a zigzag backbone and the R groups stick out from the plane of the backbone above and below, alternately.

To understand how these polypeptide chains are arranged in a protein molecule it is necessary to consider the valency of hydrogen. When hydrogen loses an electron to become a proton or hydrogen ion or when it gains an electron to form a co-valent bond, it shows its primary valency of one. But the hydrogen atom can show a weak secondary valency (exhibit a co-ordination number of two) by accommodating four "valency electrons" in its shell. This secondary valency of hydrogen can be satisfied by sharing of a "lone pair" of electrons with an oxygen or a nitrogen atom and the bond so established is called a *hydrogen bond* (Fig. 3.2). Such hydrogen bonds are important in the building up of macromolecules such as those of protein.

Now the polypeptide chain depicted above is in the extended or β-form and in protein films (and possibly in the protein monolayers of cell membranes) this form can be stabilised by hydrogen bonding between the carbonyl ($>C=O$) and imide ($HN<$) groups of the backbones of parallel orientated polypeptide chains to give a "pleated-sheet" structure (Fig. 3.3(A)). This is the structure of several fibrous proteins, e.g. β-keratin. From the "primary" polypeptide unit of structure, a "secondary" structure has arisen.

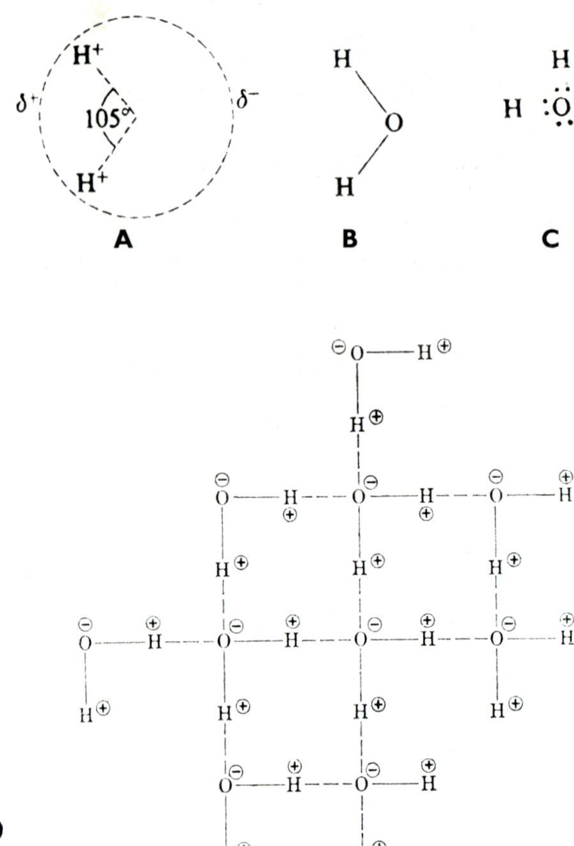

FIG. 3.2. The hydrogen bond. A. Water molecule represented as two protons embedded in an oxygen atom. B. Conventional representation of the water molecule. C. "Electronic formula" for water. D. Association between water molecules by hydrogen bonding (diagrammatic and shown as in two dimensions).

However, another kind of "secondary" structure can arise because in the polypeptide chain there can be rotation around the bonds of the α-carbon atoms of the amino acid residues so that the chain can become thrown into a complex of coils. However, this coiling in nature is apparently not entirely random but is ordered and stabilised

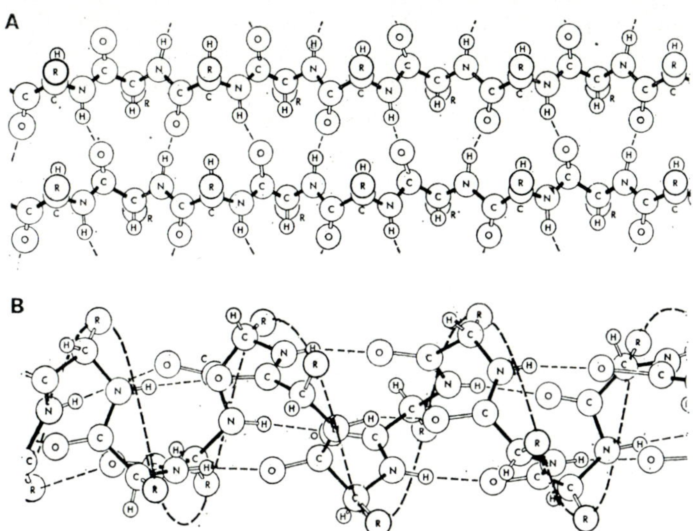

FIG. 3.3. Secondary structure of proteins. A. Polypeptide chains in the β-form associated to give a "pleated sheet" structure. B. Right handed α-helix structure of the polypeptide chain; side chains of amino acid residues labelled R. Bold broken line traces the turns of the helix. Hydrogen bonds involved in associating β-chains and in stabilising the α-helix as less bold broken lines (between oxygen and hydrogen atoms). (From Paul Doty, Scientific American Reprint, No. 7, Sept. 1957.)

by hydrogen bonding. Thus the polypeptide backbone may be thrown into a regular spiral held rigid by hydrogen bonding between the imide group of one amino acid residue and the carbonyl group three residue distances along the chain, and this occurs regularly down the spiral. The spiral or helix could, theoretically, be either right-handed or left-handed (Fig. 3.3(B)) but the L-configuration of the amino acids favours and leads to the *right-handed* helix being that present in protein molecules. The amino acid proline does not fit into this helix and where it occurs there is a break in the regular structure. It is this kind of "secondary" structure, involving folding of the polypeptide chains which seems to be of particular importance in the specific physical and biological properties of many proteins.

It is hoped that the structure of the protein α-helix can be visualised from the description of its origin which has been given and from Fig. 3.3(B). The value of using a modern set of "atom models" to build up such a structure cannot, however, be overemphasised. This is the only really satisfactory introduction to thinking about molecular architecture in three dimensions.

Now the α-helix is an elongated structure; a single helix containing say 300 amino acid residues would have a length of about 400 Å and a diameter of 10 Å. A number of fibrous proteins have α-helix structure (e.g. α-keratin). Since, however, there is evidence of α-helix structure in proteins whose molecules are globular it follows that short α-helices must be linked together by intervening folds of the polypeptide chains. Often 20–50% of the total structure is in α-helix configuration. Stabilisation of structure seems then to depend upon an association between the R groups of separate helices, and it is from such linkages between R groups that there arises the "tertiary" structure of protein molecules. The R group of the amino acid, cysteine, is $—CH_2—SH$ and two such groups can combine together, with the elimination of hydrogen to form the co-valent bond known as a disulphide linkage (—S—S—). Such disulphide linkages are very important in stabilising "tertiary" structure and the destruction of such bonds (for instance, by reduction) in enzyme molecules usually leads to loss of biological activity. This is so, for instance, in the enzyme, *ribonuclease*. "Salt" linkages can also arise by association between acidic R groups (those of aspartic and glutamic acids) and basic R groups (those of arginine and lysine) and "hydrogen bonds" can form between oxygen- and nitrogen-containing R groups. Metal ions may also form co-ordinative complexes with R groups and thereby link the α-helices together. The calcium present in the enzyme *amylase* may serve this function.

Where the protein molecule is built up of more than one distinctive polypeptide chain, its polypeptide composition is referred to as its quaternary structure (as exemplified by lactate dehydrogenase, p. 68).

The structure of a few enzymes has now been more or less completely worked out, their active centres identified and new insight obtained into the mechanism of their catalytic activity (e.g. ribonuclease, lysozyme). The enzyme *lysozyme* is present in tears, nasal mucus and in egg white (a rich source). It is so named because it lyses (dissolves) certain air-borne cocci (bacteria) by hydrolysing the characteristic polysaccharide of their walls. It has, compared with many enzymes, a low molecular weight (14,400) and consists of a single polypeptide chain of 129 amino acids cross-linked at four points by disulphide bridges and additionally stabilised in its folding by numerous hydrogen bonds. By determining the sequence of amino acids along the polypeptide chain (20 different amino acids are involved) and by X-ray crystallography of the pure enzyme, of the enzyme modified by the introduction of heavy atoms and of the enzyme in combination with a trisaccharide (a tri-N-acetylglucosamine which resembles the enzyme's substrate sufficiently to combine with it) it has been possible to construct a complete model of the enzyme molecule. This model has a number of features which are probably generally descriptive of globular enzyme proteins. Certain regions of the polypeptide chain show α-helix configuration (Fig. 3.3, B) although the helices are slightly distorted so that the carbonyl (CO) groups of the peptide bonds point outwards from the helix and the amino (NH) groups point inwards. In other sections, lengths of the polypeptide are associated in the "pleated-sheet" structure (Fig. 3.3, A). The "polar" side chains of the amino acids (side chains which are hydrophilic by nature of their O and N atoms and which hydrogen bond with water) are at the surface of the molecule. The prominent "polar" side chains are those of the acidic (aspartic and glutamic acid) and basic (lysine, arginine, histidine) amino acids. By contrast the "non-polar", hydrophobic amino acid side chains (those composed entirely or predominantly of C and H atoms) are directed towards the centre of the molecule. Although the overall shape of the molecule is globular it has two "wings" separated by a deep cleft running up one side and from the chemical groups present and their spacing in this region, it is clear that this is the active site of the molecule, the region where in a specific manner it combines with its substrate.

Certain enzymes exist in several different forms termed *isoenzymes* which are distinguishable by electrophoresis (mobility in an electric current) and by their immunochemistry. The different forms of the enzyme may occur in different tissues or alter in their relative proportions within the same tissue with time. The case which has been most studied is that of the five isoenzymes of *lactate dehydrogenase* which occur in mammalian tissues. Each isoenzyme has the same molecular weight (135,000) and is built up of four polypeptide chains of two kinds (α and β). The isoenzymes differ in their α, β make up thus: 4α, 4β, $1\alpha + 3\beta$, $2\alpha + 2\beta$, $3\alpha + 1\beta$. At least 33 different enzymes have been reported to exist as isoenzymes in plant tissues.

Isoenzymes of enzymes involved in the first step of a branched biosynthetic pathway may differ in their sensitivity to inhibitors. Thus the enzyme *aspartate kinase* catalyses in *Escherichia coli* the first step in the synthesis of lysine, methionine and threonine. Three isoenzymes occur, the synthesis and activity of one is suppressed by L-lysine, the activity of the second is depressed by L-threonine and the activity of the third by homoserine (an intermediate in methionine biosynthesis). Thus accumulation of L-lysine or L-threonine suppresses their own further synthesis but does not prevent the activity of the aspartokinase isoenzyme involved in methionine synthesis. Again peroxidase isoenzymes differ in their activity in destroying indol-3yl-acetic acid (IAA) and the pattern of peroxidase isoenzymes can be altered by feeding IAA or gibberellic acid (GA) to plant tissues.

The relative activities of isoenzymes often show marked changes during development and in certain cases particular isoenzymes may appear and disappear as growth and development proceed so that the complement of isoenzymes may be used to characterise a particular state of cell differentiation. Despite such observations the significance of isoenzymes is still in many instances not clear.

This discussion of protein structure reveals that the primary structure is built up from strong covalent peptide bonds but that weaker linkages, particularly "hydrogen bonds" are involved in the development of secondary, tertiary and quaternary structure. The actual destruction of protein molecules usually requires strong chemical action such as the application of strong acids or alkalis and a high temperature. However, aqueous solutions of proteins are very unstable in their physical properties and important changes in "native" proteins (proteins in the form in which they occur in living cells) usually occur at temperatures above 40°C or when the pH is decreased below 3 or raised above 9. The changed protein ("denatured" protein) has lost its characteristic biological properties; in the case of enzyme protein it has lost its catalytic activity. The process of "denaturation" involves changes in secondary and tertiary structure. The techniques which have been gradually developed by trial and error for the isolation of enzymes are designed particularly to obtain the enzyme in its "native" rather than its "denatured" condition.

PROSTHETIC GROUPS AND COENZYMES

The enzyme, urease, crystallised by Sumner, is a "simple" protein in the sense that its molecules are entirely built up from amino acid residues. Many enzymes are, however, "compound" proteins in that the active enzyme molecule is made up of a protein moiety combined with some other molecule or atom and it is to this non-protein moiety that we give the name *prosthetic group*. The simplest prosthetic group is a metallic atom as exemplified by certain oxidising enzymes which are copper proteins (e.g. *ascorbic acid oxidase*). Other prosthetic groups are much more complex. For example, the *cytochromes*, so important in respiration, are iron porphyrin proteins; the *flavoproteins*, also involved in respiration, have as their prosthetic groups riboflavin (vitamin B_2), mono- or dinucleotides; and the *transaminases* and *amino acid decarboxylases* have pyridoxal phosphate (a derivative of vitamin B_6), as prosthetic group. These prosthetic groups are certainly associated with the active centre of the enzyme molecule and in some cases we know that the prosthetic group undergoes a reversible chemical change as part of the mechanism of catalysis (see Chapter 5, p. 170). It is often very valuable in attempting to isolate an enzyme to submit the initial extract to dialysis (diffusion through a membrane of restricted pore size). The dialysis serves to separate the large protein molecules (too large to pass through the dialysis membrane) from small soluble molecules and ions. This often leads to inactivation of the enzyme, not by causing protein denaturation but by removing some dialysable co-enzyme or co-factor essential for enzyme activity. The term *co-factor* is usually reserved for inorganic ions such as the chloride ions essential for the action of the α-amylases and the Mg^{++} required for the enzymes promoting phosphate-transfer to and from adenosine polyphosphate. The term *co-enzyme* is used to describe organic substances which, like prosthetic groups, undergo chemical change as part of the catalytic process; such co-enzymes differ from prosthetic groups only in being readily separated from the protein moiety by dialysis. Enzyme proteins freed from their co-enzymes (or prosthetic groups) are termed apoenzymes. The classical example of a co-enzyme is co-zymase (also referred to as nicotinamide adenine dinucleotide (**NAD**), co-enzyme I or diphosphopyridine nucleotide, see Fig. 4.6), the detection of which by the biochemists, Harden and Young, will be discussed in Chapter 4. Another important example is co-enzyme A to which reference will again be made, both in regard to the respiratory breakdown of sugars and in regard to fat synthesis and degradation.

The cell constituent (metabolite) acted upon by an enzyme is usually termed its *substrate*, or where two cell constituents are caused to interact we talk of their being substrates. Enzymes are named either after the particular substrate upon which they act, e.g. *urease* whose specific substrate is urea, or according to the nature of the reaction they promote, e.g. for phosphate-transferring enzymes the term *transphosphorylases*, or by a combination of both substrate and reaction-type, e.g. *triose phosphate dehydrogenase* (glyceraldehyde-3-phosphate dehydrogenase) which catalyses the removal of hydrogen from D-3-glyceraldehyde phosphate. The distinction between substrate and co-enzyme is often not immediately clear. Thus, *2-oxoglutarate dehydrogenase* oxidises this acid in presence of the two co-enzymes, NAD and co-enzyme A (represented in the equation as HS—CoA: see Fig. 4.6) thus:

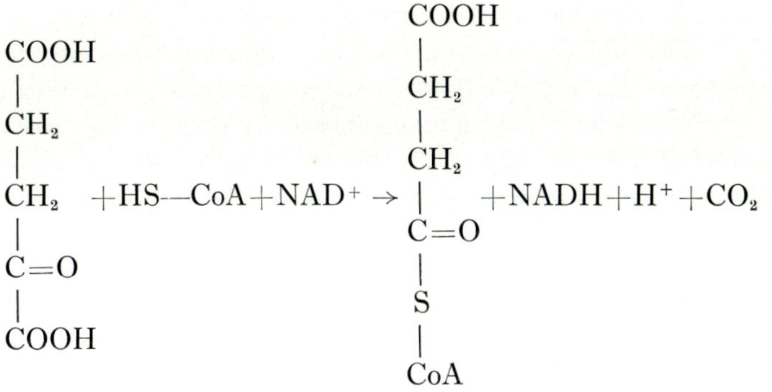

$$
\begin{array}{l}
\text{COOH} \\
| \\
\text{CH}_2 \\
| \\
\text{CH}_2 \\
| \\
\text{C=O} \\
| \\
\text{COOH}
\end{array}
\quad + \text{HS—CoA} + \text{NAD}^+ \rightarrow
\begin{array}{l}
\text{COOH} \\
| \\
\text{CH}_2 \\
| \\
\text{CH}_2 \\
| \\
\text{C=O} \\
| \\
\text{S} \\
| \\
\text{CoA}
\end{array}
\quad + \text{NADH} + \text{H}^+ + \text{CO}_2 \qquad (3)
$$

2-oxoglutarate succinyl–CoA

The NAD is reduced, carbon dioxide is liberated and the product of the reaction is a compound between succinic acid and co-enzyme A. Now this succinyl–CoA is further transformed by a second enzyme system, the overall reaction of which is

$$\text{succinyl-S.CoA} + \text{ADP} + \text{H}_3\text{PO}_4 \rightarrow \text{ATP} + \begin{array}{c}\text{succinic}\\ \text{acid}\end{array} + \text{CoA-SH} \qquad (4)$$

Here then the distinction between substrates and co-enzymes can only be made in terms of the quantities involved. These enzyme systems oxidatively decarboxylate a large amount of 2-oxoglutaric acid to succinic acid. The small amount of co-enzyme A acts over and over again and the small amount of co-enzyme I (NAD) is constantly being re-oxidised by another set of enzymes and so acts over and over again to remove hydrogen from the 2-oxoglutarate.

ENZYME ACTIVITY

Enzymes are catalysts: they accelerate the reactions of living cells. Our understanding, imperfect as it is, of this catalytic activity of enzymes is founded upon the study of the kinetics of the reactions *in vitro*. The isolated enzyme and its substrate(s) are allowed to interact under controlled conditions.

Firstly, the rate of an enzyme catalysed reaction is influenced by the concentration of the substrate. With a fixed amount of enzyme

(E) there is, with increase in substrate (S) concentration, an increase in the velocity of the reaction until a "saturating" concentration of substrate is reached when the reaction proceeds with maximum velocity (V). From a quantitative study of the relationship between S and reaction rate (v) it has been concluded that many enzymic reactions involving a single substrate can be represented thus:

$$E + S \rightleftharpoons ES \xrightarrow{k} E + \text{reaction product} \qquad (5)$$

where ES is an enzyme–substrate complex or compound and k is the velocity constant of the reaction whereby the complex breaks up to give free enzyme and reaction product. The theoretical concept of ES is supported by independent direct evidence particularly from absorption data in the ultra-violet to indicate the formation of a compound between enzyme and substrate. In the reaction sequence above it is the velocity constant (k) which determines the maximum velocity (V) and this is achieved when all the enzyme is maintained as ES by the saturating concentration of S. The affinity of the substrate for its enzyme can be expressed as a dissociation constant of the ES complex and, in the simple case, this constant—the Michaelis constant (K_m)—can be determined because it equals the substrate concentration at which the rate of reaction is $\frac{1}{2}V$ (half the maximum rate) (Fig. 3.4).

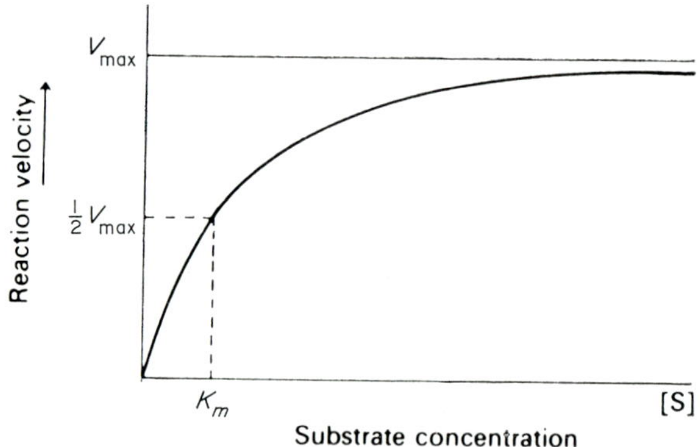

FIG. 3.4. Relationship between the rate of an enzyme catalysed reaction and the concentration of the substrate of the enzyme.

Enzyme-catalysed reactions like all thermochemical reactions increase in rate with rise in temperature. Below temperatures where denaturation of the enzyme begins to occur, enzymic reactions usually have a temperature coefficient (Q_{10}) between 2 and 3, i.e. the rate rather more than doubles for each 10°C rise in temperature.

The rate of all enzymic reactions is influenced by pH; most enzymes have a well defined pH optimum (pH at which they are most active). This presumably occurs because pH determines the ionisation of the oxygen and nitrogen-containing R groups of the amino acids at the surface of the protein molecule and a particular state of ionisation of such groups at the active centre of the enzyme is necessary for peak activity. pH may have two other effects. If sufficiently high (alkaline) or low (acid) it may denature the enzyme protein. It may also alter the position of equilibrium of the reaction. This can be exemplified by the reactions catalysed by the dehydrogenase enzymes which have a pyridine nucleotide as co-enzyme. The oxidised form of the co-enzyme, nicotinamide adenine dinucleotide, can be represented as NAD^+, the reduced form as NADH. The specific dehydrogenase which acts upon ethanol (the *alcohol dehydrogenase*) catalyses the following reversible reaction:

$$\underset{\text{ethanol}}{CH_3.CH_2OH} + NAD^+ \underset{}{\overset{\text{dehydrogenase}}{\rightleftharpoons}} \underset{\text{acetaldehyde}}{CH_3.CHO} + NADH + H^+ \quad (6)$$

H^+ is thus a product of the forward reaction and a change of pH of one unit is equivalent to a tenfold change in H^+ and will, therefore, alter the position of equilibrium since the equilibrium constant (K) is given by the equation

$$K = \frac{(CH_3CHO)\,(NADH)\,(H^+)}{(CH_3CH_2OH)\,(NAD^+)}$$

The study of the inhibitors of enzymes has also contributed to our understanding of how enzymes act. Certain enzyme poisons such as carbon monoxide, cyanide and diethyl dithiocarbamate have been shown to inactivate enzymes by virtue of their high affinity for metals. Thus the inhibition of *cytochrome oxidase* by CO and CN^- is because these combine with the iron of the haem prosthetic group. Similarly, it is the affinity of dithiocarbamate for copper which makes it a powerful inhibitor of *ascorbate oxidase*. A second group of inhibitors combine with or oxidise sulphydryl groups (—SH in cysteine) and the activity of many enzymes (the protein-degrading enzyme *papain*, the dehydrogenases and many others) depends upon some of the —SH groups in the protein being unsubstituted and reduced. The inhibitors which act in this way include iodoacetate, *p*-chloromercuribenzoic acid and the heavy metals (Cu^{++}, Hg^{++}).

The action of some inhibitors can be prevented or reversed by addition of more substrate; the extent of inhibition depends upon the ratio of inhibitor to substrate concentration. The classical example of such a relationship is the inhibition of the enzyme *succinate dehydrogenase* (this enzyme oxidises succinic acid) by malonic acid. The chemical similarity between substrate and inhibitor is clear from the formula of succinic and malonic acids (see page 104).

In this case we know that the malonic acid combines with the enzyme protein at the site normally occupied by succinic acid (to give an enzyme–inhibitor complex, *EI*) but is not oxidised. Further, the concentration of succinic acid required to displace the malonate depends upon the relative affinities of the enzyme for substrate and inhibitor. If K_s and K_i are the dissociation constants, respectively, of *ES* and *EI*, then in the presence of a given concentration of inhibitor (I), the "apparent K_s", in practical terms the substrate concentration required to give half the maximum velocity ($\frac{1}{2}V$) (the K_m or Michaelis Constant), is increased to the value $K_s(1 + (I)/K_i)$. Now this forms the basis of a practical test for this kind of competitive inhibition, a test of which we shall refer again when considering competition between ions in their uptake by plant cells (p. 222). This test depends upon the quantitative relationship between substrate concentration (S) and the velocity of the reaction (v). This is such that if we plot $1/v$ against $1/S$ (a double reciprocal plot) then we get a straight line with the intercept at $1/V$ (reciprocal of the maximum velocity) and of slope K_s/V. Then, if we had a competitive inhibitor the slope is altered to

$$K_s \frac{1 + (I)/K_i}{V}$$

but the intercept remains at $1/V$. This is in contrast to the action of a non-competitive inhibitor which does not alter the slope but alters the maximum velocity and hence the value of $1/V$ as given by the point of intercept (Fig. 3.5).

The activity of many enzymes has now been shown to be modified by interaction with molecules (effectors) often structurally unrelated

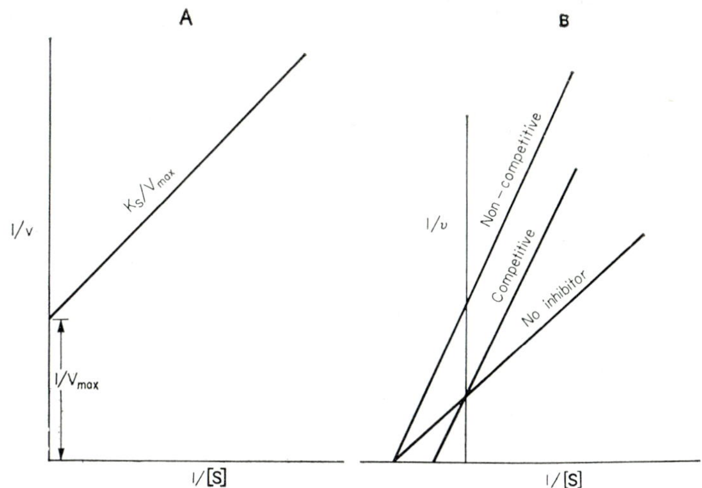

FIG. 3.5. Enzyme reaction kinetics. A. Plot of the reciprocal of the velocity of reaction ($1/v$) against the reciprocal of the substrate concentration ($1/[S]$). The value for the reciprocal of the maximum velocity shown. The slope corresponding to K_s/V_{max} where K_s is the dissociation constant of the enzyme-substrate complex. B. Similar plots to that in A but showing effect upon slope and value of the reciprocal of V_{max} of adding a competitive and non-competitive inhibitor.

to their substrates, the interactions occurring not at the active site of the enzyme but at some distant secondary *allosteric* site. It is because the effectors are not steric analogues of the substrate that this form of regulation of enzyme activity was termed allosteric by Monod and Jacob. When the effector activates the enzyme it is said to be positive in action, when it inhibits activity to be negative in action. The reversible binding of the effector to the enzyme alters the conformation of the enzyme molecule in particular at the active site, and in consequence either facilitates or impedes the formation of the enzyme-substrate complex. Further references to the allosteric regulation of enzyme activity will be found in Chapter 5, p. 164 and Chapter 8, p. 255.

HOW ENZYMES ACT

Most of the reactions which proceed in living cells are reversible reactions. A position of equilibrium is reached at which the forward and back reactions proceed at equal rate. Enzymes do not alter this position of equilibrium. What enzymes do is to speed up the reaction, they speed up equally both the forward and the back reaction.

In order to react, molecules must change from the passive to the "activated" state; the rate of reaction is determined by the number of molecules whose energy content renders them sufficiently activated. In a population of molecules all do not have the same energy content and by collisions between them this distribution is constantly changing. In a collision one molecule may have its energy increased at the expense of another molecule. If the temperature is raised the thermal agitation of the molecules is enhanced and a higher proportion will become "activated", in consequence the rate of reaction will increase. Some reactions have a high energy of activation, there is a high energy barrier impeding the reaction. Such reactions will only proceed at a measurable rate at high temperature. Many reactions known to occur in living cells will only proceed at high temperature in the absence of the appropriate enzyme. In the presence of the enzyme they proceed at temperatures compatible with life. What catalysts do, including enzymes, is to lower the energy of activation of reactions. The extra energy above the mean energy which the molecules have to acquire in order to react is reduced. Thus, it has been calculated that at 25°C the energy of activation of the acid hydrolysis of sucrose has the value of 25,560 cal/mol., whereas the energy of activation of sucrose hydrolysis in presence of the enzyme *β-fructofuranosidase* is at 25°C only 8700 cal/mol.

Any explanation, therefore, of how enzymes act must explain how they lower the energy of activation of the reaction, how they decrease the stability of their substrates or increase their potentiality to react with other molecules. Explanation of the remarkable specificity and activating properties of enzymes depends upon elucidation of the primary, secondary and tertiary structure of enzyme proteins and of enzyme-substrate compounds. This is a formidable task so far only achieved for a very few enzymes, such as lysozyme (see p. 67). The substrate of this is a polysaccharide the main chains of which are built up of amino-sugar residues; N-acetylglucosamine residues alternating regularly with N-acetylmuramic acid residues along the chains. The cleft in the lysozyme molecule can accommodate a length of six amino-sugar residues of the substrate and it has now been shown that a particular sequence of six residues is involved in the substrate enzyme interaction and that the enzyme catalyses a hydrolytic cleavage of one of the five glycosidic links between carbon 1 on a N-acetylmuramic acid residue and the oxygen of the link to carbon 4 of the adjacent N-acetylglucosamine residue. Combination of enzyme and substrate causes movement of parts of the enzyme molecule relative to one another in the region adjacent to the cleft and this change results in a distortion of one of the amino-sugar rings, the N-acetylmuramic acid ring attached to the glycosidic bond which will be cleaved. This indicates that the substrate chain is brought by its combination with the enzyme into a position of strain (its stability is decreased, it is activated). Two particular amino groups which will be involved in the hydrolytic reactions are immediately adjacent to the glycosidic bond which will be ruptured. Random encounter is replaced by "planned" encounter. From the nature of these amino groups (the R groups of aspartic acid and glutamic acid), from their proximity to the glycosidic bond, and from their immediate environments within the enzyme molecule it can be postulated that the glutamic acid group will be able to transfer its terminal hydrogen atom to the glycosidic oxygen and thus cleave the bond between the oxygen and the carbon 1 on the N-acetylmuramic acid unit. In so doing it will produce a positively charged carbonium ion (C^+) and this ion will be stabilised, by the aspartic acid group ionised (—ly charged) by its particular environment, long enough for it to await a hydroxyl ion (OH^-) diffusing into the position from the surrounding water. This will complete the hydrolysis and lead to release of the enzyme ready to attach to another exposed 6-amino-sugar unit length of the polysaccharide substrate.

A number of hydrolytic enzymes, for example the plant proteolytic enzymes *papain* and *ficin*, are also able to permit transfer reactions, in these cases transamidations. The evidence in such cases is that there occurs co-valent bonding of an intermediate to the enzyme and that this enzyme-intermediate compound retains much of the bond energy of the peptide bond. The intermediate compound in these cases is an enzyme bound thioester involving the sulphydryl group of a cysteine residue of the enzyme.

$$\text{Enz—SH} + \underset{\text{Peptide}}{\text{R—C(=O)—NH—R}'} \longrightarrow \underset{\text{Intermediate}}{\text{Enz—S—C(R)=O}} + \text{NH}_2\text{R}'$$

Intermediate $+\ \text{R}''\text{NH}_2$ — Transamidation → $\text{Enz—SH} + \text{R—C(=O)—NH—R}''$

Intermediate $+\ \text{H}_2\text{O}$ — Hydrolysis → $\text{Enz—SH} + \text{R—C(=O)—OH}$ (7)

Similar covalent bonding is also involved in the action of the enzyme *sucrose glucosyltransferase* found in *Pseudomonas saccharophila* grown on sucrose medium. The normal reaction catalysed is

$$\text{sucrose} + \begin{matrix}\text{inorganic}\\\text{phosphate}\end{matrix} \rightleftharpoons \beta\text{-D-glucose-1-phosphate} + \text{fructose} \qquad (8)$$

However, the enzyme also catalyses the following transfer reaction:

$$\text{sucrose} + \underset{\text{(a monosaccharide)}}{\text{L-sorbose}} \rightleftharpoons \text{glucose-sorboside} + \text{fructose} \qquad (9)$$

and an interchange reaction (revealed by using radioactive ^{32}P labelled phosphate):

$$\text{glucose-1-PO}_4 + {}^{32}\text{P} \begin{array}{c}\text{inorganic}\\ \text{phosphate}\end{array} \rightleftharpoons \text{glucose-1-}{}^{32}\text{PO}_4 + \begin{array}{c}\text{inorganic}\\ \text{phosphate}\end{array} \qquad (10)$$

To explain these reactions it must be postulated that a glucosyl-enzyme is formed as intermediate. Since in the initial reaction α-glucose-1-phosphate (and not the β compound) is formed the mechanism must involve two consecutive double displacements. Thus, the combining group in the enzyme being postulated to be an oxygen function, and R—OH = fructose, the mechanism can be depicted thus:

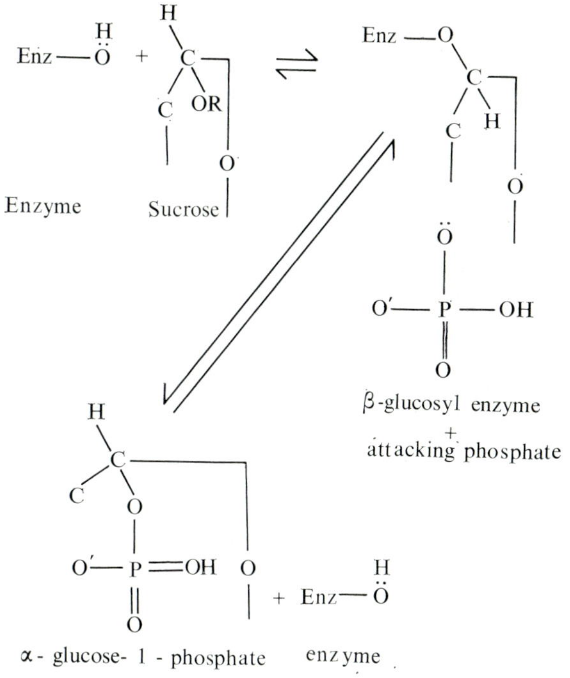

Enz—Ö(H) + H—C(—C)(—OR) ⇌ Enz—O—C(H)(—C) + R—OH

Enzyme Sucrose

Ö(—P(—O′)(—OH)(=O))

β-glucosyl enzyme
+
attacking phosphate (11)

H—C(—C)—O—P(—O′)(=OH)(=O) O + Enz—Ö(H)

α- glucose- 1 - phosphate enzyme

In other instances there is no evidence of a covalently bonded intermediate. Thus the kinases, which catalyse transfer of a phosphate group from ATP to a specific acceptor, do not appear to form intermediate phosphate–enzyme compounds.

ENZYMES IN THE LIVING CELL

The purified enzymes of the biochemist are not artefacts but the separate catalytic agents of living cells. However, we must remember that within a cell each enzyme functions in the presence of a multiplicity of other enzymes and other protein molecules. Further, we do not know the precise location within the cell of any enzyme nor the concentrations of its substrate(s) or of interfering substances which are present at that location. This means that while the particular catalytic activities of many proteins are now known their quantitative role in metabolism is much less easy to assess. The importance of "enzyme teams" is emphasised by the discovery that cellular organelles or "particles" like plastids, mitochondria and microsomes have each a characteristic population of enzymes and that the biochemical activities of such particles can be interpreted in terms of their enzymic complements. We shall have further occasion to stress the significance in metabolism of organised "multi-enzyme" units of this type.

There are, however, many enzymes which, because they do not seem to be contained in such cellular particles, are referred to as "soluble enzymes". Thus, the chemical reactions whereby the yeast cell converts glucose into ethanol and which seem to be identical with the first sequence of reactions in the normal respiration of plant cells are catalysed by such "soluble" enzymes. In the living cell such enzymes may, however, function as an "organised multi-enzyme" system which forms part of the endoplasmic reticulum. The fact that they cannot be obtained in association may merely reflect our inability to preserve the endoplasmic reticulum in "cell-free" preparations. Thus, in alcoholic fermentation by the yeast cell, which occurs in the absence of oxygen, there is an oxidative reaction which is balanced by a reduction. These both involve dehydrogenase enzymes having the same co-enzyme, nicotinamide adenine dinucleotide. One of these reactions, that catalysed by *alcohol dehydrogenase*, has already been mentioned (p. 73, eqn. 6). The other reaction involves the oxidation of a triosephosphate by the enzyme *glyceraldehyde-3-phosphate dehydrogenase:*

$$\text{glyceraldehyde-3-phosphate} + H_3PO_4 + NAD^+ \rightleftharpoons \text{1,3-diphosphoglyceric acid} + NADH + H^+ \quad (12)$$

The NADH generated in this oxidation is utilised in the reduction of acetaldehyde to ethanol by the alcohol dehydrogenase. It is possible, since the co-enzyme is diffusable, that co-enzyme reduced by triosephosphate dehydrogenase has to migrate in solution to a "remote" site where the alcohol enzyme is located. From the speed of the reaction in the cell it is, however, much more likely that these two "soluble" enzymes do *in vivo* work at the same site. In support of this is the evidence of a difference in the steric specificity of the two enzymes for the NAD. It is the nicotinamide part of the co-enzyme molecule which suffers oxidation and reduction by the removal or addition of hydrogen:

```
                               β     α
                               H     H
       H                         \ /
   H /   \\ CONH2                H / \ CONH2                  (13)
   ||       |      +2H →     ||     ||          +H+
   H \\    // H               H       H
       N                         \ N /
       |+                          |
       R                           R
```

If one of the H atoms at the para position of the nicotinamide ring is replaced by deuterium (heavy hydrogen) then the C atom at this position becomes asymmetric. When this is done it is found that hydrogen is removed from reduced NAD from one side of the nicotinamide ring by some dehydrogenases and from the other side by other dehydrogenases. The alcohol enzyme by virtue of the side of the nicotinamide ring to which the hydrogen is added is defined as being α-specific. The *lactate dehydrogenase* which takes the place of the alcohol enzyme in mammalian muscle has the same specificity, whereas the glyceraldehyde-3-phosphate dehydrogenase is β-specific (Fig. 3.6).

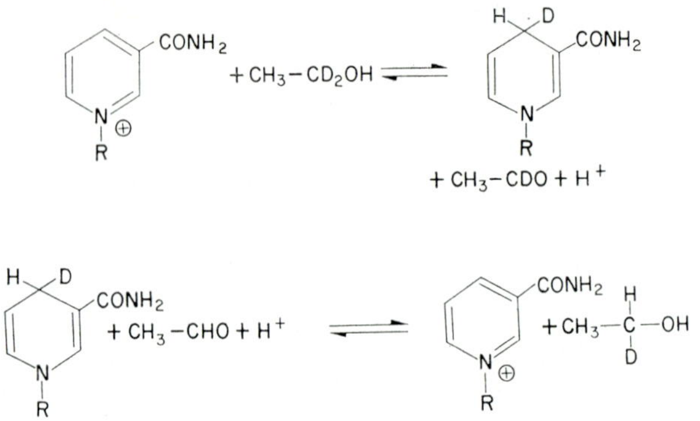

$CONH_2$ + CH_3-CD_2OH ⇌ H D $CONH_2$ N R + $CH_3-CDO + H^+$
N⊕ R

H D $CONH_2$ + $CH_3-CHO + H^+$ ⇌ $CONH_2$ + $CH_3-\overset{H}{\underset{D}{C}}-OH$
N R N⊕ R

FIG. 3.6. The α-specificity of the alcohol dehydrogenase as demonstrated by using deuterium labelled alcohol as substrate and then recovering the deuterium by the reverse reaction. D, deuterium; R, ribose-pyrophosphate-ribose-adenine moiety of NAD (see Fig. 4.6).

Another very important aspect of enzymes in living cells is that cells which are performing different functions have different enzyme complements and activities. The contrasts between the metabolic activities of the absorbing cells of the root, the photosynthetic cells of the leaves, the living cells of the xylem and phloem, the secretory cells of nectaries and other glands, and the actively dividing cells of meristems reflect differences in enzymic activity within the different cell types. Similarly, the synthesis of alkaloids in some plants, of gums and mucilages in others, of tannins and resins in others and of particular polysaccharides (e.g. inulin) in still others, reflect differences in enzymic composition and activities as between different species. These considerations indicate that genetic differences can be reflected in enzymic differences and raise the possibility that it is entirely through the control of enzyme synthesis and activation that the genes control physiology and development. They also raise the possibility that specialisation of structure and function within the organism is initiated by changes in the number and relative activity of enzymes within the incipient tissues. Great interest clearly attaches to the mechanism of protein synthesis, including the synthesis of enzymic proteins and of how the genes exert a control over this vital process.

In subsequent chapters we shall not only consider the action of further individual enzymes and of the multienzyme systems of chloroplasts, mitochondria and microsomes but the regulation of metabolism and differentiation through the controlled synthesis and activation and inhibition of enzymes.

FURTHER READING

D. C. Phillips. The three-dimensional structure of an enzyme molecule. *Scientific American*, **215**: 78–93, 1966.

W. H. Stein and S. Moore. *The Structure of Proteins*. Scientific American Reprint No. 80 (Feb. 1961). W. H. Freeman & Co., San Francisco.

J. Wesley. *Enzymic Catalysis*. Harper Revue, New York, 1969.

MORE ADVANCED READING

B. Vennesland. *Proteins, Enzymes and the Mechanism of Enzyme Action* in *Plant Physiology*, vol. IA, pp. 131–205, edited by F. C. Steward. Academic Press, New York, 1960.

M. Dixon and E. C. Webb. *Enzymes*. Academic Press, New York, 1964.

L. M. Shannon. Plant isoenzymes. *Annual Review of Plant Physiology*, **19**: 187–210, 1968.

Enzyme Nomenclature. Elsevier Publ. Co., Amsterdam, 1965.

CHAPTER 4

Catabolism

"The existence of such anaerobic organisms is hardly in agreement with the old dogma 'no life without respiration' unless we assume, as is here done, that respiration includes all metabolic processes that involve a liberation of energy."

W. Pfeffer, in *The Physiology of Plants*, vol. 1. Translated by E. J. Ewart, Oxford, 1900.

"Respiration in plants is taken to include all the phenomenon of dissimilation, the characteristics of which are the breaking down of complex substances into simpler ones with a consequent release of energy."

W. Stiles and W. Leach, in *Respiration in Plants*. Methuen & Co. Ltd., London, 1932.

INTRODUCTION

CATABOLISM embraces those processes which involve degradation of the chemical architecture of the complex organic molecules of cells. These processes result in the release of the potential energy of organic compounds and the major catabolic systems of living cells play an essential role in metabolism in that they release a substantial part of this energy in forms utilisable by the cell. Further the intermediate compounds arising during catabolism are often the starting materials for the synthesis of essential cell constituents.

Embraced by Pfeffer's definition that respiration includes all metabolic processes that involve a liberation of energy are two broad classes of respiratory reactions, aerobic respiration in which oxygen is the terminal acceptor of protons and electrons released in the course of catabolic reactions and anaerobic respiration in which terminal acceptors other than oxygen are utilized.

In aerobic respiration oxygen is reduced to water and carbon dioxide is released—this is the significance of the continuous uptake of oxygen and release of carbon dioxide in green plants in the dark experimentally demonstrated by de Saussure in 1804.

When cells of higher plants are under anaerobic conditions (deprived of oxygen) carbon dioxide release persists for some time and at least part of this carbon dioxide arises from a type of anaerobic respiration known as fermentation in which the electrons and protons are transferred to organic compounds such as pyruvic acid and acetaldehyde. The reduction of these compounds results in the appearance in the cell of compounds like lactic acid and ethanol. Clearly, this fermentation which can occur in the cells of higher plants closely resembles the alcoholic fermentation of yeast first studied in detail by Pasteur in 1870 and the production of lactic acid in mammalian muscles under conditions of an oxygen deficit, a process studied in detail by Fletcher and Hopkins in 1907.

The hexose sugar, glucose, can serve as the starting point for either respiration or fermentation and the following equations are often used to contrast these two processes:

Aerobic respiration

$$C_6H_{12}O_6 + 6O_2 \rightarrow 6CO_2 + 6H_2O \qquad \Delta G^\circ = -686 \text{ kcal} \tag{14}$$

Fermentation

$$C_6H_{12}O_6 \rightarrow 2C_2H_5OH + 2CO_2 \qquad \Delta G^\circ = -54 \text{ kcal} \tag{15}$$

ΔG° represents the change in Gibbs free energy per mole of substrate (glucose) and the negative sign of this value indicates that the reaction involves a *decrease* in free energy, i.e. a release of energy from the system.

The equations (eqns. 14 and 15) indicate that both processes involve carbon dioxide evolution and decrease in dry weight of the cell. They also draw attention to the involvement of molecular oxygen in aerobic respiration and to the much greater amount of energy released per molecule of glucose in this process.

These introductory paragraphs clearly prompt a number of interesting questions which will form the basis of our developing discussion of catabolism. At what rates do these catabolic processes proceed in plant cells and what are the factors which determine these rates? What substances other than glucose can and do regularly act as substrates for respiration? What are the chemical reactions involved and whereabouts in the cell do these reactions take place? How far do the reactions of fermentation still proceed in the cell when oxygen is available? How are these catabolic processes linked to the synthetic (anabolic) reactions of the cell and to other cellular processes which are energy-requiring?

RATES OF RESPIRATION

Rates of respiration can be calculated from measurements of carbon dioxide evolution and/or oxygen uptake, from recording rates of liberation of heat, or from measuring, in replicates of the experimental material, decreases in dry weight, in calorific value or in the content of specific food substances. The rate of respiration when calculated will be expressed by reference to unit time (e.g. per hour) and to unit amount of plant material (e.g. per unit fresh or dry weight, per single cell of the tissue, or per unit weight of cellular protein). The experimental conditions under which the determinations were made (temperature, oxygen supply, light or darkness, etc.) and the species, age and exact nature and previous nutritional history of the experimental plant material must also be precisely described.

Studies of respiration rate well illustrate the importance of expressing the rates of physiological processes in as many different ways as possible. Figures 4.1 and 4.2 show the oxygen uptake of young roots of maize expressed in different ways. When the respiration rate is expressed by reference to *fresh weight* the peak rate is in the region of most active cell division (0·5–2·2 mm behind the apex). When, however, rate is expressed *per cell* or *per unit of protein nitrogen* the region of most active cell division has a minimal rate of respiration while a high rate of respiration is recorded in the region of active cell elongation (2·2–7·0 mm behind the apex).

In the classical researches of Kidd, West and Briggs (1921) on

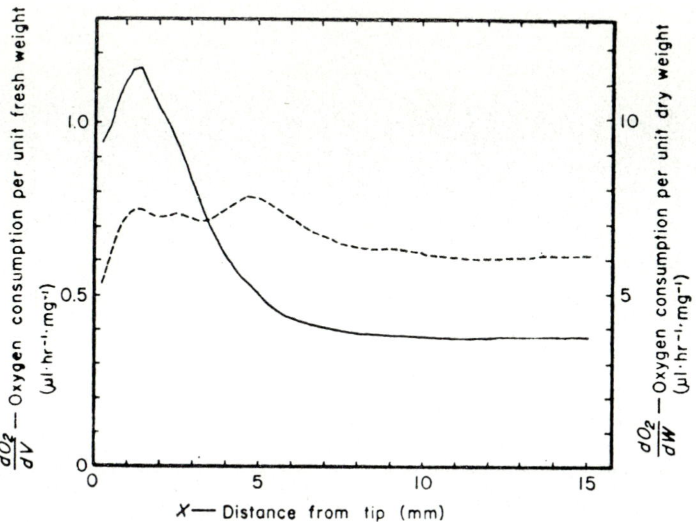

FIG. 4.1. Oxygen consumption (µl hr) of seedling *Zea mays* (corn) roots at different distances from the tip. Solid curve per mg fresh weight; broken curve per mg dry weight. (From D. R. Goddard and W. D. Bonner, in *Plant Physiology*, vol. 1A, edited by F. C. Steward, Academic Press, New York, 1960.)

the respiration of the sunflower, *Helianthus annuus*, it was found that while during the first 60 days of growth the total carbon dioxide evolved per plant per hour rose, there was from the onset of seedling growth and throughout the subsequent period a fall in the intensity of respiration expressed per unit of plant dry weight largely due to the increasing accumulation of inert dry matter in the form of cell wall materials. However, when the intensity of respiration is expressed per unit of protein nitrogen the rate is seen to increase with age probably indicating an increase in the rate of respiration per unit of cytoplasm as cells mature and senesce. Goodwin and Goddard's studies of the oxygen uptake of the different tissues of the stem of *Fraxinus* again illustrate the importance of the units in which respiration rate is expressed. Per unit of fresh weight the cambium is the tissue with the most active respiration; however, when the rate is

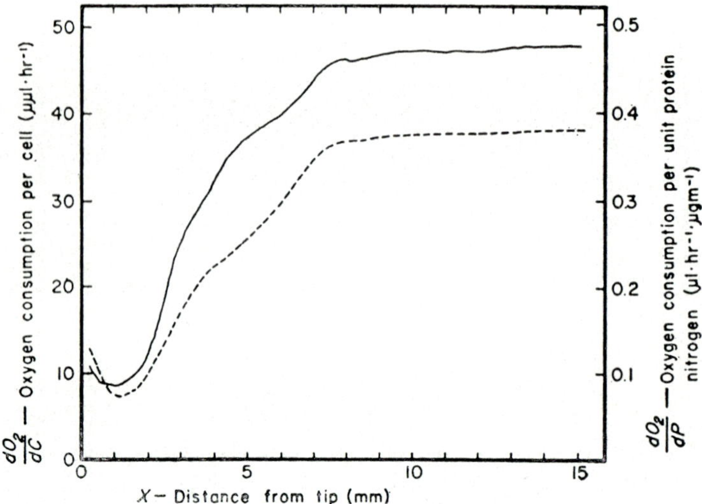

FIG. 4.2. Data as in Fig. 4.1 plotted by reference to cell number (solid curve) and to protein nitrogen (broken curve). Mitosis occurs from 0·7 to 2·2 mm from tip, elongation is maximal at 4–5 mm and ceases 9·0 mm from root tip.

expressed per unit of nitrogen the recently differentiated xylem is more active than the cambial tissue. Presumably, on a fresh weight basis, the respiratory activity of the xylem parenchyma and vascular ray cells is obscured by the presence in the xylem tissue of the mature and metabolically inert conducting cells.

Certain environmental factors markedly influence the rate of respiration. It is well known, for instance, that most tissues respire very slowly at or close to 0°C and that with rise in temperature there is a steady rise in respiration rate until we reach the optimum temperature (temperature at which the maximum rate is recorded) somewhere between 30–40°C. Within the range 0–30°C the temperature coefficient may be fairly constant, and such temperature coefficients

$$\left(Q_{10} \text{ value} = \frac{\text{rate at } t+10°\text{C}}{\text{rate at } t°\text{C}}\right)$$

frequently fall within the range 2·1–2·6. This means that the respiration process as a whole is responding like a thermo-chemical reaction to change in temperature (Chapter 3, p. 73). Above the optimum temperature the rate of respiration declines, particularly when the duration of exposure is prolonged. The deleterious effects of high temperatures are usually interpreted in terms of the denaturation of the essential respiratory enzymes.

A number of investigators have reported an enhanced rate of respiration in non-chlorophyllous tissues, e.g. *Vicia faba* roots, following transfer from darkness to light. Activity was reported as being in the blue region of the spectrum (wavelengths below 560 mμ) suggesting absorption of the light by carotenoid or flavonoid pigments. Further, even in chlorophyllous tissue, e.g. leaves, there appear to be stimulatory carry-over effects of illumination on the rate of respiration which cannot be explained in terms of enhanced levels of respiratory substrate produced by photosynthesis. In calculating rates of photosynthesis from measurements of carbon-dioxide uptake or oxygen evolution it is necessary to take into account

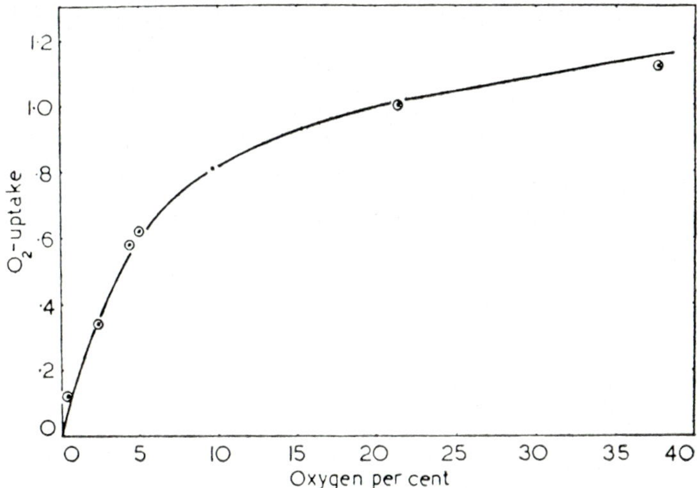

FIG. 4.3. Relation of oxygen uptake by barley seedlings to concentration of oxygen. Uptake in air = 1·0. (From D. F. Forward, *New Phytologist*, **50**, 297, 1951.)

that the respiration rate may be directly influenced by light and uncertainty on this question (see p. 143) has complicated attempts to determine the optimum efficiency of light energy utilisation in the photosynthesis of sugars.

The rate of oxygen uptake increases progressively with increase in the external concentration of oxygen and the curve relating uptake to external concentration frequently takes the form of a rectangular

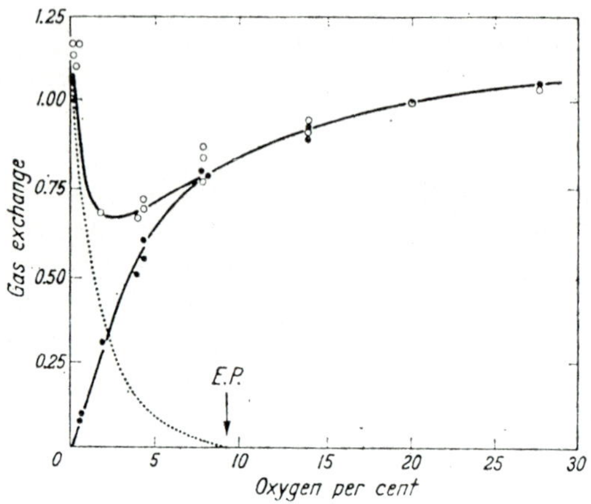

FIG. 4.4. The influence of the per cent oxygen upon the oxygen uptake (●) and the carbon dioxide evolution (○) of Bramley's Seedling apples. Dotted line shows the carbon dioxide evolution due to fermentation. E.P., the oxygen tension at which fermentation is completely suppressed. (After Watson, from W. O. James, *Plant Respiration*, Clarendon Press, Oxford, 1953.)

hyperbola (Fig. 4.3). When respiration rate is measured by determining the rate of carbon dioxide evolution, a minimum rate is recorded at a low oxygen concentration (usually at some point within the range 1–9% oxygen depending on the tissue used) and below this oxygen tension the carbon dioxide evolution rises steeply (Fig. 4.4) due to fermentation. The form of the carbon dioxide evolution curve illustrates the suppression of fermentation by oxygen but does not indicate the oxygen tension at which fermentation is completely suppressed (the extinction point = EP); this can only be determined by simultaneous determinations of alcohol and lactic acid formation at these low levels of oxygen supply.

The rate of respiration may be controlled by the concentrations of respiratory substrates in the cells. A response curve similar to that shown in Fig. 4.3, where respiration rate is plotted against concentration of oxygen, can be obtained by inducing different levels of soluble sugar in potato tubers by storage at temperatures within the range 1–10°C and then measuring their oxygen uptake at constant temperature. Further study of the soluble sugars present in the tubers shows that it is the content in the cells of the disaccharide, sucrose, which controls the rate of respiration. The control of respiration rate by carbohydrate supply is also illustrated by measurements on the oxygen uptake of root tips derived from excised tomato roots, depleted of carbohydrate reserves by a period of culture in sugar-free solution and then supplied with sucrose at different concentrations (Fig. 4.5).

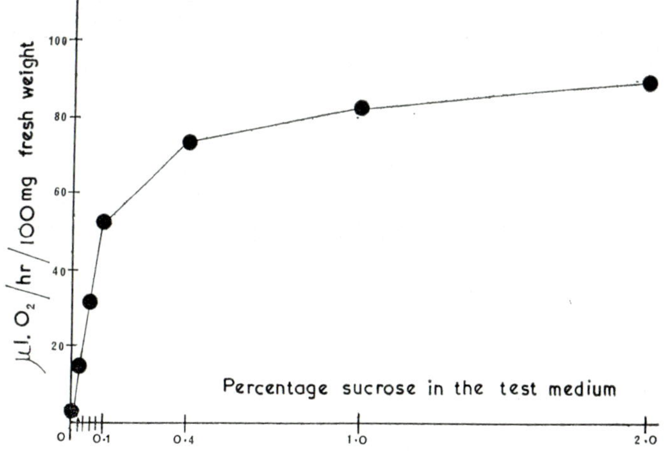

FIG. 4.5. Influence of the sucrose concentration in the external medium upon the rate of respiration of cultured tomato roots. The roots were grown in presence of 2% sucrose, maintained in a sugar-free solution for 48 hr and then transferred to the test media and their oxygen uptake measured for 1 hr commencing 67 hr after transference to the test medium.

Experiments on the respiration of detached leaves not only illustrate the control of respiration rate by availability of respiratory substrates but illustrate very clearly that not only carbohydrates but also proteins may be respired. This protein respiration involves protein hydrolysis to yield amino acids and amides, deamination of these amino acids and subsequent decarboxylation (leading to carbon dioxide release) and oxidation of the resulting organic acids.

In the overall equation (eqn. 14, p. 85) used to express the respiration of glucose, the volume of carbon dioxide evolved is equal to the volume of oxygen absorbed. The fraction

$$\frac{\text{volume of } CO_2 \text{ evolved}}{\text{volume of } O_2 \text{ absorbed}}$$

is referred to as the respiratory quotient (R.Q.) and in the case of the respiration of glucose is unity. When fats are the respiratory substrates the R.Q. is 0·7, when proteins, the R.Q. is 0·8. When organic acids are oxidised R.Q. values greater than unity are obtained. For instance, the complete oxidation of malic acid according to the equation

$$\underset{\text{malic acid}}{\begin{array}{l} CH_2.COOH \\ | \\ CHOH.COOH \end{array}} + 3O_2 \rightarrow 3H_2O + 4CO_2 \qquad (16)$$

corresponds to an R.Q. $= 1 \cdot 33$. Values of R.Q. above unity, particularly at low oxygen tensions, may also be indicative of the occurrence of fermentation.

Although measurement of the R.Q. may give some guide as to the nature of the substrate being respired by a particular tissue the simultaneous respiration of a combination of substrates such as for example fats and organic acids can result in R.Q.s which give no indication of the substrates being consumed. Only quantitative analysis of the changing contents of the respiring tissue allied to careful measurements of concomitant gaseous exchanges is likely to yield useful information as to the identity of the respiratory substrate(s).

CHEMICAL PATHWAYS OF RESPIRATION

One of the outstanding achievements of biochemistry has been the elucidation of the sequence of chemical reactions and the isolation of the enzymes which are involved in the alcoholic fermentation of sugars by yeast. This story had its beginning in one of the historic experiments in plant science, the preparation of "yeast press juice" by Hans and Edward Buchner in 1897 and the demonstration that this juice could promote an active fermentation of sucrose.

W. Pfeffer, influenced by the earlier studies of Pasteur, advanced in 1900 the view that respiration might proceed in two stages, a first stage corresponding to fermentation and a second stage involving the oxidation by oxygen of the products of fermentation. This concept was developed by two distinguished Russian plant physiologists, V. I. Palladin (1922) and S. Kostychew (1927), who emphasised that the enzymes of fermentation occur in the cells of higher plants and that there is no evidence to suggest that their activity is suppressed by oxygen. Further, knowing that ethyl alcohol is not itself readily oxidised by plant cells, they postulated that some precursor of alcohol in the fermentation sequence of reactions suffered rapid oxidation in the presence of oxygen. In support of Kostychew's hypothesis it is now known that (1) low concentrations of cyanide and of hydrogen sulphide by inhibiting the oxidative reactions of respiration lead to the accumulation of alcohol and acetaldehyde in plant tissue supplied with oxygen; (2) substances known to inhibit fermentation have been shown to inhibit respiration; (3) all the compounds known to be intermediates in fermentation can be detected in respiring cells. The "precursor of alcohol" which suffers complete oxidation to carbon dioxide and water in respiration is now known to be *pyruvic acid*. This contention is supported by a large body of evidence starting from the pioneer discovery by von Grab in 1921, using β-naphthylamine as a trapping agent, that pyruvic acid is formed during yeast fermentation prior to acetaldehyde and alcohol. This was followed by the demonstration that the addition of 1-naphthyl-2-sulphonic acid to higher plant cells causes the accumulation of pyruvate by inhibiting its conversion to acetaldehyde by the enzyme *pyruvic carboxylase* and by numerous experiments in which the cells of higher plants and animals have been shown to effect rapid oxidation of added pyruvic acid. We can now, therefore, develop our consideration of the chemical pathways of respiration by asking what reactions are involved in the formation of pyruvic acid and by what reactions this is oxidised by molecular oxygen.

The unravelling of the sequence of chemical steps involved in the formation and oxidation of pyruvic acid has followed from studies of the chemical activities of plant extracts and of the individual enzymes isolated from such extracts. Such studies have now yielded a very detailed picture of the biochemical reactions involved in respiration; have provided us with the essential biochemical foundation for a critical study of the process of respiration as it proceeds in the living cell. In this volume where our emphasis is on the metabolism of living cells a detailed consideration of the biochemistry of respiration or of any of the other vital aspects of metabolism would be inappropriate. Nevertheless, and just in so far as interpretation of cellular metabolism requires appreciation of the nature of the underlying chemical events we can properly draw upon the findings of biochemistry.

In 1908 the British biochemists, Harden and Young, when studying the alcoholic fermentation of glucose by "yeast press juice" found that the fermentation was augmented by the addition of inorganic phosphate. This enhancement was short-lived and if when the rate began to decline the fermentation liquor was boiled and then treated with uranium acetate, a sugar phosphate was precipitated. This was shown to be a diphosphate of fructofuranose and, since the phosphate groups are attached to carbon atoms 1 and 6 of the sugar, is known as fructose-1-6-diphosphate. Subsequent biochemical work has shown that this compound arises whenever the primary substrate of respiration is glucose, fructose, some other simple sugar or a polysaccharide such as starch. The first stage of carbohydrate respiration involves phosphorylation and interconversion of sugars to give this fructose-1-6-diphosphate.

In 1934 it was shown that yeast and mammalian muscle contain an enzyme (*aldolase*) which splits this fructose-1-6-diphosphate between carbon atoms 3 and 4 to give triose phosphates (glyceraldehyde-3-phosphate and dihydroxyacetone phosphate). The same enzyme was, by 1949, shown to be universally distributed in higher plants. Further, the two isomeric triosephosphates formed are interconvertible by means of a second enzyme, phosphotriose isomerase. This is important because it is only the glyceraldehyde-3-phosphate which is normally further metabolised in respiration.

Harden and Young, who revealed the importance of phosphorylation in fermentation, also showed that the "yeast press juice" (containing the mixture of enzymes known as zymase) could be separated by passage through a porous porcelain candle impregnated with gelatin into (1) a colloidal protein fraction which did not pass the filter and (2) a crystalloidal fraction stable to heat. Neither fraction alone promoted glucose fermentation but the activity was restored by mixing together the two fractions. Following the concept introduced by Bertrand as early as 1897, Harden and Young recognised the crystalloidal fraction as a co-ferment or *co-enzyme* which came, therefore, to be referred to as co-zymase. This co-enzyme was subsequently isolated in pure form and named diphosphopyridine nucleotide (or, for short, as DPN). This name has been discarded in favour of the term nicotinamide adenine dinucleotide (NAD) which more adequately indicates the nature of its two component nucleotides (Fig. 4.6). Now NAD is a vital component of those enzyme systems called dehydrogenases which catalyse the oxidation of substances by removal from them of pairs of hydrogen atoms. The hydrogen removed from the molecule of the metabolite reduces the co-enzyme. NAD functions as the co-enzyme for a number of dehydrogenases each of which acts on a specific metabolite by virtue of the specificity of the catalytic enzyme protein. The glyceraldehyde-3-phosphate formed during fermentation is metabolised by being acted upon by one such specific dehydrogenase, the enzyme *triosephosphate dehydrogenase*. The action of this particular dehydrogenase is rather unique as the oxidation of the free aldehyde group of the triosephosphate to an acidic carboxyl group is coupled with combination of the carboxyl group with

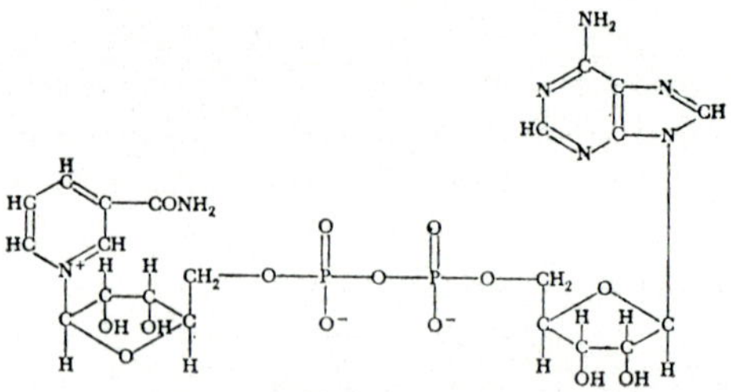

NAD Nicotinamide adenine dinucleotide

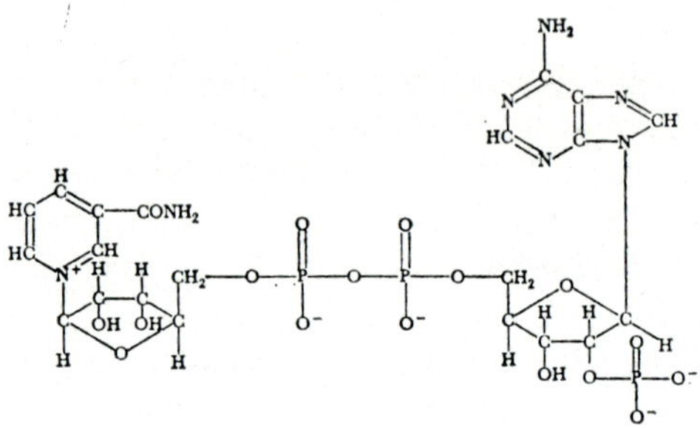

NADP Nicotinamide adenine dinucleotide phosphate

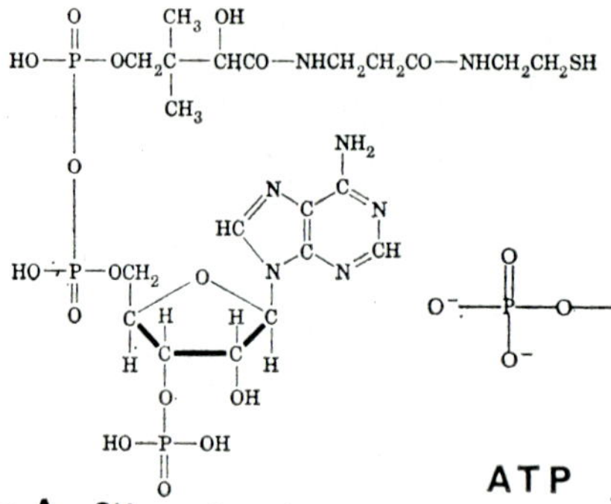

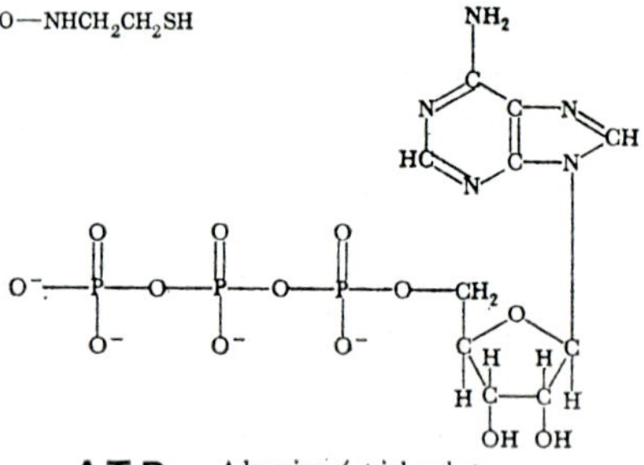

$HO-P(=O)(-O-)-OCH_2C(CH_3)_2-CH(OH)CO-NHCH_2CH_2CO-NHCH_2CH_2SH$

Co A –SH Coenzyme A

ATP Adenosine-5′-triphosphate

FIG. 4.6. The chemical structure of co-enzymes and of adenosine triphosphate.

inorganic phosphate to give a diphosphoglyceric acid. The overall reaction can be represented thus (as Chapter 3, p. 81):

$$\begin{array}{c} CHO \\ | \\ HCOH + P_i + NAD^+ \\ | \\ CH_2O(P) \\ \text{glyceraldehyde-3-phosphate} \end{array} \underset{\substack{\text{glyceraldehyde} \\ \text{phosphate} \\ \text{dehydrogenase}}}{\rightleftharpoons} \begin{array}{c} COO(P) \\ | \\ HCOH + NADH + H^+ \\ | \\ CH_2O(P) \\ \text{1,3-diphosphoglyceric acid} \end{array} \quad (17)$$

In fermentation this oxidative step is equated with a balancing reduction in which the co-enzyme is re-oxidised. In yeast fermentation it is acetaldehyde which is reduced to ethyl alcohol by a second specific dehydrogenase (*alcohol dehydrogenase*), and the same co-enzyme (see Chapter 3, p. 80). These oxidation reductions are reversible and here the dehydrogenase is acting "in reverse":

$$\begin{array}{c} CH_3 \\ | \\ CHO \\ \text{acetaldehyde} \end{array} + NADH + H^+ \rightleftharpoons \begin{array}{c} CH_3 \\ | \\ CH_2OH \\ \text{ethanol} \end{array} + NAD^+ \quad (18)$$

In respiration the reduced co-enzyme is oxidised by the intervention of molecular oxygen as will be discussed below.

Now the chemical steps between diphosphoglyceric acid and pyruvic acid are of interest in another direction. The diphosphoglyceric acid is first acted upon by one of a group of enzymes known as *phosphotransferases* or *kinases*, enzymes which catalyse the transfer of chemical groups from one molecule to another. The action of the kinase here is to produce mono (3)-phosophoglyceric acid but the phosphate group removed does not appear as inorganic phosphate but is transferred to an organic acceptor molecule, a molecule of adenosine diphosphate (ADP) which is converted to the triphosphate (ATP) thus:

1:3-diphosphoglyceric acid + ADP
→ 3-phosphoglyceric acid + ATP (19)

The 3-phosphoglyceric acid then undergoes an internal chemical change and releases a molecule of water to give phosphoenol pyruvic acid (PEP). The phosphoenol pyruvic acid now acted upon by a second kinase, transfers its phospho-group to a molecule of ADP (thereby converted to ATP) and pyruvic acid is formed. From a molecule of the hexose diphosphate, two molecules of the 3 carbon keto-acid, pyruvic acid are formed.

Figure 4.7 correlates the reactions leading to the formation of pyruvic acid from glucose—often referred to as the *Embden–Meyerhof–Parnas* (*EMP*) *pathway* in recognition of the particular importance of the work of these three German scientists in its elucidation. Originally established in yeast and muscle extracts evidence quickly accumulated for the widespread occurrence of the EMP pathway in a great variety of organisms including all plant tissues investigated.

The pyruvic acid produced by the EMP pathway has now to be completely oxidised and on the assumption that the oxidative steps, like the oxidation of triosephosphate, each involve the removal of two hydrogen atoms, the overall reaction can be represented thus:

$$CH_3.CO.COOH + 5O \rightarrow 3CO_2 + 2H_2O \qquad (20)$$

or

$$\underset{\text{pyruvic acid}}{CH_3.CO.COOH} + 3H_2O \rightarrow 5 \times 2H + 3CO_2 \qquad (21)$$

This latter method of representation (eqn. 21) is justified because we know that five specific dehydrogenases each acting on a separate intermediate are involved and that the carbon of pyruvic acid is released as carbon dioxide by three reactions in each of which a specific decarboxylase enzyme splits off carbon dioxide from a carboxyl group. To initiate the oxidation the pyruvic acid has to be converted to "active acetate" and there has to be present in the cells a catalytic amount of another acid (oxaloacetic acid) with which this can combine to give citric acid. The citric acid then undergoes the series of chemical changes which result in the reduction of NAD (dehydrogenase steps), the release of carbon dioxide (decarboxylase steps) and the regeneration of oxaloacetic acid ready to combine with further "active acetate". It is this series of reactions which is known as the citric acid, tricarboxylic acid (TCA) or *Krebs cycle* (Fig. 4.8).

"Active acetate" is important; not only is it an essential intermediate in the oxidation of pyruvate arising from carbohydrate but also in many other facets of metabolism. The chemical nature of this "active acetate" (highly reactive acetate) was, for a time, an outstand-

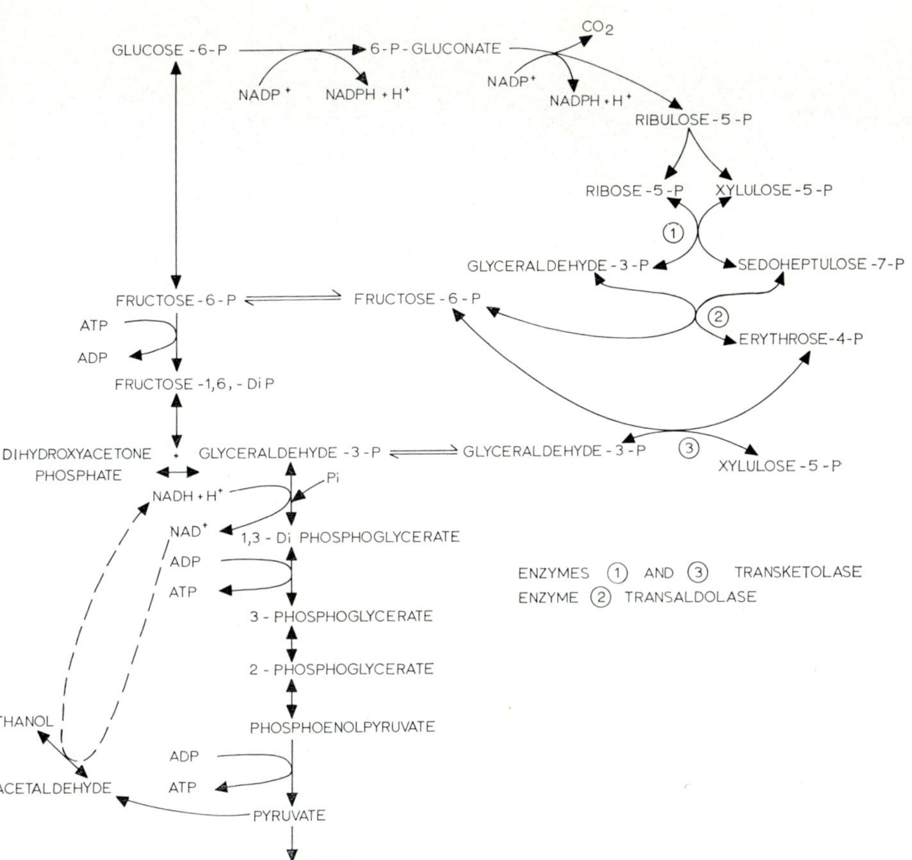

FIG. 4.7. Pathways of glucose metabolism and their interrelationships. The reactions of the EMP pathway are to the left of the figure and those of the Pentose Phosphate pathway on the right. Broken arrow illustrates the NAD mediated coupling of the reduction of glyceraldehyde-3-P to the oxidation of acetaldehyde which takes place under anaerobic conditions. (After M. W. Fowler.)

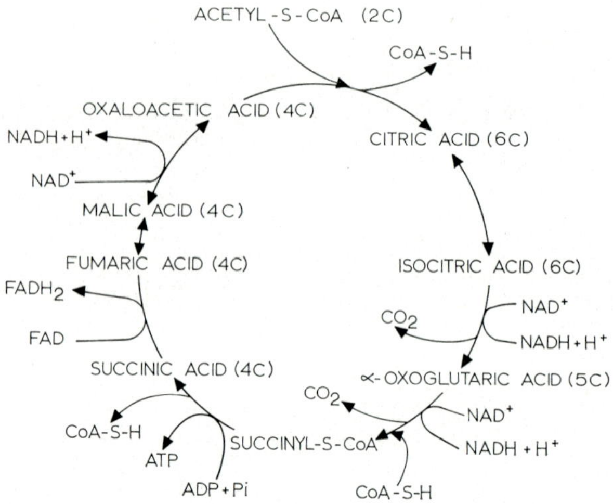

FIG. 4.8. The tricarboxylic acid (Krebs) cycle.

ing biochemical puzzle until Lipmann in 1946 obtained evidence for and later isolated a co-enzyme of acylation now known as *co-enzyme A* (Fig. 4.6) and which because it reacts through its sulphydryl group is usually abbreviated in biochemical equations to CoA-SH. When, in 1951, Lynen demonstrated that the "active acetate" of yeast was acetyl-co-enzyme A the problem was posed of how pyruvate is converted to this compound. This has also proved a difficult problem and its unravelling has shown it to be a complex process in which still other co-factors are involved. However, the overall equation for the oxidative decarboxylation which occurs can be represented:

$$\underset{\text{pyruvic acid}}{\begin{array}{l} CH_3 \\ | \\ C{=}O \\ | \\ COOH \end{array}} + CoASH + NAD^+ \rightarrow \underset{\text{acetyl-CoA ("active acetate")}}{\begin{array}{c} CH_3 \\ | \\ C{=}O \\ | \\ S \\ | \\ CoA \end{array}} + CO_2 + NADH + H^+ \quad (22)$$

This acetyl-CoA then transfers a "2C fragment" to oxaloacetic acid to form the citric acid whose metabolism back to oxaloacetate is effected by the Krebs cycle enzymes.

The cyclic nature of the process, as postulated by Sir Hans Krebs in 1937, is illustrated in experiments initially using animal tissues in which the addition of a cycle acid—say oxaloacetate to a tissue homogenate results in an increase in the rate of pyruvate utilisation and the release of much more carbon dioxide than can be accounted for by the oxidation of the oxaloacetate added. Further the consumption of citrate is specifically inhibited by fluoroacetate and, in an inhibited preparation, the addition of oxaloacetate results in the release of a stoichiometrically equivalent amount of CO_2—one molecule of CO_2 per molecule of oxaloacetate added.

As noted above the operation of the TCA cycle results in the conversion of acetyl-CoA to carbon dioxide and reduced co-enzymes, the oxidation of which may be coupled to the synthesis of ATP (see "terminal oxidation" below). However the TCA cycle must not be thought of simply as an incinerator—it is also important as a means whereby acetyl-CoA from a variety of sources is converted to compounds involved in a number of biosynthetic pathways. For example oxoglutarate is involved in amino acid synthesis and succinyl-CoA is a precursor of porphyrins. These and other intermediates are withdrawn from the cycle—however, a moment's thought will indicate that the carbon withdrawn must be replaced if the rate of utilization of acetyl-CoA by the cycle is not to be reduced. The reactions shown below result in the replenishment of the cycle with malic or oxaloacetic acid from phosphoenolpyruvate (PEP) or pyruvate and are sometimes referred to as anaplerotic reactions.

$$\text{PEP}+\text{ADP}+\text{CO}_2 \xrightarrow[\text{carboxykinase}]{\text{PEP}} \text{oxaloacetate}+\text{ATP} \qquad (23)$$

$$\text{PEP}+\text{CO}_2 \xrightarrow[\text{carboxylase}]{\text{PEP}} \text{oxaloacetate}+\text{Pi} \qquad (24)$$

$$\text{pyruvate}+\text{CO}_2+\text{NADPH}+\text{H}^+ \xrightarrow[\text{enzyme}]{\text{malic}} \text{malate}+\text{NADP}^+ \qquad (25)$$

The isolation of active mitochondrial preparations from plants (mung beans and cauliflowers being particular favourites for this purpose) confirmed the importance of the TCA cycle in plant respiration and demonstrated clearly that the site of this biochemical pathway is within the mitochondria.

Several of the TCA cycle acids are so named because they accumulate to high concentrations in certain plant tissues—malic acid in apple (*Pyrus malus*), citric acid in citrus fruits and fumaric acid in fumitory (*Fumaria officinalis*). Accumulations of this sort (there are a number of other examples) were initially taken as evidence that the TCA cycle was not operating rapidly in these plants. However when ^{14}C-labelled TCA acids are fed to such tissues and then the specific activities (radioactivities per unit amount of carbon) of the evolved carbon dioxide and of the TCA acids within the tissue are measured it is found that whereas the specific activity of the carbon dioxide may be the same from two different acids, the specific activities of these acids in the cells may be very different indeed. This indicates that the whole cellular content of the acids does not participate in the TCA cycle, part of the acid is participating in the TCA cycle, part is remote from the site of this cycle. In certain crassulacean plants which contain high concentrations of malic acid it can be shown that there is a small metabolically active "pool" associated with the TCA cycle and a large "pool" which is relatively metabolically inert. Equilibration of label between these pools is slow and it is probable that the inert pool corresponds to the vacuole and the active pool with an intramitochondrial compartment.

This concept of compartmentalization of intermediates is important and the range of potential compartments in plant cells is a complicating factor which must always be borne in mind when interpreting experiments involving measurements of amounts of metabolic intermediates in tissues, cells or even organelles. For example changes in the total phosphoglyceric acid (PGA) content of a leaf—perhaps as a result of transfer from light to dark—would incorporate the sum of the contents of at least two discrete pools (a chloroplastic photosynthetic pool and a cytoplasmic respiratory pool) the changes in which could well be opposite in sign.

The reactions of the TCA cycle coupled with those of the EMP pathway came to be regarded as *the* pathway of glucose dissimilation although since the period 1932–7 it has been known that yeast and many other plant cells contain a dehydrogenase which directly oxidises a monophosphate of glucose arising during the initial phosphorylation of sugars in respiration. Further, that the coenzyme of this system is nicotinamide adenine dinucleotide phosphate (NADP), formerly known as *tri*-phosphopyridine nucleotide (TPN). The action of this enzyme can be represented thus:

$$\text{glucose-6-phosphate} + \text{NADP}^+ \rightarrow \text{6-phosphogluconate} + \text{NADPH} + \text{H}^+ \quad (26)$$

Yeast and higher plant cells also contain a second enzyme which oxidatively decarboxylates the phosphogluconate thus:

$$\begin{array}{c} \text{COOH} \\ | \\ \text{HCOH} \\ | \\ \text{HOCH} \\ | \\ \text{HCOH} \\ | \\ \text{HCOH} \\ | \\ \text{H}_2\text{C—OPO}_3\text{H}_2 \\ \text{6-phosphogluconate} \end{array} + \text{NADP}^+ \xrightarrow[\text{dehydrogenase}]{\text{6-phosphogluconate}} \begin{array}{c} \text{CO}_2 \\ + \\ \text{CH}_2\text{OH} \\ | \\ \text{C=O} \\ | \\ \text{HCOH} \\ | \\ \text{HCOH} \\ | \\ \text{H}_2\text{COPO}_3\text{H}_2 \\ \text{D-ribulose-5-phosphate} \end{array} + \text{NADPH} + \text{H}^+ \quad (27)$$

Subsequent research has shown that the pentose phosphate so produced is further metabolized in reactions utilizing the enzymes transaldolase and transketolase and involving as intermediates pentose, heptulose, tetrose and triose (5, 7, 4 and 3 carbon sugars) phosphates which are also involved in the photosynthetic carbon reduction cycle. A version of the *pentose phosphate pathway* is shown in Fig. 4.7.

The elucidation of the biochemistry of the pentose pathway raises the question of its importance and significance in carbohydrate breakdown relative to the EMP pathway. Evidence on this question comes from two approaches. The first arises from the finding that one of the reactions of the Krebs cycle, the oxidation of succinic acid, is very specifically inhibited by another organic acid of similar structure, malonic acid.

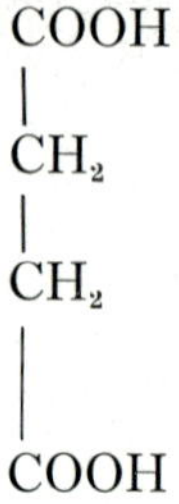

$$\begin{array}{c} COOH \\ | \\ CH_2 \\ | \\ CH_2 \\ | \\ COOH \end{array} \qquad \qquad \begin{array}{c} COOH \\ | \\ CH_2 \\ | \\ COOH \end{array}$$

succinic acid 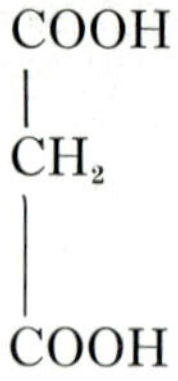   malonic acid

The inhibiting action of the malonic acid depends not only upon its own concentration but also in the concentration of the natural substrate of the enzyme, the succinic acid. The inhibition is, therefore, described as competitive and is known to involve competition between the two acids for combination with the enzyme protein; the protein in combination with the malonic acid being inactive as a catalyst of the oxidation of succinate (see Chapter 3, p. 74). Malonic acid, by inhibiting this reaction, inhibits the Krebs cycle sequence of reactions and blocks the oxidation of pyruvate by this pathway. Malonate is, in consequence, an inhibitor of respiration and it has been argued that the malonate-sensitive respiration is mediated by the Krebs cycle and the malonate-insensitive by an alternative pathway. However, this conclusion may be an over-simplification because the permeability of cells to malonate varies and is very sensitive to factors like pH and also the shutting down of the Krebs cycle may itself activate a pathway not normally operative at an appreciable rate.

There is, however, another and rather interesting way of approaching this problem by the use of radio-active carbon (^{14}C). This technique involves the use as respiratory substrates of samples of glucose phosphate in which only the first carbon atom or only the 6th carbon atom of the sugar is enriched with radio-active carbon (the glucose molecule is not uniformly labelled but *specifically labelled in particular carbon atoms*). The carbon dioxide which is released immediately after supplying the glucose phosphate will come, in the pentose pathway, only from the carbon atom 1 of the sugar and only when this is radioactive should the carbon dioxide be labelled. By contrast, the inter-conversion of the triosephosphates (p. 95) in the EMP pathway should result in equally labelled carbon dioxide, whether C1 or C6 is labelled in the glucose sample. Thus, the extent to which the radioactivity of the carbon dioxide is reduced when the ${C^{14}}_6$ instead of the ${C^{14}}_1$ sugar is provided as substrate will be a measure of the contribution of the pentose pathway to the carbon dioxide evolved in respiration.

The application of this technique to plant tissues under various experimental conditions has made possible the generalization that the pentose phosphate pathway commonly operates in plants and is likely to be most in evidence in tissues in which there is much synthetic activity, for example in mature cells synthesising primary metabolites as opposed to differentiating cells which are utilising products transported to them from such mature tissue cells. This when allied to the generalization that synthetic pathways usually employ NADP and not NAD has led to the suggestion that the primary function of the pentose phosphate pathway may be to supply reducing power for synthetic reactions. Another function may be to transform storage carbohydrate (via glucose-6-P) to carbon skeletons required by particular synthetic pathways, e.g. ribose-5-P for nucleic acid synthesis. Similarly the tetrose, erythrose-4-P, an intermediate of the pentose phosphate pathway is also involved in the synthesis of shikimic acid, a key intermediate in the pathways leading to the aromatic amino acids and other cyclic compounds (see p. 187).

Thus although the operation of the pentose phosphate pathway alone or in conjunction with the TCA cycle can cause the complete dissimilation of glucose to CO_2 it is certain that this pathway is more than simply an alternative respiratory route. The evidence suggests that it is deeply involved in the utilisation of carbohydrates to generate reduced coenzymes and intermediates destined for synthetic reactions.

RESPIRATORY SUBSTRATES

The TCA cycle is central to respiratory reactions and may be supplied with acetyl-CoA by a number of metabolic pathways leading from the variety of storage compounds found in plants. These storage compounds represent the accumulated products of photosynthesis and are drawn upon when the requirements of the plant for energy and carbon skeletons are greater than are supplied currently by photosynthetic reactions. Such a situation arises during the hours of darkness, in unsatisfactory climatic conditions (e.g. winter in temperate climates) and also in seeds before and during germination until an operational photosynthetic apparatus is established and the seedling achieves a positive carbon balance.

The reactions of the EMP and pentose phosphate pathway can account for the conversion of hexose phosphates to acetyl-CoA (p. 99) but hexose phosphates are not found in high concentrations in plants. The storage compounds commonly encountered in plants are sucrose—not common in ungerminated seeds but found in such plants as sugar beet and sugar cane; starch—found almost universally in photosynthetic tissue as assimilation starch (p. 161) and as a "true" storage product in many plant parts including for example potato tubers and many seeds such as the cereals; fats—found in many seeds, a classical example being the castor bean, and protein—found as a storage product in some seeds, especially some leguminous seeds such as lupin. The metabolic pathways involved in the conversion of these compounds (and the list above is not by any means comprehensive) to acetyl-CoA or to intermediates of the EMP or pentose phosphate pathway are outlined below.

The consumption of the disaccharide sucrose (see p. 160) involves its hydrolysis resulting in the formation of free glucose and fructose. The enzyme catalysing this reaction is *β-fructofuranosidase* (also known as invertase):

$$H_2O + \text{sucrose} \rightarrow \text{fructose} + \text{glucose} \qquad (28)$$

The free sugars are then phosphorylated under the influence of *hexokinase*:

$$\text{fructose} + \text{ATP} \rightleftharpoons \text{fructose-6-phosphate} + \text{ADP} \qquad (29)$$

$$\text{glucose} + \text{ATP} \rightleftharpoons \text{glucose-6-phosphate} + \text{ADP} \qquad (30)$$

The glucose-6-phosphate may enter the pentose phosphate pathway

(p. 103) or may be converted by the action of the enzyme *phosphohexoseisomerase* to fructose-6-phosphate:

$$\text{glucose-6-phosphate} \rightleftharpoons \text{fructose-6-phosphate} \qquad (31)$$

Fructose-6-phosphate may then be phosphorylated in a reaction catalysed by *phosphofructokinase* which results in the production of fructose-1-6-diphosphate which may enter the EMP pathway (p. 94 and Fig. 4.7):

$$\text{fructose-6-phosphate} + \text{ATP} \rightleftharpoons \text{fructose-1-6-disphosphate} + \text{ADP} \qquad (32)$$

The utilisation of starch (see Fig. 5.11 for structure) appears to involve two distinct mechanisms—one hydrolytic and the other involving phosphorolysis.

The hydrolytic reactions are catalysed by the enzymes, α, β and iso-amylase. *α-Amylase* is a calcium requiring enzyme which attacks 1:4α-glucosidic linkages randomly throughout amylose and amylopectin molecules resulting in almost complete hydrolysis to maltose. The 1:6α-linkage branch points of amylopectin are not affected. *β-Amylase* also attacks only the 1:4 linkages but does so from the non-reducing ends of chains cleaving off maltose units until a 1:6α-linkage is approached in amylopectin and until, theoretically at least, amylose is completely converted to maltose (anomalous linkages in the amylose usually prevent this). The third enzyme, *iso-amylase*, specifically attacks the 1:6α-linkage branch points converting amylopectin into straight chain fragments. Co-operation of all three enzymes results in the rapid hydrolysis of starch preparations to maltose which yields glucose units under the influence of the enzyme maltase. The glucose may be phosphorylated and enter the EMP or pentose phosphate pathway as described above for sucrose. During germination of starch containing seeds, cereals having been most carefully studied, amylase activities increase, sometimes, as in wheat, through activation of latent enzymes or through *de novo* synthesis of enzymes as is the case with α-amylase in the barley seed. Evidence such as this clearly indicates that the role of the amylases is in the mobilisation of starch.

The second enzymic mechanism of starch degradation involves the enzyme *phosphorylase* which in the presence of inorganic phosphate attacks 1:4α-linkages proceeding from the non-reducing end of the molecule:

$$\text{starch } (n \text{ glucose units}) + P_i \rightleftharpoons \text{glucose-1-phosphate} + \text{starch } (n\text{-1 glucose units}) \quad (33)$$

By the action of the enzyme *phosphoglucomutase* the glucose-1-phosphate so produced may be converted to glucose-6-phosphate which can enter the EMP or pentosephosphate pathway. Unlike the hydrolysis of starch, its phosphorolysis is readily reversible (the energy level of the glucosidic linkages being similar to that of the glucose-1-phosphate ester linkage). Phosphorylase can therefore under suitable conditions catalyse the synthesis of starch from glucose-1-phosphate and a suitable primer. However, it appears that in developing pea seeds and a number of other tissues that the level of inorganic phosphate is such as to ensure that the reaction will proceed in the direction of starch breakdown. However, the inorganic phosphate measured here is the total cellular phosphate and the possible effects of compartmentation should be borne in mind. Nevertheless, on the basis of this evidence and from the observation that in animal tissues phosphorylase deficiency affects breakdown but not synthesis of glycogen it has been suggested that plant phosphorylase functions to degrade and not to synthesise starch. Correlations between photosynthetic starch synthesis and phosphorylase activities reported in some plant tissues and the increase in phosphorylase activity in *developing* rice grains caution, however, that a verdict on whether or not phosphorylase is involved in starch synthesis should be regarded as "not proven" until more critical experimental evidence is available (see p. 162 for an alternative mechanism for starch synthesis).

The first step in the respiratory consumption of fats is a hydrolysis, catalysed by the enzyme *neutral lipase* which results in the release of glycerol and free fatty acids. The glycerol is phosphorylated to α-glycerolphosphate, and then converted to dihydroxyacetonephosphate, a point of entry into the EMP and pentosephosphate pathways.

The free fatty acids are partially oxidised and converted to two carbon units in a remarkable series of reactions known as the β-oxidation pathway or fatty acid spiral (Fig. 4.9). A detailed considera-

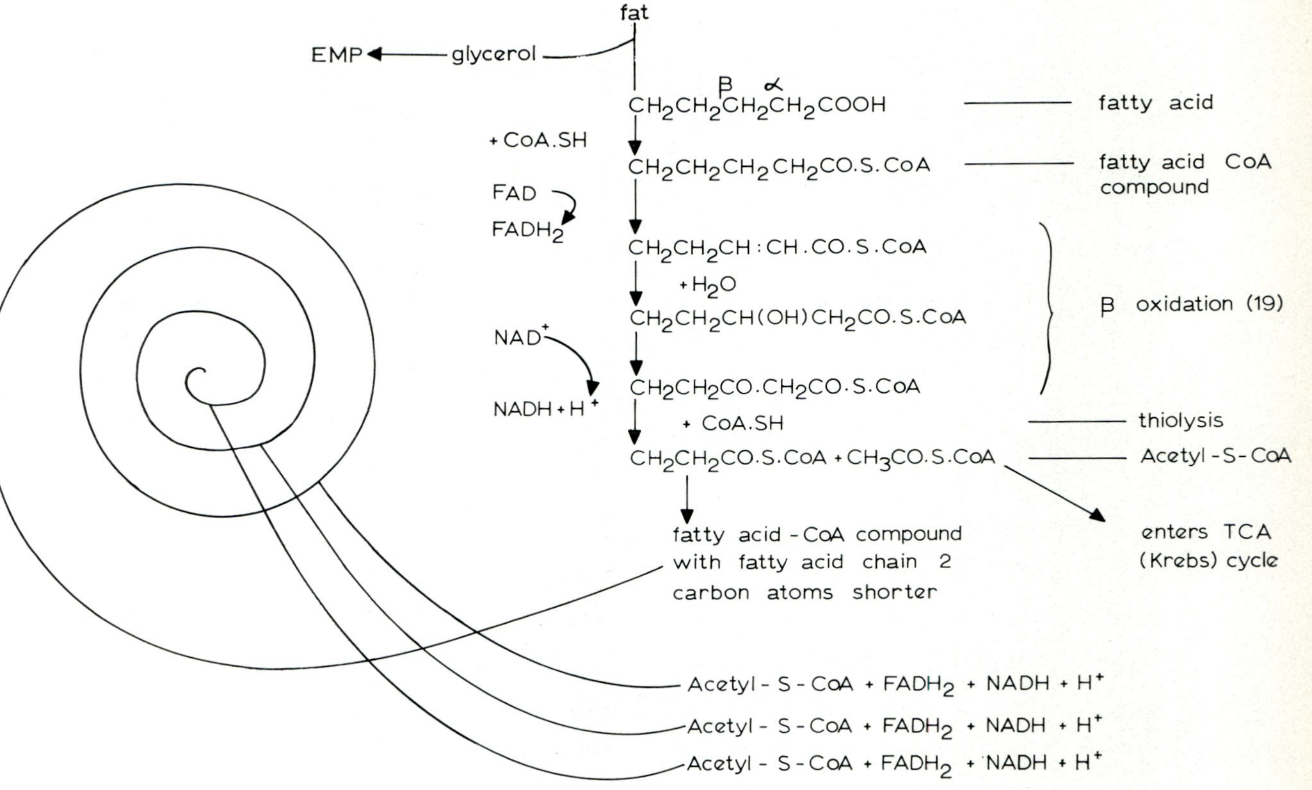

FIG. 4.9. The fatty acid (β-oxidation) spiral. Fatty acids are degraded to reduced NAD and FAD (flavin adenine dinucleotide) which can transfer electrons to the respiratory electron transport system, and acetyl-CoA which can enter the TCA cycle or the glyoxylate cycle.

tion of this pathway is beyond the scope of this text but in essence the sequence proceeds as follows. The free fatty acid is converted to the fatty acyl-CoA derivative which is oxidised to the α, β unsaturated fatty acyl-CoA by a *fatty acyl-CoA dehydrogenase* which requires flavin adenine dinucleotide (FAD) as co-enzyme. This unsaturated fatty acyl-CoA is hydrated and then oxidised, in a reaction resulting in the reduction of NAD, to the β keto-derivative which undergoes thiolysis by reaction with reduced acetyl-CoA resulting in the production of a molecule of acetyl-CoA and a fatty acyl-CoA containing two less carbon atoms than the original acid. Repetition of this process, depicted by the spiral in Fig. 4.9 results eventually in the complete conversion of the fatty acid to acetyl-CoA units.

The reduced co-enzymes may donate electrons to the electron transport system (p. 118) although in the glyoxysomes (see below) the reduced flavoprotein co-enzyme is oxidised directly.

The acetyl-CoA may be consumed by way of the TCA cycle thus: a combination of the β-oxidation spiral, the TCA cycle and associated electron transport systems can result in the complete conversion of fatty acids to CO_2 and water. The overall reaction if palmitic acid were respired in this way would be:

$$C_{16}H_{32}O_2+23O_2 \rightarrow 16H_2O+16CO_2 \qquad \text{R.Q.} = \frac{16}{23} \simeq 0.7 \quad (34)$$

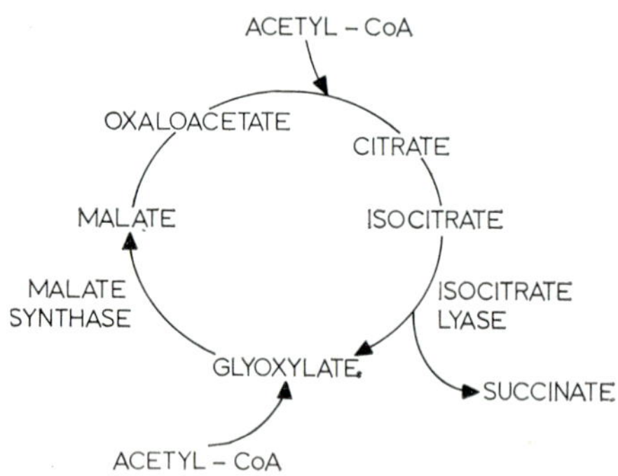

FIG. 4.10. The glyoxylate cycle.

As well as being involved in the oxidative consumption of fats it has been shown, predominantly by the work of Beevers and collaborators, that the β-oxidation spiral participates in the biochemical reactions involved in the conversion of fats to sucrose during the germination of fatty seeds such as the castor bean. The mobilization of the fatty reserves involves hydrolysis and β-oxidation as described above but the acetyl-CoA produced enters not the TCA cycle but a modification of it known as the *glyoxylate cycle* (Fig. 4.10). In this cycle the gap between isocitrate and malate is bridged by the enzymes *isocitrate lyase* and *malate synthase* which catalyse the reactions shown below.

$$\underset{\text{Isocitrate}}{\begin{array}{c} \text{OH} \\ | \\ \text{H}-\text{C}-\text{COOH} \\ | \\ \text{H}-\text{C}-\text{COOH} \\ | \\ \text{CH}_2\text{COOH} \end{array}} \xrightarrow[\text{lyase}]{\text{Isocitrate}} \underset{\text{Succinate}}{\begin{array}{c} \text{CH}_2\text{COOH} \\ | \\ \text{CH}_2\text{COOH} \end{array}} + \underset{\text{Glyoxylate}}{\begin{array}{c} \text{CHO} \\ | \\ \text{COOH} \end{array}} \tag{35}$$

$$\underset{\text{Acetyl-CoA}}{\text{CH}_3-\overset{\overset{\displaystyle \text{O}}{\|}}{\text{C}}-\text{S}-\text{CoA}} + \underset{\text{Glyoxylate}}{\begin{array}{c} \text{CHO} \\ | \\ \text{COOH} \end{array}} \underset{\text{Synthase}}{\overset{\text{Malate}}{\rightleftharpoons}} \underset{\text{Malate}}{\begin{array}{c} \text{COOH} \\ | \\ \text{CH}_2 \\ | \\ \text{CHOH} \\ | \\ \text{COOH} \end{array}} + \underset{\text{CoA}}{\text{CoASH}} \tag{36}$$

A second molecule of acetyl-CoA enters by eq. 36 so that one complete turn of the cycle synthesizes one molecule of succinate from two molecules of acetyl-CoA.

The two enzymes isocitrate lyase and malate synthase appear to occur together only in plants converting fats to sugars and have not been detected in animal tissue although the cycle operates (and was in fact first discovered) in certain micro-organisms during the assimilation of acetate.

The succinate produced by the glyoxylate cycle is converted via malate and oxaloacetate to phosphoenolpyruvate this last reaction being accompanied by the loss of CO_2—the only carbon lost during the conversion of fats to sucrose. Phosphoenolpyruvate is converted, via intermediates of the EMP pathway, to hexose phosphate which, by the operation of the reaction involving UDP-glucose described on p. 160 is finally converted to sucrose.

Authentication of the overall reaction scheme shown in Fig. 4.11 comes from a variety of sources, notably experiments demonstrating correlations between the activities of the glyoxylate cycle enzymes and most convincingly in experiments in which the labelling pattern in sucrose synthesized by castor bean endosperm from specifically labelled acetate is as predicted by the scheme.

The intracellular sites of the various sections of this reaction scheme have been partially located. The reactions of the β-oxidation spiral

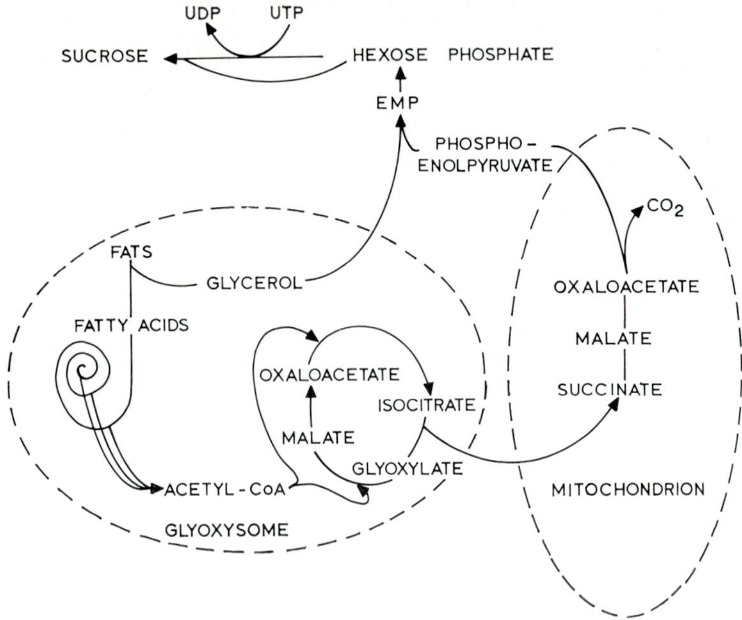

FIG. 4.11. The conversion of fats to sucrose in germinating fatty seeds. The reactions of the fatty-acid spiral and glyoxylate cycle are shown to occur in the glyoxysome, the conversion of succinate to phosphoenolpyruvate in the mitochondrion and the remaining reactions at sites other than these.

and the glyoxylate cycle have all been shown to take place in discrete, approximately-spherical, single-membrane bound organelles of about 0·8 μ diameter known as the *glyoxysomes*. Isolated glyoxysomes from castor bean endosperm can convert fatty acids to succinate (Fig. 4.11). They do not contain the enzymes of the TCA cycle necessary to convert succinate to oxaloacetate and it is believed that these reactions and the conversion of oxaloacetate to phosphoenolpyruvate occurs within the mitochondria. The remaining reactions between phosphoenolpyruvate and sucrose may take place in the cytoplasm although it is tempting to speculate that the sucrose synthesizing system may reside within the proplastids which are present in the castor bean endosperm.

The respiration of fats can result in low R.Q.s—but the conversion of fats to sucrose, in effect an incomplete respiration of the fats, can result in even lower R.Q.s. Consider, for example, the conversion of palmitic acid to sucrose:

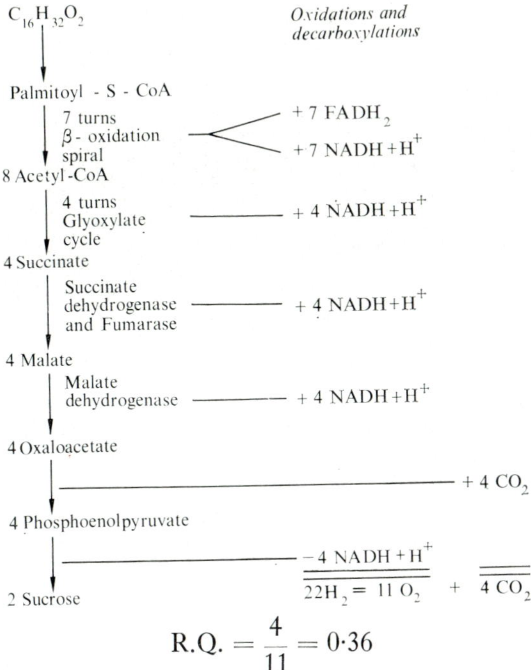

$$\text{R.Q.} = \frac{4}{11} = 0{\cdot}36$$

Neglecting the contribution of reactions concerned in the consumption of glycerol and assuming that all electrons, apart from those involved in the reduction of triose phosphate, find their way to oxygen the overall R.Q. for the process would be 0·36. However, the simultaneous operation in living tissues of respiratory pathways consuming carbohydrates usually raises observed R.Q.s to around 0·7.

The conversion of fats to sugars appears to be regulated by the activity of whichever of the enzymes, isocitrate lyase or malate synthase, is rate-limiting and the involvement of a special organelle in the process may be necessary to maintain the integrity of the pathway against possible consumption of acetyl-CoA by the TCA cycle.

Protein respiration almost certainly involves first a hydrolytic breakdown by proteinases and peptidases to give free amino acids. Of the amino acids of plant cells, only glutamic acid is known to be actively oxidised by an enzyme universally distributed, the *glutamic acid dehydrogenase*:

$$\text{glutamic acid} + NAD^+ + H_2O \rightleftharpoons \text{oxoglutaric acid} + NH_3 + NADH + H^+ \quad (37)$$

The oxoglutaric acid is an intermediate in the oxidation of pyruvate by the Krebs cycle and, therefore, can enter the main pathway of carbohydrate degradation. Plant cells also contain transaminase enzymes which catalyse the transfer of the amino groups of amino acids to keto acids (see also p. 169). One of the most active of these enzymes in plant cells is that transferring amino groups to oxoglutaric acid (which is thus converted back to glutamic acid). Such a reaction can be represented thus:

$$\begin{matrix}\alpha\text{-oxoglutaric} & & \text{amino} & & \text{glutamic} & & \text{keto acid} \\ \text{acid (amino} & + & \text{acid} & \rightleftharpoons & \text{acid} & + & \text{corresponding} \\ \text{group acceptor)} & & \text{(R)} & & & & \text{to R}\end{matrix} \quad (38)$$

It could, therefore, be that all the nitrogen of respired protein can, in this way, be canalised into the formation of glutamic acid and then be released as ammonia by the glutamic acid dehydrogenase enzyme. We are not certain that this is the case; α-oxoglutaric acid, and oxalacetic acid (corresponding to the amino acid aspartic acid) and pyruvic acid (corresponding to alanine) are acids of the Krebs cycle and can, therefore, be oxidised by this pathway. However, we do not yet know how the other keto acids arising by transamination can be oxidised nor even, in many cases, whether they are normal cell constituents. There is, therefore, as yet no certain knowledge of the reactions by which many of the known amino acids of plant cells are oxidised during protein respiration.

TERMINAL OXIDATION IN PLANT CELLS

The oxidations involved in the respiration of carbohydrates, fats and proteins involve the transfer of hydrogen ions and electrons and result in the reduction of the co-enzymes. Terminal oxidation involves transport of these hydrogen ions and electrons to oxygen and a coupling of this with the creation of new chemical bonds in which most of the conserved energy of respiration is held in utilisable form.

Our present understanding, still very incomplete as far as plant cells are concerned, of the enzymic reactions which effect this transport has a long and interesting history. In 1886, MacMunn detected in many animal tissues pigments chemically related to haemoglobin, capable of existing in both an oxidised and a reduced form and to which he assigned the role of enabling the tissues in which they occur to take up oxygen. This work of MacMunn was not at the time accepted but was later vindicated in a classical paper published in 1925 by Keilin under the title *On Cytochrome, a Respiratory Pigment Common to Animals, Yeast and Higher Plants*. In this paper Keilin not only described the characteristics of these pigments but obtained evidence that they formed a link between the dehydrogenases and oxygen. He was able to recognise three distinct cytochromes which he termed cytochromes *a*, *b* and *c*. These were iron-porphyrins combined with protein (hemoproteins) and in the oxidised form contained ferric (Fe^{+++}) and, in the reduced form, ferrous (Fe^{++}) iron. The reduced forms of the cytochromes have distinct absorption bands in the visible spectrum; the oxidised forms diffuse poorly defined absorption spectra. If we take a flat-sided vessel containing a suspension of yeast between a bright source of light and a hand spectroscope and either keep the suspension under anaerobic conditions or add inhibitors like cyanide or sulphide or a reducing agent like hydrosulphite these absorption bands of the reduced cytochromes are clearly visible. If we admit oxygen to the anaerobic suspension these bands disappear as the cytochromes are oxidised by molecular oxygen.

When Keilin succeeded in isolating cytochrome *c* from baker's yeast he found that it could be reduced by mild reducing agents but was then not reoxidised by shaking a solution in oxygen. Its oxidation immediately occurred, however, in the presence of living cells. This suggested the existence in living cells of an enzyme catalysing the oxidation of cytochrome *c* by oxygen. Now some years earlier the German biochemist, Otto Warburg, as a result of a series of brilliant researches started in 1918, had detected in all the living cells he had examined, an iron-containing oxygen-activating respiratory enzyme (*Atmungsferment*) and shown that the activity of this enzyme was inhibited by cyanide and carbon monoxide and that the carbon monoxide inhibition was reversed by illuminating the cells. In this connection it is interesting to recall that as early as 1897, Haldane had shown that oxygen is displaced from haemoglobin (also a hemoprotein) by carbon monoxide and that the haemoglobin-carbon monoxide complex is unstable in light.

Warburg went on to demonstrate that the absorption spectrum of his respiratory enzyme indicated it to be a hemoprotein and in 1939 Keilin and Hartree separated from a cytochrome *a* preparation, a new cytochrome (cytochrome a_3 or *cytochrome oxidase*) which reacted directly with oxygen and was in its properties identical in every way with Warburg's respiratory enzyme. Cytochrome oxidase is unique in the rapidity of its reaction with oxygen and in its high affinity for this gas. This affinity for and reactivity with oxygen points to cytochrome oxidase as the last step in the transport of electrons and hydrogen ions to unite with oxygen and form water. The name cytochrome oxidase indicates that the enzyme catalyses a direct oxidation of its substrate (a cytochrome) by molecular oxygen and in this respect resembles the copper-protein enzymes, *catechol oxidase* first described by Bourquelot and Bertrand in 1895 and ascorbic acid oxidase discovered by Szent-Györgyi in 1931. These copper-protein enzymes have also been considered as possible terminal oxidases in plant respiration but the evidence that they function in this role is very controversial particularly in view of their relatively low affinity for oxygen. Further, while cytochrome oxidase appears to be universally found in respiring cells, these other enzymes have a wide but not universal distribution.

Whereas in animal cells cyanide can virtually completely inhibit oxygen uptake, the respiration of many plant tissues is partially cyanide resistant. The *Arum* spadix is a classical example and at a particular stage of development may be 90% resistant to the effects of cyanide. This is an extreme example perhaps associated with the ability of the spadix to maintain a temperature above ambient (a factor in the attraction of pollinating flies). However less dramatic examples of cyanide-resistant respiration are common in plants but at present neither the electron pathway nor the metabolic functions of this process are properly understood.

Substances which can exist in an oxidised or reduced form and which can interact can be arranged in order of their ability to oxidise or reduce one another. The highly reducing substances at one extreme of such a series have increasingly negative oxidation-reduction potentials (the standard hydrogen electrode having by definition a zero potential), the highly oxidising substances at the other end of the series have increasingly positive oxidation-reduction potentials. The span of oxidation-reduction potential involved in the electron transport system of respiration ranges from $-0{\cdot}32$ volts for the NAD^+/NADH system to $+0{\cdot}82$ volts for the oxygen electrode. The path of hydrogen and electrons from the dehydrogenases to oxygen will be through a sequence of "carriers" each with a more positive oxidation-reduction potential. This enables us to say that electrons will flow from cytochrome *b* → cyt. *c* → cyt. *a* → cytochrome oxidase → oxygen. This poses the question of whether in the respiring cell this cytochrome complex directly oxidises the reduced co-enzymes. The evidence is that this does not occur, thereby suggesting that some additional hydrogen carrier is interposed between the reduced co-enzymes and the cytochromes. In 1932, Warburg and Christian isolated a "yellow enzyme" (a flavoprotein) from yeast and such flavoproteins are now known to be universally distributed in living cells. These flavoproteins (FP) consist of protein in combination with flavin nucleotides and the nucleus of these nucleotides can undergo reversible oxidation and reduction; the molecule is colourless in the oxidised form and yellow in the reduced form. The reduced molecules are either slowly oxidised by oxygen or are quite stable in oxygen but they are very rapidly oxidised by the cytochrome complex provided that the flavoprotein has been purified by a method which does not remove the iron or other heavy metal with which it is apparently associated in nature. These flavoproteins are, as might be anticipated, intermediate in oxidation-reduction potential between $NAD^+/NADH$ and the cytochromes. One of the enzymes involved in electron transport, *succinic dehydrogenase* (see p. 74) is itself a flavoprotein in which the prosthetic group is flavin adenine dinucleotide (FAD) and which therefore directly transfers electrons to the cytochrome complex. Other flavoproteins function to receive hydrogen from the reduced (NADH) co-enzymemolecules generated by the other dehydrogenases of the Krebs cycle.

A simplified version of the complex chain of electron transport from the Krebs cycle reactions to oxygen is shown below (Fig. 4.12). This also indicates the sites of oxidative phosphorylation. CoQ in this diagram represents a quinone termed *co-enzyme Q* which is now thought to be interposed in the chain between the metalloflavoproteins (FP) and the cytochrome complex.

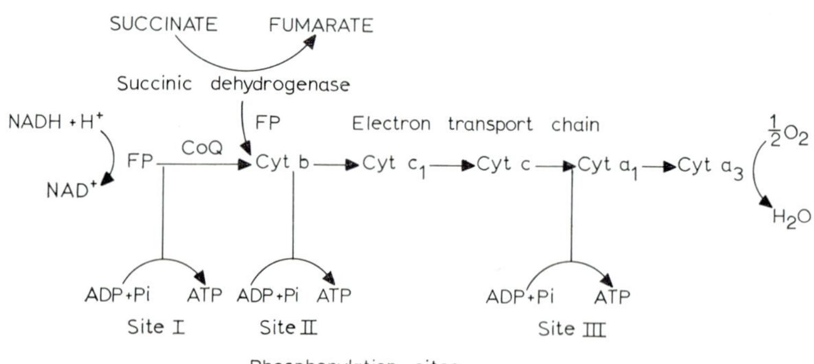

FIG. 4.12. The respiratory electron transport chain and phosphorylation sites (Sites I to III).
Key: CoQ, co-enzyme Q; FP, flavoprotein; Cyt, cytochrome.

The elucidation of this sequence of electron carriers has been achieved, for the most part, by the use of mitochondria isolated from cells in such a way as to maintain for a time much of their *in vivo* activity.

As is the case with studies in photosynthesis, isolation of organelles has permitted the application of sophisticated biochemical techniques which were difficult or impossible to use with whole tissues. Techniques of particular importance in this instance and which have been used singly or more usually in combination, involve observations of the effects on electron transport of additions of Krebs cycle intermediates and inhibitors and of physical fragmentation and reconstitution of the chain. Electron transport is monitored by measurements of substrate oxidation, oxygen uptake or by following the changes in absorption spectra associated with change in the oxidation-reduction status of certain components—the cytochromes being particularly suited to this latter type of investigation.

The isolation of sub-cellular organelles for such studies involves disrupting the cell in such a way as to leave intact the organelle of interest; mechanical, ultrasonic and enzymic destruction of the cell wall each having its particular advantages. The cells are broken in a medium which is cold, adjusted in osmotic potential, and buffered at a suitable pH. It may also contain other factors, e.g. metal ions such as magnesium, which have been empirically shown to be necessary for the maintainance of the activity of a particular organelle. The organelles are separated from the crude homogenate by the technique of differential centrifugation. A scheme for mitochondrial preparation would involve for example, disruption of the tissue in a suitable medium followed by filtration to remove coarse debris, centrifugation at low speed to remove plastids, nuclei and cell wall fragments, and finally centrifugation of the supernatant at a speed sufficient to sediment the mitochondria. Properly isolated mitochondria contain, as well as numerous other systems, all the enzymes of the TCA cycle and the mechanism for the transfer of electrons from reduced pyridine nucleotides to oxygen with the concomitant synthesis of ATP. They are, therefore, centres of respiratory activity.

The measurement of these activities can be made under steady-state conditions of electron flow, or kinetic measurements may be made during transitional conditions—a technique used to good effect by Chance and collaborators to determine the sequence of components in particular parts of the chain. It would be inappropriate to give extensive examples of the information obtained by these methods but a hypothetical example will help illustrate one of the most frequent kinds of deduction made. Suppose, upon the addition of an inhibitor of electron transport to a suspension of mitochondria actively transferring electrons from some substrate to oxygen, spectroscopic investigation indicates that the components of the electron transport chain have been divided into two groups—those of one group becoming more oxidised following the addition of the inhibitor and those of the other group more reduced. The point in the chain at which the direction of change alters (the crossover point) is the site of action of the inhibitor and compounds becoming more oxidised must be on the "oxygen side" of this site and compounds becoming more reduced on the side nearest the source of electrons. Thus a start has been made in determining the sequence of carriers in the chain.

The real metabolic importance of the electron transport system lies of course in the synthesis of ATP associated with the passage of electrons along the chain—a subject best discussed in relation to the other energy conservation reactions of the plant cell.

CONSERVATION OF THE ENERGY LIBERATED IN RESPIRATION

Reference has previously been made to the formation of adenosine triphosphate from adenosine diphosphate at two steps in the sequence of reactions by which triosephosphate is converted to pyruvic acid.

This substance, adenosine triphosphate (ATP) (see Fig. 4.6), is a universal constituent in living cells and plays a central role in the conservation of energy and in the transfer of energy from catabolic to anabolic (synthetic) reactions.

The phosphate bonds of ATP can be broken hydrolytically under appropriate conditions and the free energy changes associated with the cleavage of each of the three phosphate bonds is as follows:

$$\text{ATP}+\text{H}_2\text{O} \rightarrow \text{ADP}+\text{phosphate} \qquad \Delta G' = -8{\cdot}5 \text{ kcal.} \qquad (39)$$

$$ADP + H_2O \rightarrow AMP + \text{phosphate} \qquad \Delta G' = -7{\cdot}8 \text{ kcal.} \quad (40)$$

$$AMP + H_2O \rightarrow \text{adenosine} + \text{phosphate} \qquad \Delta G' = -3 \text{ kcal.} \quad (41)$$

The molecule of ATP therefore contains two phosphate bonds which are "energy-rich" and one which is relatively poor in energy. We use the symbol ~P to represent a high energy phosphate radical so that the ATP molecule can be represented A–P~P~P where A represents the nucleoside, adenosine.

The synthesis of ATP from ADP in transfer reactions catalysed by specific kinases conserves energy released in the metabolism of triosephosphate to pyruvate. The reaction

$$\begin{matrix}\text{glyceraldehyde} \\ \text{3-phosphate}\end{matrix} + NAD^+ \rightarrow \begin{matrix}\text{3-phosphoglyceric} \\ \text{acid}\end{matrix} + NADH + H^+ \quad (42)$$

involves a large decrease in free energy and most of this would be lost as a positive heat of reaction. However, this reaction should be compared with the reaction catalysed by triosephosphate dehydrogenase (p. 97) which has a $\Delta G' = +0{\cdot}4$ kcal. Part of the energy released in the oxidation of triosephosphate is conserved by the simultaneous synthesis of the high energy phosphate bond in the carboxyl group of C atom 1 of the diphosphoglyceric acid (the instability of this 1 carboxyl phosphate group is indicative of its energy-rich nature). However, despite the instability of this carboxyl phosphate bond, its energy is conserved because in the presence of ADP and of the specific kinase, ATP is synthesised (eqn. 19, p. 97). A similar conservation of the high energy content of the phosphate bond of phosphoenolpyruvic acid is achieved by the formation of pyruvic acid being linked to the synthesis of ATP from ADP.

It thus follows that in the conversion of a molecule of fructose 1-6-disphosphate to 2 molecules of pyruvic acid, four of the terminal phosphate bonds of ATP are synthesised. However, ATP is, as already mentioned, also involved in synthetic reactions and when glucose is the respiratory substrate two molecules of ATP are degraded to ADP to form the fructose-diphosphate. The net gain in ~P during the fermentation of a gram mol. of glucose is, therefore, $2 \times 8{\cdot}5 = 17$ kcal as phosphate bond energy. The equation (eqn. 15) on p. 85) shows that the ΔG° of fermentation is —54 kcal from which we can calculate that of the total energy released 17/54 $\times$ 100 = 31% is conserved as energy in ~P of ATP. Comparison of the equations for Respiration and Fermentation (eqns. 9 and 10) leads us now to enquire whether the very much greater energy release in respiration is also conserved by ATP synthesis. Thus we are led to consider the reactions involved in pyruvate oxidation in which the energy released is conserved by simultaneous synthesis of ATP from ADP.

Using pigeon brain dispersions and heart muscle extracts. Ochoa, in 1944, obtained evidence that 6 molecules of inorganic phosphate were taken into organic combination per *molecule* of oxygen absorbed in respiration. Later work with mitochondria showed that a similar phosphate assimilation accompanied the oxidation of organic acids, including pyruvate. Mitochondrial preparations are particularly suited to the study of the quantitative synthesis of ATP during the oxidation of pyruvate by oxygen. To do this a rather interesting technique can be employed. Although the adenosine phosphates are so important in the energy relationships of the living cell they are present in cells in small amounts. Consequently, if ATP synthesis from ADP proceeds actively the small amount of ADP present is quickly depleted unless some mechanism by which the terminal phosphate bond of ATP is transferred in a further phosphorylation reaction is proceeding simultaneously and thereby constantly regenerating ADP. Reference has already been made (p. 94) to the phosphorylations of sugar which precedes the formation of triose-phosphate and to the involvement of ATP in these reactions. One of these phosphorylation reactions is catalysed by the enzyme, hexokinase and can be represented thus:

$$\text{glucose} + \text{ATP} \rightarrow \text{glucose-6-phosphate} + \text{ADP} \qquad (43)$$

If we take some recently isolated mitochondria, add some ADP to reinforce the natural content, glucose and hexokinase to trap the phosphate of newly formed ATP as glucose-6-phosphate, and an oxidisable substrate such as pyruvate or other organic acid of the Krebs cycle, it is possible to demonstrate active phosphorylation. Further, by measuring glucose-6-phosphate formation and oxygen uptake by the mitochondria one can work out the P:O ratio, the molecules of phosphate incorporated into ATP per oxygen *atom* absorbed.

An alternative method of measuring the amount of phosphate esterified per oxygen atom consumed depends upon the "coupling" of the phosphorylation process to electron transport. In properly isolated mitochondrial preparations electron transport will not proceed in the absence of the substrates of phosphorylation ADP and P_i. In the intact cell the limiting substrate is usually ADP which is produced as a result of the consumption of ATP. Thus the rate of electron transport and therefore substrate utilization can be geared to the cell's requirements for ATP, a phenomenon referred to as respiratory control. Phosphorylation and therefore electron transport and oxygen uptake in a "coupled" mitochondrial preparation will commence only upon the addition of ADP and will cease when the ADP is used up. Assuming that the ADP is consumed only in the synthesis of ATP and that the breakdown of ATP to ADP and P_i does not proceed at a significant rate then the ADP:O ratio, which is equivalent to the P:O ratio, can be calculated from a knowledge of the amount of ADP added and the amount of oxygen consumed. The first successful measurement of P:O ratios were carried out using animal mitochondria, but in work since 1953 by Laties and many others, active and efficient phosphorylation has been obtained with mitochondrial suspension from higher plants. The most efficient suspensions of mitochondria have given P:O ratios above 3·0 and approaching 4·0 which corresponds well with Ochoa's estimates made earlier and implies that at least some 15 $(5 \times 3) \sim$ P can be synthesised per molecule of pyruvate oxidised (see eqns. 20 and 21 for pyruvate oxidation on p. 98). It is interesting that many mitochondrial preparations have much lower efficiencies than those recorded above, the P:O ratios often approximating to 1·0, and that as mitochondrial suspensions "age" they lose their ability to effect phosphorylation while their oxidative activity as measured by oxygen uptake and carbon dioxide release is unimpaired. A similar effect follows addition of dinitrophenol to cells or isolated mitochondria; this substance "uncouples" oxidation from phosphorylation, oxygen uptake is usually stimulated but no ATP synthesis takes place.

This brings us to the difficult question of the reactions involved in ATP synthesis in respiration. The reactions of the Krebs cycle have been very fully elucidated, both as regards the chemical changes involved and the changes in free energy occurring at each step. Only at one step in this cycle is ATP synthesised. This is in the oxidation of oxoglutaric acid to succinic acid. In this reaction a compound between succinic acid and Co-A is formed (succinyl-S.CoA) and this reacts with ADP in the presence of inorganic phosphate thus (see Chapter 3, p. 71):

$$\text{succinyl-S.CoA} + \text{ADP} + H_3PO_4 \rightarrow \text{ATP} + \begin{array}{c}\text{succinic}\\ \text{acid}\end{array} + \text{CoA-SH} \quad (44)$$

Incidentally, this is an interesting reaction because it shows us that the S-linkage binding CoA to metabolites is also an "energy-rich" bond; that it contains sufficient energy to power the synthesis of ATP from ADP and inorganic phosphate. Now, in the conversion of triosephosphate to pyruvate there are also two reactions (p. 97) involving ATP synthesis. Thus, we know of three *substrate phosphorylations* occurring in the complete oxidation of triosephosphate, i.e. 6 molecules of ATP are synthesised by such reactions per molecule of glucose and, since two of these are consumed in the formation of fructose-1-6-diphosphate, there is a gain of $4 \times \sim P$ per glucose molecule by substrate phosphorylation.

Clearly, most of the ATP molecules formed in respiration are synthesised at some part of the respiratory sequence other than in either pyruvic acid formation or its oxidation in the Krebs cycle. Now there is a very large release of energy (decrease in free energy) involved in the oxidation of reduced NAD and this, calculated from the gap in oxidation-reduction potential between the $NAD^+/NADH$ system and oxygen, comes out at some 52 kcal/g-mol. In line with this is the experimental observation that when reduced NAD is fed to mitochondria it is possible to obtain P:O ratios close to but usually slightly below 3·0. This means that for every two hydrogen atoms from reduced NAD transported along the terminal oxidation pathway to combine with one atom of oxygen there are up to three molecules of ATP synthesised. The three sites of phosphorylation have been located in the electron transport chain at or near the positions shown in the diagram on Fig. 4.12. The methods used to locate these are basically similar to those used in the investigation of the electron transport chain itself. For example when reduced NAD is supplied as substrate to a mitochondrial preparation the P:O ratio as noted above is 3·0. When succinate is substrate, however, the P:O ratio is 2·0, the inference being that one site of phosphorylation (site I), is located at a point on the electron transport chain further away from the terminal oxidase than the point at which electrons donated by succinate enter the chain. By other experiments involving the use of inhibitors and various substrates sites II and III have been located (see Fig. 4.12). Since, to oxidise a molecule of glucose $12 \times 2H$ are transported and 12 atoms of oxygen absorbed, this indicates the synthesis of 36 ~ P of ATP per molecule of glucose respired. It is this phosphorylation occurring in mitochondria and linked to the oxidation of reduced co-enzymes via the flavoprotein-cytochrome pathway which is termed *oxidative phosphorylation*.

From the above discussion it will be seen that some 40 ~ P (36 by "oxidative" and 4 by "substrate" phosphorylation), each with an energy content of $\simeq$ 8·5 kcal/g-mol, are synthesised per g-mol of glucose respired and some 340 kcal therefore conserved as utilisable energy. Since the overall decrease in free energy involved in the complete oxidation of a g-mol of glucose is 686 kcal (eqn. 14, p. 85) this means that approximately 50% of this energy is trapped during respiration in a form which can be used in cellular metabolism for the synthesis of cell constituents or to perform work.

A problem of great and general importance not so far discussed is the mechanism by which phosphorylation of ADP is linked to electron transport. Although the present chapter is concerned with catabolism and therefore with oxidative phosphorylation the mechanism of photophosphorylation is believed to be basically similar—it will therefore be convenient to discuss the two together.

If a chloroplast suspension capable of photophosphorylation is illuminated in the absence of phosphate and then after a period of darkness phosphate is supplied, measurable synthesis of ATP takes place. This implies the existence of an appreciable amount of a relatively stable, non-phosphorylated, high-energy intermediate.

However, although much effort has been expended no chemical compound of this kind has been identified. This apparent contradiction can be solved in terms of the *Chemiosmotic hypothesis* of Mitchell which suggests that electron transport—whether respiratory or photosynthetic—results in the transport of protons and the establishment of a proton gradient. The energy represented by this proton gradient (or secondary changes resulting from it) may be utilized in ATP synthesis. If this is so then it is no surprise that efforts to isolate the non-phosphorylated high energy intermediates did not succeed!

Two basic questions may be asked: (1) How does the transport of electrons result in the transport of protons? and (2) How is the energy represented by a proton gradient utilized in the synthesis of ATP from ADP and P_i? Detailed consideration of both of these questions is beyond the scope of this book but the basic principles whereby the chemiosmotic hypothesis can account for these phenomena can be described. The transport of protons as a result of the passage of electrons along the electron transport chain depends upon the fact that some electron transport components accept only electrons during their reduction (e.g. cytochromes) whereas others such as FAD accept both electrons and protons. We must expect, therefore, the uptake and release of protons at various points along the chain and if the chain is oriented within a membrane in such a way that protons are taken up on side one and protons released on the other we have a mechanism for proton transport. Such a situation is depicted in Fig. 4.13.

A light dependent pH rise has in fact been observed in the bathing solution of a suspension of isolated chloroplast thylakoids indicating that protons have been taken into the thylakoids. This is in contrast to the situation in mitochondria in which it has been shown that protons are extruded as a result of electron transport. That there is in fact a relationship between a proton (pH) gradient and phosphorylation has been elegantly demonstrated by Jagendorf and Hind in experiments in which the proton gradient normally established by photosynthetic electron transport was induced by manipulation of the pH inside and outside chloroplast thylakoids. In complete darkness an appreciable amount of ATP was formed—clearly demonstrat-

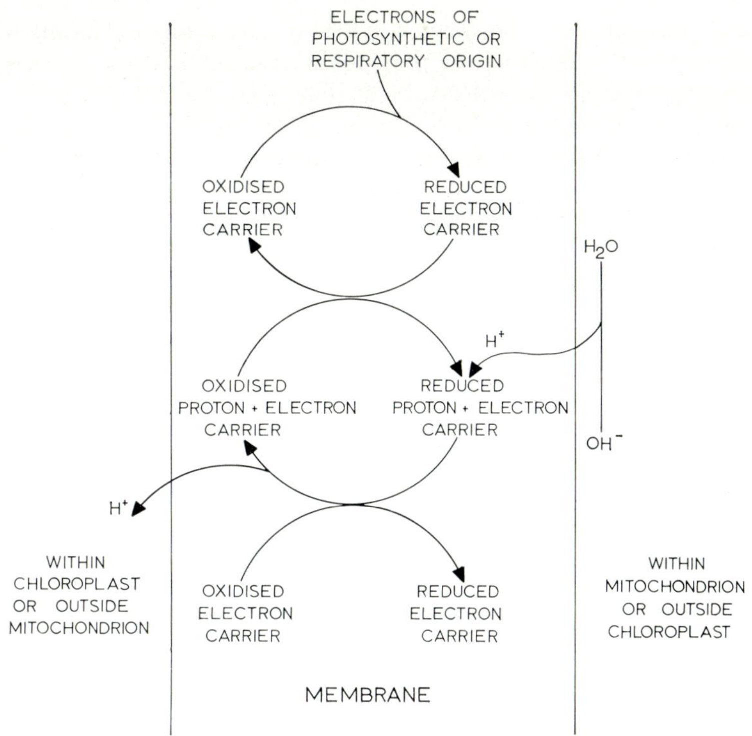

FIG. 4.13. The linkage of proton transport across a cell membrane to electron transport as envisaged by the chemiosmotic hypothesis of Mitchell.

ing that energy represented as a pH gradient can be utilized in phosphorylation.

The answer to the second question above which concerns the actual mechanism by which phosphorylation occurs at the expense of a proton gradient lies, so the Mitchell hypothesis suggests, in another membrane phenomenon. The formation of ATP can be represented as the removal of the elements of water from ADP + P_i. It is suggested that this may be achieved by an enzyme system (ATPase) situated within the hydrophobic interior of the membrane which has a proton gradient across it. The removal of H^+ to the negative (OH^-) side and OH^- to the positive (H^+) side of the membrane will result in the dissipation of the proton gradient and an associated synthesis of ATP from ADP and P_i (Fig. 4.14).

Evidence consistent with a mechanism of this kind has come from experiments with chloroplasts and mitochondria in which it has been shown that compounds which uncouple phosphorylation from electron transport often also result in the dissipation of the proton gradient (or of gradients derived from the proton gradient which for

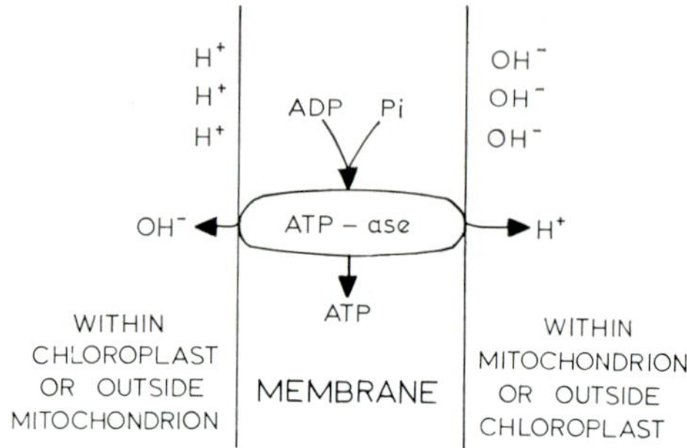

FIG. 4.14. Synthesis of ATP by dissipation of a proton gradient.

the sake of simplicity have been neglected in the foregoing discussion). The possibility of ion movements associated with the chemiosmotic hypothesis is of interest in relation to active ion transport and the subject is pursued further in Chapter 7.

THE CONTROL OF RESPIRATION

Presumably, at full oxygen tension and optimum temperature and in the presence of abundant respiratory substrate, the rate of respiration is controlled by those chemical reactions whose rate puts a "brake" on the overall process, by reactions which act as "pacemakers". Isolated enzyme systems may be self-regulatory in their activity as instanced by the dehydrogenase which oxidises malic to oxaloacetic acid. When in this system the oxaloacetic acid concentration reaches 10^{-4} M it inhibits its own synthesis. Further, oxaloacetic acid also inhibits another enzyme of the Krebs cycle, the enzyme succinic acid dehydrogenase. The concentration of oxaloacetic acid, itself regulated by the rate of formation of acetyl-CoA with which it condenses, may, therefore, control the rate of cycling of the Krebs cycle. Reference has already been made to the phenomenon of respiratory control in mitochondria in which phosphorylation is coupled to electron transport. By influencing the availability of oxidised pyridine nucleotides, electron transport may indirectly control substrate utilization by the TCA cycle or β-oxidation pathway. Thus the dependence of electron transport upon a supply of ADP and P_i is clearly important in the regulation of respiration according with the rate of utilization of ATP by the cell.

One of the most dramatic, longest known, and most difficult to explain examples of the regulation of carbohydrate catabolism is the observation, first made by Pasteur, that at low oxygen tensions the rate of aerobic respiration is strongly suppressed and the rate of fermentation increased and that under these conditions, and particularly at zero oxygen pressure, the rate of glucose breakdown can be four times that in air. This pronounced "sparing" action of oxygen upon the breakdown of sugars is known as the Pasteur effect. In this case, although the effect of oxygen upon the relative levels of ATP, ADP and inorganic phosphate is clearly involved, it is also necessary to consider the cellular sites of ATP synthesis and utilisation to develop a satisfactory explanation of the Pasteur effect (see Chapter 8, p. 254).

Synthetic processes proceeding in respiring cells not only consume the energy stored in ATP and other phosphorylated compounds but draw upon the intermediates of respiration. For instance, the enhanced rate of respiration of cells actively synthesising proteins is, in part, due to the utilisation in amino acid synthesis of organic acids which are Krebs cycle intermediates. It is clearly a major objective in the study of plant metabolism to understand the quantitative aspects of such inter-relationships between the metabolic processes of the cell.

FURTHER READING

W. O. James. *Plant Respiration.* Oxford University Press, London, 1953.

H. Beevers. *Respiratory Metabolism in Plants.* Row, Paterson & Co., New York, 1961.

V. S. Butt and H. Beevers. The plant lipids. Chapter 7 in *Plant Physiology*, vol. 4B, edited by F. C. Steward. Academic Press, New York, 1966.

MORE ADVANCED READING

F. Kidd, C. West and G. E. Briggs, A quantitative analysis of the growth of *Helianthus annuus*. Part I. The respiration of the plant and of its parts throughout the life cycle. *Proceedings of the Royal Society*, **B92**: 368–384, 1921.

W. O. James. Reaction paths in the respiration of the higher plants. *Advances in Enzymology*, **18**: 281–318, 1957.

F. F. Blackman. *Analytic Studies in Plant Respiration*. Cambridge University Press, 1954.

E. F. Hartree. Cytochrome in higher plants. *Advances in Enzymology*, **18**: 1–64, 1957.

D. R. Goddard and W. D. Bonner. Cellular Respiration, in *Plant Physiology*, vol. 1A, pp. 209–312, edited by F. C. Steward. Academic Press, New York, 1960.

J. Bonner and J. E. Varner (Editors). *Plant Biochemistry*. Academic Press, New York, 1965 (particularly chapters 6 and 10).

D. E. Griffiths. Oxidative phosphorylation. *Essays in Biochemistry* **1**: 91–120, 1965.

A. T. Jagendorf and E. Uribe. Photophosphorylation and the chemiosmotic hypothesis. *Brookhaven Symposia in Biology*, **19**: 215–245, 1966.

P. Mitchell. Translocation through natural membranes. *Advances in Enzymology*, **29**: 33–87, 1967.

CHAPTER 5

Anabolism

"*. . . that this operation . . . begins only after the sun has for some time made his appearance above the horizon . . . that this operation of the plants is more or less brisk in proportion to the clearness of the day and the exposition of the plants . . . that this operation of plants diminishes towards the close of day, and ceases entirely at sunset; that this office is not performed by the whole plant, but only by the leaves and the green stalks; that even the most poisonous plants perform this office in common with the mildest and most salutary; that the most part of leaves pour out the greatest quantity of this dephlogisticated air from their under surface . . .*"

Jan Ingen-Housz, in *Experiments upon Vegetables, Discovering Their Great Power of Purifying the Common Air in Sunshine and Injuring it in the Shade and at Night*. Elmsly & Payne, London, 1779.

"*ATP is the main fuel of life produced in photosynthesis and oxidative phosphorylation. In both cases it is produced by an electric current, that is, the energy released by a 'dropping' electron.*"

A. Szent-Györgi, in *Introduction to a Submolecular Biology*, Academic Press, New York, 1960.

INTRODUCTION

THE term *anabolism* covers all those aspects of metabolism which involve the development of complex molecules from less complex molecules, of larger molecules from smaller molecules. These are the synthetic reactions of metabolism, the reactions whereby the primary nutrients, the inorganic ions, water and carbon dioxide required by green plants, are built up into a myriad of organic molecules.

The primary process of anabolism is *photosynthesis*, the process whereby sugars are synthesised, in green cells exposed to sunlight, from carbon dioxide and water. As earlier emphasised the unique aspect of this form of carbon assimilation is the conversion of light energy into the chemical energy of newly synthesised sugar molecules. It is photosynthesis which maintains life in all its abundance on this planet, it is this process which has reduced the content of carbon dioxide in the earth's atmosphere to 0·03% and raised its oxygen content to 21%.

It may well be that the first forms of life were colourless anaerobic micro-organisms dependent upon being bathed by a sea containing a great variety of complex organic compounds; that the first organisms were extreme *heterotrophs*. However, continued evolution of life resulted from an event perhaps second only in improbability to the origin of life, the acquisition by some organism of the ability to utilise as its primary source of energy the sun's radiation, the "invention" of photosynthesis. Thereby arose a new type of organism, which had the potential to become independent of external organic matter and exist autotrophically requiring only simple organic nutrients. This led to a separation of oxygen from its union with hydrogen; led to the first appearance of oxygen in the previously anaerobic atmosphere. The process of photosynthesis provided the raw materials (reduced organic compounds and oxygen) for a new mechanism of energy release, the aerobic respiration of organic cell constituents. The molecules elaborated by photosynthesis were at one and the same time the starting molecules (precursor molecules) for the synthesis of other organic molecules essential to life and a source of energy for such synthesis.

It is therefore appropriate to open any discussion of anabolism with a consideration of the process of photosynthesis.

THE DISCOVERY AND GENERAL NATURE OF PHOTOSYNTHESIS

As early as 1772, Joseph Priestley described the power of plants to restore air vitiated by the burning of candles or animal life. Then in 1779 in the book *Experiments upon Vegetables, discovering their*

Great Power of Purifying the Common Air in Sunshine and Injuring it in the Shade and at Night, Jan Ingen-Housz clearly established that the purifying activity of plants depended upon their exposure to light, that it involved the absorption of some constituent of the air and that during this process "dephlogisticated air" was evolved. By 1782 came the recognition by Jean Senebier that the constituent absorbed by plants from the air in light was "fixed air" and by 1796 the discovery of the nature of photosynthesis was almost complete for Ingen-Housz in his book *Food of Plants and Renovation of the Soil* published in that year clearly enunciated the view that plants acquire their carbon as "fixed air" which he recognised as carbonic acid, that this carbon is elaborated into organic matter in the light and that during the processes oxygen (previously referred to as "dephlogisticated air") is evolved. It remained for N. Théodore de Saussure in his book *Recherches chimiques sur la vegetation* published in 1804 to describe the quantitative experiments which had led him to the conclusion that both *water* and carbon dioxide are involved in the synthesis of organic matter by green plants in light.

Ingen-Housz recognised that photosynthesis was a property of the green parts of the plant, its leaves and green stems. The pigment complex conferring this green colour was termed chlorophyll in 1818 by Pelletier and Caventou and the importance of this complex was stressed by Dutrochet in 1837 who regarded it as a veritable "elixir of life". Julius Sachs in 1862 first postulated starch as a direct product of photosynthesis and described the famous experiment in which if half of a leaf is effectively shaded and the other half exposed to light, then starch appears only in the illuminated half. Since the starch grains arose in the plastids, he regarded these as the centres of photosynthesis. That the chloroplasts are also the centres of oxygen release in photosynthesis was elegantly demonstrated by Englemann (1882) by using oxygen-sensitive motile bacteria which migrated to the position of the chloroplasts. Another famous nineteenth-century German plant physiologist, W. Pfeffer (1873), showed that starch formation only occurred in illuminated leaves if the atmosphere contained carbon dioxide.

The general nature and importance of photosynthesis was established by the end of the nineteenth century. During the first 40 years of the twentieth century the chemistry of the chlorophyll pigments and their absorption spectra were worked out beginning with the classical studies of Willstätter and Stoll (1913–1918). During this period, the quantitative relationships between photosynthesis and temperature, light intensity and carbon dioxide concentration were also worked out, beginning with the classical studies of F. F. Blackman in 1905. These studies of Blackman showed that when photosynthesis is proceeding in well illuminated leaves receiving an ample supply of carbon dioxide then its Q_{10} is always above 2·0 whereas at low light intensities the Q_{10} approaches 1·0. Q_{10} is the temperature coefficient, the ratio of the rate of a "reaction" at $t + 10°C$ to its rate at $t°C$. For ordinary chemical reactions (thermochemical reactions) the Q_{10} is 2·0 or higher (see Chapter 3, p. 73; Chapter 4, p. 88), whereas photochemical reactions (reactions involving the absorption of radiant energy) are almost insensitive to temperature and, therefore, have Q_{10} values close to unity. From these observations, Blackman suggested that at low light intensities photosynthesis was limited by reactions involving light energy (photochemical or "light" reactions), whereas the response of photosynthesis to temperature at high light intensities indicated the occurrence of thermochemical reactions, which he designated, for contrast with the photochemical reactions, as "dark" reactions. In support of this concept the renowned German biochemist, Otto Warburg, showed that cyanide (a powerful inhibitor of enzymes with metal ions as prosthetic groups) was a powerful inhibitor of photosynthesis at high and a relatively weak inhibitor at low light intensities. The cyanide inhibition of a "dark" reaction was, as expected, most marked at high light intensities when "dark" reactions would be limiting the rate of photosynthesis. Warburg also showed that at high light intensities, the amount of photosynthesis resulting from a given number of flashes of light is increased if the flashes are separated by a dark period rather than being given contiguously. Emerson and Arnold (1932) found in their experiments that, using intense flashes of 10^{-5} sec, the minimum dark interval permitting maximum oxygen evolution per light flash was 0·03 sec at 25°C and 0·4 sec at 1°C. These flashing light experiments again pointed to "dark" reactions which could be limiting and which, after a period of intense illumination, could "catch" up in a dark interval to permit again of the maximum rate of oxygen evolution in light. These observations formed the background for the generalisation of the nature of photosynthesis advanced by van Niel in 1930–1 in a survey of photosynthesis both in green plants and in certain purple sulphur bacteria. Van Niel suggested that all forms of photoreduction of carbon dioxide can be formulated:

$$CO_2 + 2H_2A \rightarrow \{CH_2O\} + H_2O + 2A \qquad (45)$$

In the case of the sulphur bacteria, A = sulphur, in the case of green plants, A = oxygen. Photosynthesis is represented as an oxidation-reduction reaction in which carbon dioxide is reduced to the level of carbohydrate, represented as $\{CH_2O\}$, by 4 hydrogens from the reductant H_2A which, therefore, also yields 2A (2 atoms of sulphur or a molecule of oxygen according as to whether the reductant is hydrogen sulphide or water). Since clearly the sulphur formed in bacterial photosynthesis must come from the H_2S, then if this general formulation is correct the oxygen of photosynthesis must come entirely from water in green plants. Van Niel's equation for photosynthesis in the green plant is, therefore,

$$CO_2 + 2H_2O \rightarrow \{CH_2O\} + H_2O + O_2 \qquad (46)$$

and the introduction of the second molecule of water in the left-hand side of the equation is meaningful since there must be no less than two water molecules involved to release the molecule of oxygen.

Van Niel further postulated that the photochemical reactions of photosynthesis are concerned with the decomposition of the reductant, water. In its simplest form this would presumably involve four identical photochemical reactions

$$4H_2O \xrightarrow{4h\nu} 4H^+ + 4e + 4[OH] \qquad (47)$$

In this equation $h\nu$ represents a unit (a photon or quantum) of light energy,

$$4[OH] \rightarrow 2H_2O + O_2 \qquad (48)$$

Recent work suggests that the decomposition of water may not be achieved directly by a photochemical reaction but may involve the utilization in enzymic "dark" reactions of a product of the photochemical reactions. The work of Joliot suggests that four molecules of this product (which may well contain manganese), synthesised by four separate light reactions, are involved in the liberation of one molecule of oxygen. The concept that the photochemical reactions are concerned with the oxidation of water received support from the demonstration in 1937 by R. Hill that he could isolate chloroplasts which although they had lost the ability to reduce CO_2 could, when and only when illuminated, both evolve oxygen and reduce ferric compounds or quinones. The Hill reaction was interpreted as a partial photosynthetic system, one in which the linkage between the primary reducing activity generated photochemically and the "dark" reduction of CO_2 had been broken. The concept that the reactions involved in carbon dioxide assimilation and reduction were the "dark" reactions of photosynthesis was supported by the demonstration, first made in 1935 by Wood and Werkman with bacteria, that CO_2 can be assimilated into organic form using the energy and "reducing activity" generated by respiration. The unique feature of photosynthesis was not the assimilation of CO_2 but the utilisation of light as the source of energy for this assimilation.

Despite these tentative steps towards an understanding of the mechanisms involved in photosynthesis no real breakthrough in our understanding of its biochemistry came until immediately after the Second World War which brought with it the availability for biological research of isotopes including the radioactive isotope of carbon (^{14}C) and the mass isotope of oxygen (^{18}O). One of the first of the post-war experiments was the demonstration by S. Ruben in 1941 using, alternatively, CO_2 and H_2O labelled with ^{18}O, that the oxygen evolved in photosynthesis comes from the H_2O, as postulated by van Niel, and not from the CO_2. But this is leading us on to the biochemistry of photosynthesis and it would be appropriate if first we considered further certain aspects of photosynthesis as it proceeds in the living leaf or algal cell.

INFLUENCE OF EXTERNAL FACTORS ON THE RATE OF PHOTOSYNTHESIS

It was a result of his studies of the influence, at constant temperature, of carbon dioxide concentration and light intensity that

Blackman (1905) was led to enunciate the principles of *limiting factors* thus, "when a process is conditioned as to its rapidity by a number of separate factors the rate of the process is limited by the pace of the slowest factor". Later and more accurate determinations showed that the curves relating rate of photosynthesis to light intensity or carbon dioxide concentration are of the form shown in Fig. 5.1. With fixed

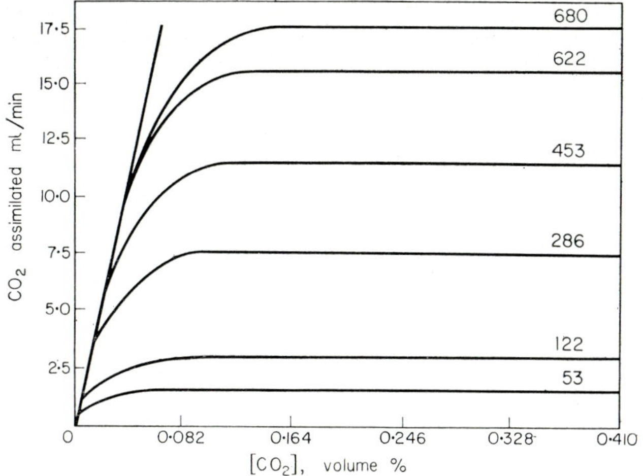

FIG. 5.1. Relationship of carbon dioxide concentration to the rate of carbon dioxide assimilation at various light intensities (shown against each curve in kerg/cm^2/sec). (Data for whole plants of *Triticum sativa* (wheat) after W. H. Hoover, E. S. Johnston and F. S. Brackett, *Smithsonian Inst. Publ. Misc. Collections*, **87:** No. 16, 1933.) (From E. I. Rabinowitch, *Photosynthesis*, vol. 2, pt. 1, Interscience Pub. Inc., New York, 1951.)

CO_2 concentration this means that there is a region where the rate of photosynthesis rises in a linear fashion with rise in light intensity (where light is a "limiting factor") but that with further increase in light intensity the curve begins to bend (some other factor is beginning to limit the rate) and beyond a certain point becomes parallel to the abscissa. At this point the system is light-saturated and the rate is determined by limiting "dark" reactions. One very interesting aspect of this light curve is that it seems from the very beginning to be linear; light at very low intensities is used with an *efficiency* similar to that at all intensities up to the point when other factors begin to limit the rate. This suggests that the amount of light being absorbed by the "chlorophyll complex" is linearly related to the intensity and that a fixed amount of carbon dioxide is assimilated (or oxygen evolved) per unit of light energy absorbed even though, at low intensities, only a small proportion of the chlorophyll molecules will, at any one time, be in an "activated" state (their energy enhanced by absorption of energy) (see p. 153).

These studies have involved the use of artificial "white" light having a spectral composition similar to natural sunlight. Other workers have used monochromatic light (literally light of a single wavelength but in practice light within a narrow band of wavelength) from different regions of the visible spectrum in an attempt to assess which regions of the spectrum are utilised in photosynthesis. A typical curve derived from such an experiment which may be termed an action spectrum is shown in Fig. 5.2. In such studies it is necessary to know the extent to which the particular wavelength is absorbed and, if possible, to compensate for non-specific absorption (absorption other than by the pigments of the chlorophyll complex). It is also necessary to express the absorbed radiation in similar units irrespective of wavelength. One way of expressing "intensity" or absorbed radiation would be in the conventional terms of energy, such as ergs per cm^2 per sec. However, the radiant energy concerned in photosynthesis is involved in photochemical reactions and in such reactions radiation is absorbed in units or "packets" of energy termed *quanta* (usually denoted hv†) and the energy value of these quanta varies with the wavelength (long-wave radiation having a lower energy per quantum than short-wave radiation). It is, therefore, preferable to express the amount of energy absorbed as a number of quanta and to express the efficiency of each wavelength in terms of the number of quanta absorbed for each molecule of carbon dioxide assimilated (as a *quantum efficiency*). This reveals, for instance, whether

† E (energy of a quantum in ergs) $= hv$, where $h =$ Planck's constant ($6{\cdot}61 \times 10^{-27}$) and v (the frequency of the radiation) $= C/\lambda$, where $C =$ velocity of light (3×10^{-10} cm/sec) and $\lambda =$ wavelength in cm.

a quantum of red light is used with equal efficiency to a quantum of green or yellow or blue light in photosynthesis. We may regard these action spectra as evidence of differences in photosynthetic efficiencies of different pigments. The trough which occurs between 640 and 460 mμ corresponds to the trough in chlorophyll *a* absorption (Fig. 5.4). However the significant utilisation of quanta between these wavelengths is taken as evidence that other pigments, which absorb

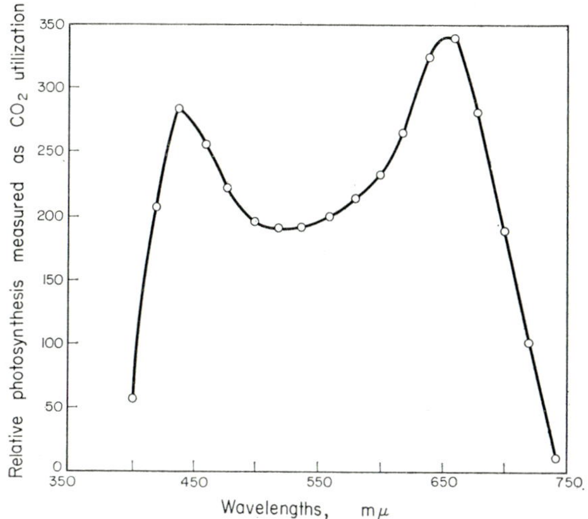

FIG. 5.2. Relative rates of photosynthesis of wheat in different wavelengths of light of equal intensity. (Data of W. H. Hoover, *Smithsonian Inst. Publ. Misc. Collections,* **95:** No. 21, 1937; 11.)

in this range, can contribute to photosynthesis. In higher plants, chlorophyll *b* and the carotenoids are the accessory pigments which absorb in this region and transmit the energy to chlorophyll *a*. In red algae a similar function is performed by phycobilins, and in brown algae the carotenoid fucoxanthol spans the gap. That energy transfer between accessory pigments and chlorophyll *a* (which for reasons discussed below is believed to be the only plant pigment involved directly in photosynthetic reactions) occurs can be shown by illuminating an organism with light absorbed only by the accessory pigment. If this is done under suitable experimental conditions a fluorescence characteristic of chlorophyll *a* can be detected. This fluorescence represents a release of energy from chlorophyll *a* which must have received this energy from the accessory pigments. The drop in quantum efficiency corresponding to the troughs of the chlorophyll absorption spectrum does not indicate that quanta absorbed by chlorophyll in this region are less efficiently used in photosynthe-

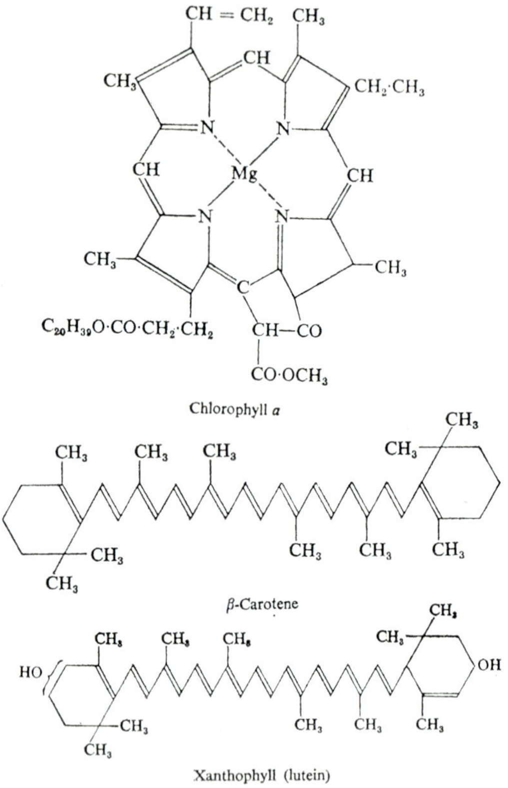

Chlorophyll *a*

β-Carotene

Xanthophyll (lutein)

FIG. 5.3. Chemical structure of chlorophyll *a* (note the long phytyl side chain) and skeleton formulae of β-carotene and the carotenol lutein.

sis but rather that energy absorbed in this region by other pigments is less efficiently transferred to chlorophyll *a*. By making certain assumptions about the structure of the photosynthetic apparatus and the values of the absorption coefficient *in vivo* it is possible to estimate for example that fucoxanthol in brown algae transfers approximately 80% of the energy it absorbs to chlorophyll *a*.

As already indicated, temperature has a marked effect upon the rate of photosynthesis under conditions where light intensity and carbon dioxide concentration are not limiting, and this observation led Blackman to postulate the occurrence of "dark reactions". For most plants, photosynthesis increases from just above 0°C to 35°C, although often at temperatures above 30°C the initial high rate is not maintained and this is markedly so above 35°C. One major factor in the decline in rate of photosynthesis at these high temperatures is the denaturation of enzyme proteins. It is, therefore, very interesting that some tropical plants can photosynthesise at temperatures above 40°C and that some algae, indigenous to hot springs, can grow at temperatures up to 89°C.

Further consideration of Fig. 5.1 indicates that under conditions of illumination approximating to those of a bright summer's day the rate of photosynthesis is limited by the atmospheric CO_2 concentration of 0·03%. In fact increase in rates of photosynthesis are observed with increase in CO_2 concentration up to at least 1% and CO_2 fertilisation as it is sometimes called is a commercial proposition for some glasshouse crops.

If a plant is illuminated in a closed container the carbon dioxide concentration inside the container eventually reaches a steady level which is known as the carbon dioxide compensation point. The final level represents the balance attained between carbon dioxide fixing and carbon dioxide releasing reactions. That carbon dioxide releasing reactions do occur is readily demonstrated if leaves which have previously been photosynthesising in labelled carbon dioxide are transferred to unlabelled carbon dioxide—labelled carbon dioxide can be detected in the atmosphere surrounding the leaves. This sort of experiment underlines the fact that when photosynthesis is measured as CO_2 uptake or oxygen evolution the value obtained represents apparent photosynthesis—the balance between reactions resulting in

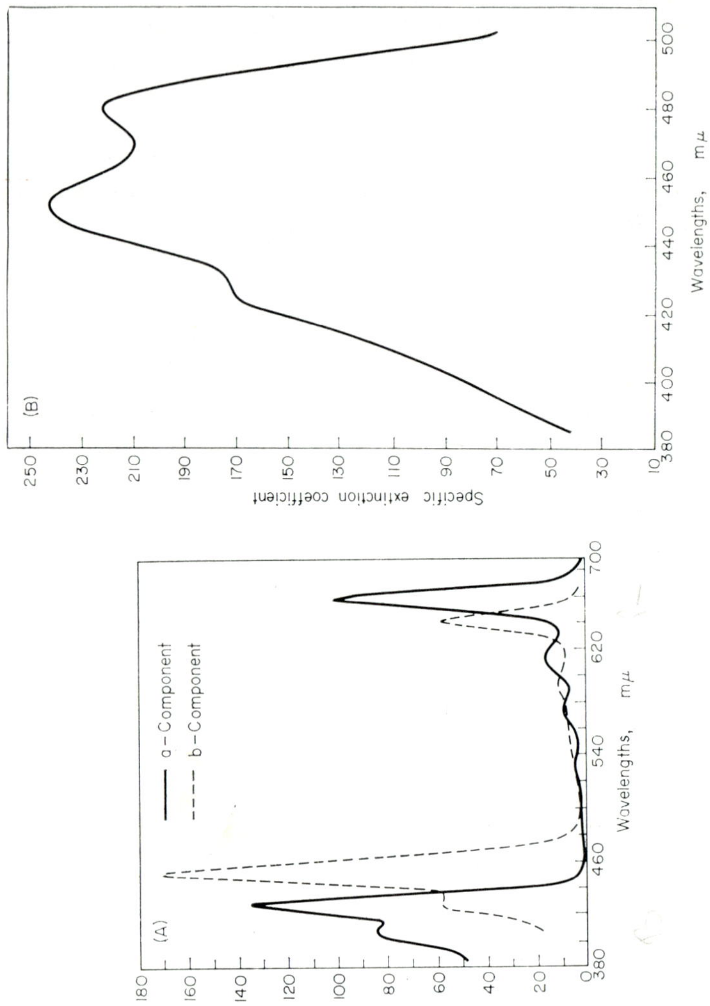

FIG. 5.4. Absorption spectra. (A) of chlorophyll *a* (solid line) and chlorophyll *b* (broken line). (B) of β-carotene.

CO_2 and oxygen uptake and evolution. The reactions involved in oxygen uptake and carbon dioxide evolution in the light have attracted considerable attention, initially as an annoying complication in measurements of photosynthesis and more lately in their own right as reactions reducing the overall efficiency of photosynthesis and therefore worthy of study.

Early work involving monitoring carbon dioxide and oxygen levels during transition from light to darkness gave conflicting results and only with the availability of isotopic oxygen did it become possible to follow oxygen uptake in the light in an unequivocal way. The work of Brown using this isotope clearly indicated that in one of the organisms used (*Ankistrodesmus*) the rate of oxygen uptake in the light was less than in the dark, suggesting inhibition, by light, of respiratory reactions. However, in many higher plants, e.g. tobacco, it has been shown that rate of oxygen uptake and of carbon dioxide evolution is increased in the light. This phenomenon, known as *photorespiration*, decreases the overall efficiency of photosynthesis and involves a reaction sequence quite different from "normal" respiration and in which glycollate is probably an intermediate. Plants exhibiting photorespiration usually have a CO_2 compensation point of around 50 ppm. Plants which do not exhibit photorespiration have CO_2 compensation points very close to zero and are capable of photosynthesising at very high rates—possibly because they can maintain a higher CO_2 gradient between the outside air and the chloroplasts. Plants lacking photorespiration include the important crop plants sugar-cane and maize, a number of tropical grasses and some dicotyledenous plants (a mixed taxonomic group).

These cane-type plants also exhibit chloroplast dimorphism—possessing normal grana-containing chloroplasts and also, near the vascular bundles, chloroplasts in which the granal system is much reduced. This latter type of chloroplast may be more involved in temporary storage and transport of photosynthates than in photosynthesis proper. The division of labour between types of chloroplast may also be associated with the ability to photosynthesise at high rates. There are other correlations between members of this group, perhaps the most significant being their mechanism of carbon dioxide fixation which will be discussed later in this chapter (p. 149).

THE PATH OF CARBON IN PHOTOSYNTHESIS

Our knowledge of the path of carbon in photosynthesis comes mainly from a series of brilliant researches initiated in 1946 by Melvin Calvin and his associates at the Lawrence Radiation Laboratory of the University of California. These workers made use of the radioactive isotope of carbon, ^{14}C, from which they prepared radioactive carbon dioxide ($^{14}CO_2$) and then fed this to unicellular algae (*Chlorella pyrenoidosa* and *Scenedesmus obliquus*) and green leaves.

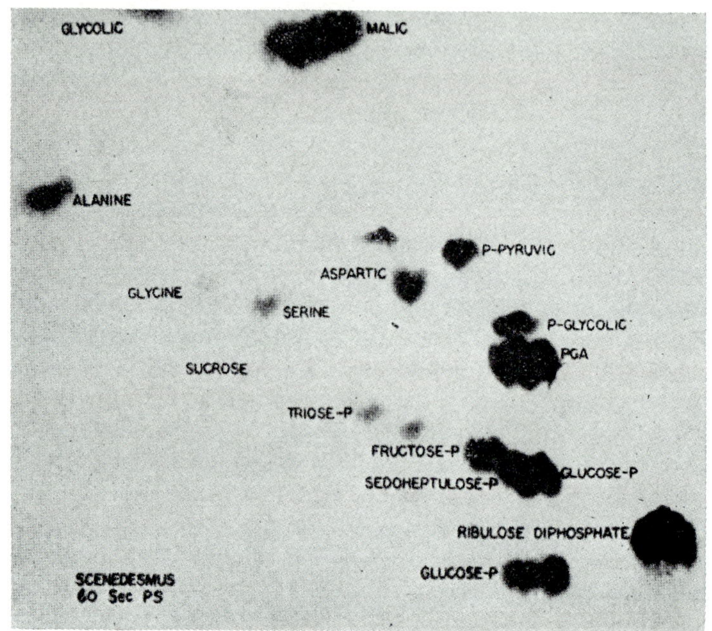

FIG. 5.5. Radio-autograph of a paper chromatogram of the alcoholic extract of cells of the alga *Scenedesmus* actively photosynthesising radioactive carbon dioxide $^{14}CO_2$. Radioactivity located by contact of the chromatogram with a photographic plate. Each radioactive spot labelled with its chemical composition. (From J. A. Bassham and M. Calvin, *The Path of Carbon in Photosynthesis*, Prentice-Hall Inc., New Jersey, 1957.)

After short periods of photosynthesis in $^{14}CO_2$ the cells were instantaneously killed and their enzymes denatured by plunging into boiling alcohol. The compounds into which the ^{14}C had been incorporated by photosynthesis were then separated from the alcoholic

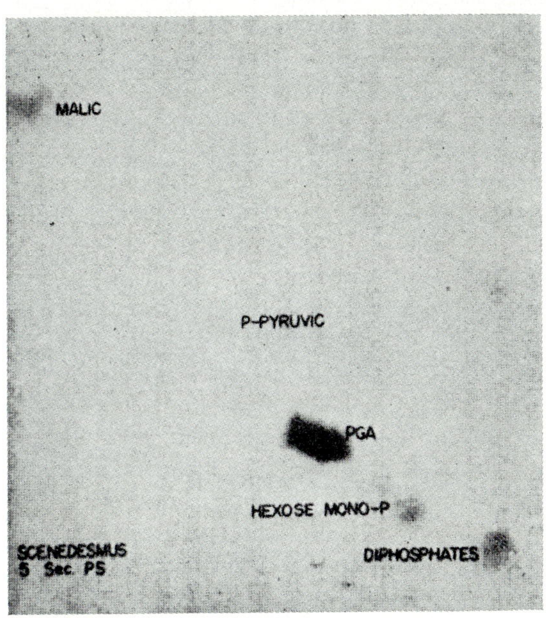

Fig. 5.6. Radio-autograph prepared as in Fig. 5.5. *Scenedesmus* cells assimilating CO_2 were, at zero time, supplied with CO_2 and extracted 5 *sec* later. Almost the whole of the detected radioactivity is in phosphoglyceric acid (PGA). (Via E. I. Rabinowitch, *Photosynthesis*, vol. 2, part 2, Interscience Pub. Inc., New York, 1956.)

extract. To achieve this separation of the radioactive carbon compounds and to identify them they used the technique of two-dimensional paper partition chromatography which had been developed by A. J. Martin and R. L. M. Synge in 1941. The compounds were located on the paper chromatograms, both by their chemical reactions and their radioactivity. The individual compounds were then washed off the chromatograms and their identity confirmed by re-chromatography with known "cold" compounds. Degradation of the isolated radioactive intermediates by micro-methods made it possible to determine the relative radioactivity of each carbon atom they contained. From such studies came first the identification of the intermediates between carbon dioxide as starting material and the end-products of photosynthesis which proved to be mainly sugars and amino acids. Then the degradation studies enable the sequence of the reactions to be confirmed. Finally, there followed the isolation of the enzymes which catalyse these "dark" reactions of photosynthesis. Appropriately the cyclic metabolic pathways thereby worked out are referred to as the *Calvin cycle.* The significance of these biochemical studies has subsequently been upheld by strong experimental evidence that in many photosynthetic organisms the greater part of the carbon dioxide assimilated in the chloroplasts passes through the Calvin cycle.

Within less than 1 min the radioactivity of carbon dioxide can be picked up in sugar phosphates, phosphoglyceraldehyde, phospho-pyruvic acid, phosphoglyceric acid (PGA), amino acids (particularly alanine and aspartic acid) and organic acids (particularly malic acid) (Fig. 5.5). By reducing the period of exposure to $^{14}CO_2$ to a few seconds it was possible to show that the first stable intermediate product of photosynthesis was PGA (Fig. 5.6) and that virtually all the ^{14}C was located in the carboxyl carbon (marked ⋆) of this compound.

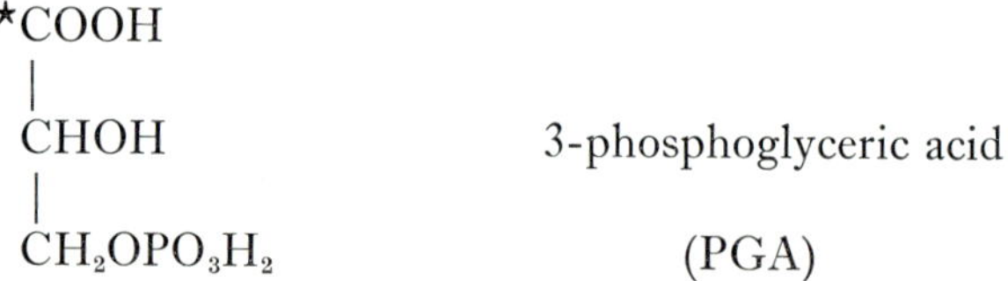

The PGA and the phosphoglyceraldehyde and the hexose phosphates which are quickly labelled in photosynthetic carbon dioxide assimilation are also intermediates in the EMP pathway of respiration (p. 98). This immediately suggested that PGA might be the precursor of the hexose phosphates and ultimately of starch by undergoing phosphorylation and reduction to give first 3-phosphoglyceraldehyde and then fructose-1-6-diphosphate by the equilibrium reactions which involve these compounds in respiration. This would require the chloroplasts to contain the necessary enzymes and to be capable of synthesising ATP and reducing the appropriate co-enzyme. This view that the hexose phosphates are thus derived from PGA was supported by the demonstration that the first-formed hexose phos-

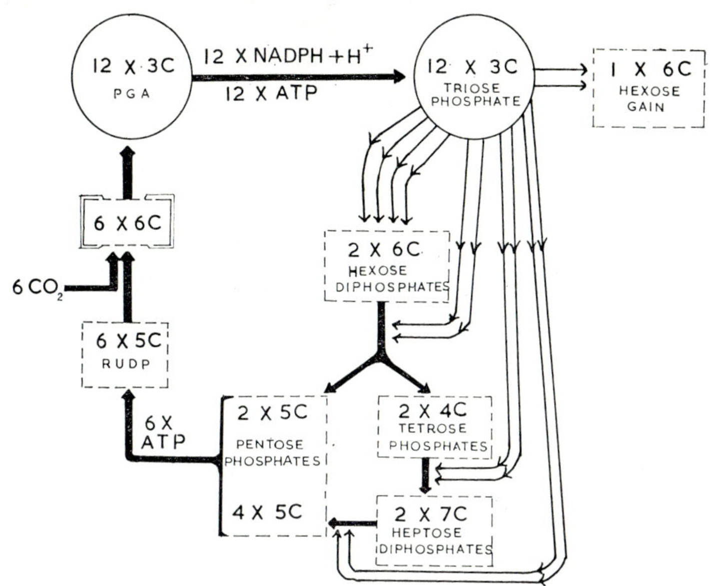

FIG. 5.7. Simplified version of the Calvin cycle showing the path of carbon dioxide assimilation in photosynthesis.

phates had their radioactivity almost entirely in carbon atoms 3 and 4 (the middle carbon atoms of the chain of 6) as would be the case if two triose units with radioactive aldehyde groups joined through these to give a hexose diphosphate.

The origin of the PGA in photosynthesis was not immediately clear. Calvin and his group first searched unsuccessfully for a 2-carbon compound which could be the acceptor of CO_2 to give PGA. Then A. A. Benson, working in Calvin's laboratory, showed that the chromatograms contained not only labelled hexose phosphates but also labelled 7-carbon and 5-carbon (pentose) phosphates including the 5-carbon ribulose-1-5-diphosphate (RuDP). This opened up the new possibility that the CO_2 combined with a pentose diphosphate to give a 6-carbon compound which then split to give 2 molecules of PGA. That this was, in fact, the case was demonstrated first by an experiment in which, after a period of $^{14}CO_2$ feeding, the light was switched off when PGA accumulated and RuDP was consumed and then by the complementary experiment in which the light was kept on but the $^{14}CO_2$ withdrawn when PGA disappeared and RuDP accumulated. RuDP is now generally accepted to be the carbon dioxide acceptor in the Calvin cycle in a reaction catalysed by the enzyme *RuDP carboxylase* (carboxydismutase). Further, the observation that PGA was the first stable intermediate product of photosynthesis indicated, and this has been confirmed by *in vitro* studies, that the 6-carbon compound formed by combination of CO_2 with RuDP is quite unstable and immediately undergoes cleavage to give two molecules of PGA.

By a patient study of the radioactive carbon labelling of the 7-carbon and 5-carbon sugar phosphates it was concluded that PGA was the starting material for their synthesis and, therefore, the precursor, not only of hexoses, but of RuDP. Further, it was possible to set out the reaction sequences by which these 7- and 5-carbon sugars must be formed. These reactions were at first hypothetical but subsequent work has led to the isolation from plant cells of enzymes which catalyse the postulated reactions. It is these sugar interconversions giving rise to RuDP together with the reactions whereby PGA gives rise to hexose that constitute the Calvin cycle.

It would be going beyond the scope of the present text to consider these reactions in detail. The Calvin cycle is, however, presented in a simplified form in Fig. 5.7 and the following is the balance sheet of the path of carbon in photosynthesis:

(a) $6RuDP + 6CO_2 \rightarrow 12PGA$

(b) $12PGA \xrightarrow[12ATP]{12 \times (2H)} 12C_3$ (phosphoglyceraldehyde)

(c) $2C_3 \rightarrow 1C_6$ (net gain of hexose)

(d) $4C_3 \rightarrow 2C_6$

(e) $2C_6+4C_3 \rightarrow 2C_5+2C_7$

(f) $2C_7+2C_3 \rightarrow 4C_5$

(g) $6C_5 \xrightarrow{6ATP} 6RuDP$ (49)

Thus the assimilation of six molecules of CO_2 leads to the net gain of a molecule of hexose and requires the energy supplied by 18 moles of ATP and the reducing activity of 12 molecules of reduced co-enzyme II (NADP) (p. 103; Fig. 4.6; reactions 49(b), (g); and Fig. 5.9).

Further study of the effects of light–dark transitions on the levels of RuDP suggested to Bassham that the affinity of RuDP carboxylase for RuDP was lower in the dark than in the light. This may be one aspect of control mechanisms which co-ordinate the catabolic and anabolic reactions of the cell. The regulation of the flow of carbon between the two by way of the several compounds which are intermediate in both chloroplastic and non-chloroplastic reaction sequences and which can traverse the chloroplast membrane may also be involved. Doubtless many subtle interactions between the photosynthetic and non-photosynthetic reactions of the green cell await discovery.

For some considerable time the Calvin cycle as described above was generally accepted as the only photosynthetic reaction sequence operating in higher plants and algae. Recently, however, evidence has accumulated that "cane-type" plants (a taxonomically amorphous group of plants which are distinguished by anatomical similarities and the absence of photorespiration) may be similar to one another also in that they may utilize phosphoenol-pyruvate (PEP) and not RuDP as the substrate of the primary carboxylation reaction. The enzyme involved, *PEP carboxylase*, is found in high concentration in cane-type plants, has a higher affinity for CO_2 than RuDP carboxylase and catalyses the synthesis of oxaloacetate from PEP and CO_2 (Chapter 4, eqn. (24)). Oxaloacetate, malate and aspartate (these latter two compounds rapidly equilibrating with oxaloacetate) are found to be the earliest products of photosynthetic CO_2 fixation in cane-type plants. This being so the metabolic sequence may be described as the C_4-dicarboxylic acid pathway—alternatively it is sometimes termed the *Hatch and Slack pathway* after the two workers predominantly concerned in its elucidation. This rapid accumulation of label in the dicarboxylic acids is not a phenomenon exclusive to cane-type plants (see Figs. 5.5 and 5.6) but in these plants

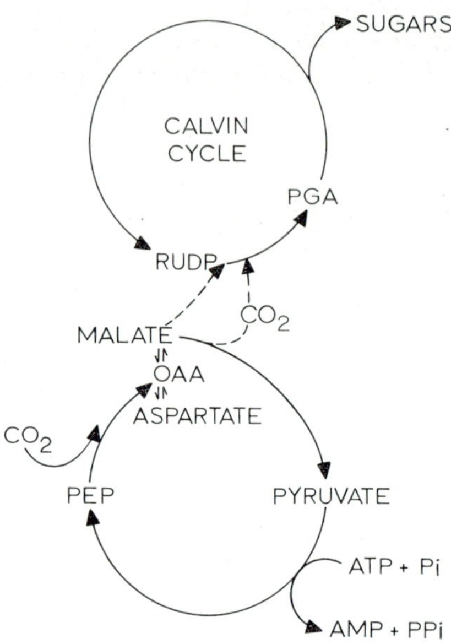

FIG. 5.8. The Hatch and Slack pathway of CO_2 fixation. The alternative routes between the primary carboxylation and the Calvin cycle are shown as dotted arrows. OAA, oxaloacetate; PEP, phosphoenolpyruvate; PP_i, pyrophosphate.

there appears to exist a light-requiring mechanism for the continued synthesis of PEP and the transfer of the fixed carbon from the dicarboxylic acids to carbohydrates via the Calvin cycle (Fig. 5.8). The mechanism resulting in this transfer of fixed carbon is controversial. It may be that there is a direct transfer of carbon from a dicarboxylic acid to a 5-carbon acceptor to produce two molecules of 3-PGA which could then participate in the reactions of the Calvin cycle. However, until a transcarboxylase enzyme which can catalyse such a reaction is isolated from cane-type plants this suggestion must remain hypothetical. An alternative possibility is that CO_2 may be released from the dicarboxylic acid and refixed by RuDP carboxylase. Providing some mechanism operates which ensures that PEP carboxylase does not compete effectively for the released CO_2 then the Hatch and Slack pathway could act as a CO_2 concentrating system for RuDP carboxylase. Such a mechanism could involve transport of a dicarboxylic acid away from the site of PEP carboxylase activity followed by the release of CO_2 at a physically separate site of RuDP carboxylase activity. Either mechanism would account for the ability of cane-type plants to photosynthesize in very low levels of carbon dioxide.

It has been suggested that the Hatch and Slack pathway is a modification of the Calvin cycle of advantage to plants growing in dense stands of tropical vegetation where the carbon dioxide concentration may be reduced to a very low level. The reduction of atmospheric CO_2 concentration which has occurred since the evolution of photosynthetic reactions may also have contributed to the selection of this reaction sequence. The discovery of the Hatch and Slack pathway, although it is not yet fully authenticated, has demonstrated the possible existence of photosynthetic reactions other than the conventional Calvin cycle and suggests that other and as yet undiscovered variations on the photosynthetic theme may exist.

THE CONVERSION OF LIGHT ENERGY INTO CHEMICAL ENERGY IN PHOTOSYNTHESIS

Carbon dioxide assimilation via the Calvin cycle requires a supply of reduced co-enzyme and of ATP. This suggests that the photochemical reactions of photosynthesis involve the conversion of electromagnetic energy into the chemical energy of the terminal pyrophosphate bond of ATP and the reduced form of the co-enzyme.

Hill's early work with isolated chloroplasts of *Stellaria media* and *Lamium album* showed that when, and only when, illuminated they evolved oxygen, that this oxygen came from water and that associated with this oxygen evolution the chloroplasts developed "reducing activity". However, this reducing activity not only failed to reduce carbon dioxide but appeared to be of very limited reducing potential. Then, in 1951, illuminated chloroplasts from other plants (particularly chloroplasts from spinach, Swiss chard and certain algae) were shown to be capable of reducing co-enzyme II (NADP) (p. 103) and,

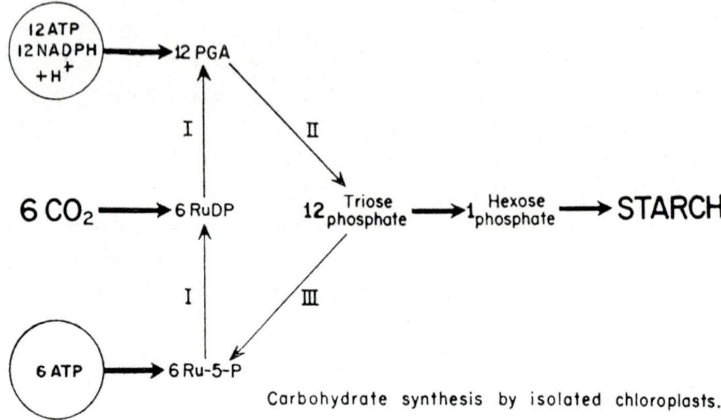

FIG. 5.9. Diagram summarising the path of carbon assimilation in chloroplasts (see also eqn. 49). The assimilation can be divided into three phases. Phase I: The phosphorylation of ribulose monophosphate (Ru-5-P) to ribulose diphosphate (RuDP) which then accepts a molecule of CO_2 and is cleaved to 2 molecules of PGA (eqn. 49 (g) and (a)). Phase II: PGA is reduced to triose phosphate (eqn. 49 (b)). Phase III: Triose phosphate acts as the precursor of both Ru-5-P and of hexose and starch. The reactions are driven by ATP and reduced NAD. The assimilation of 1 mole of CO_2 requires 3 moles of ATP and 2 moles of reduced NADP. (After D. I. Arnon, in *Biological Structure and Function*, edited by T. W. Goodwin and O. Lindberg, Academic Press, New York, 1961.)

by adding an appropriate enzyme to the system, of assimilating carbon dioxide. The reaction used by Dr. Vishniac and Professor Ochoa to effect carbon dioxide uptake was that catalysed by their *malic enzyme*:

$$\underset{\text{pyruvic acid}}{CH_3CO.COOH} + {}^{14}CO_2 + NADPH + H^+$$

$$\rightleftharpoons \underset{\text{malic acid}}{HOOC.CHOH.CH_2.{}^{14}COOH} + NADP^+ \quad (50)$$

Of course, this was still far short of full photosynthesis. However, in 1954, Professor D. I. Arnon and his colleagues at the University of California, showed that spinach chloroplasts properly prepared *could* carry out full photosynthesis, assimilating carbon dioxide and giving rise to the intermediates and end-products of the Calvin cycle. In the initial experiments only very low rates of photosynthetic CO_2 fixation were obtained but improvements in the techniques used in the isolation of chloroplasts have resulted in *in-vitro* rates closely approaching maximum *in-vivo* rates. Clearly the whole of the photosynthetic apparatus is within the chloroplast. The photochemical acts are linked to the "dark" reactions of carbon dioxide assimilation through the synthesis of ATP and the reduction of the co-enzyme of the chloroplast (NADP). Further, by illuminating a chloroplast preparation in absence of CO_2 but in presence of large amounts of ADP and inorganic phosphate and NADP, Arnon demonstrated oxygen evolution and the accumulation of ATP and reduced NADP ($NADPH + H^+$):

$$H_2O + NADP^+ + ADP + P_i \xrightarrow{\text{light}} NADPH + H^+ + ATP + \tfrac{1}{2}O_2 \qquad (51)$$

The next step must be to visualise a model for such a reaction—how can light energy be converted to chemical energy in the form of reduced NADP and ATP?

When a chlorophyll molecule absorbs a photon (quantum) it becomes an excited molecule, a molecule with more energy than the ground state energy (becomes, if you like, very hot and therefore very reactive). Such molecules have a very short lifetime and if they do not immediately participate in photochemical reactions they lose their energy partly as heat and partly as light (fluorescence). When a chlorophyll molecule absorbs a quantum of blue light it becomes raised to the "excited" state known as the "second singlet state" which has a very short half-life (10^{-11} sec approx.) and which loses energy as heat to give molecules at the "first singlet state". First singlet state molecules are also produced when chlorophyll absorbs red light. They have a half-life of approximately 10^{-9} sec and may undergo an internal conversion to the "triplet state" which is metastable and has a half-life of approximately 10^{-3} sec. It is these activated "triplet state" chlorophyll molecules which by virtue of their much longer life can most easily be involved in photochemical reactions.

Arnon has suggested that such activated chlorophyll molecules expel their energy as an electron

$$\text{Chl} \xrightarrow{\text{light}} \text{triplet state Chl} \rightarrow [\text{Chl}]^{+} + e^{-} \qquad (52)$$

and that this electron is accepted by a suitable molecule and transferred along an electron transport chain which terminates with the reduction of NADP. Some of the energy dissipated by the electron as it passes along the electron transport chain is conserved as ATP. $[\text{Chl}]^{+}$ in eqn. (52) represents electron deficient chlorophyll, a deficiency which is made good at the expense of the oxidation of water. The details of the mechanism whereby the green plant transfers electrons from water to NADP with an associated synthesis of ATP have been an area of intense research and a scheme which incorporates results of much work by many investigators is shown in Fig. 5.10. An important feature of the currently accepted scheme is

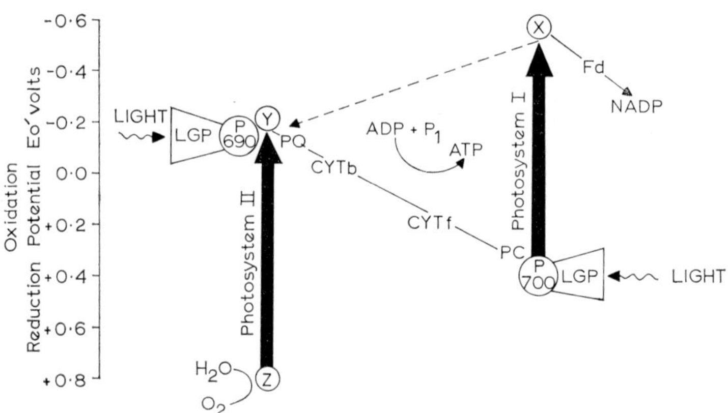

FIG. 5.10. Photosynthetic electron transport.
Key: LGP, light-gathering pigments (accessory pigments and chlorophylls other than P700 and P690); PQ, plastoquinone; CYT, cytochrome; PC, plastocyanin; FD, ferredoxin.

that two separate light reactions working in series, each with its own assortment of pigments, are involved. This concept owes much to the experiments of Emerson and co-workers on the effect of wavelength of light on the quantum efficiency of photosynthesis. It was found that, at wavelengths up to approximately 680 mμ, a minimum of 8 quanta were required per molecule of oxygen evolved. At wavelengths longer than this the efficiency dropped sharply and many more quanta were required per molecule of oxygen evolved. This fall in efficiency has become known as the "red drop". Further experiments showed that when photosynthesis was occurring in the "red drop" region additional illumination with light of wavelength shorter than 680 mμ resulted in a decrease in the number of quanta involved in the evolution of one molecule of oxygen. Furthermore, the total amount of oxygen evolved as a result of simultaneous illumination by the two wavelengths was found to be greater than the sum of the amounts of oxygen evolved as a result of separate illumination by an equal number of quanta at the two wavelengths; enhancement was occurring. These results are consistent with the operation of a system involving two separate light reactions each with a different assortment of pigments (hence the effect of different wavelengths of light) and with a need for a balance in the amounts of energy received by each light absorbing system. The "red drop" is taken to indicate that system II, the shorter wave absorbing system, is receiving insufficient energy for efficient operation of the overall system. Additional illumination by shorter wavelength light restores the balance and results in enhancement. Subsequent experiments by Bendall and Hill and Duysens on the photosynthetic cytochromes support the concept of two co-operating pigments separated by an electron transport chain involving cytochromes *b* and *f*. Further indication of the existence of two photosystems (although not necessarily serial systems) comes from the work of Boardman and Anderson who were able to effect partial physical separation of the two systems by fractionation of digitonin-treated chloroplast preparations.

The process of electron transport may be considered as being initiated by the absorption of a quantum of light by the pigments of system I. No matter which of the several species of chlorophyll *a* known to be associated with system I initially absorbs the quantum, energy transfer occurs between pigment molecules resulting eventually in the "trapping" of the energy by a special chlorophyll *a* known as P_{700}. This pigment which represents the "active centre" of system I can be reversibly bleached (oxidised) by light under suitable conditions. This bleaching can be detected by difference spectroscopy (comparison of the absorption spectra of photosynthetic systems in the light and dark) and has a maximum change in absorption at about 700 mμ. The process which results in bleaching of P_{700} involves the ejection of an electron and leaves behind an electron deficiency (a "hole") in the P_{700} molecule. The ejected electron is trapped by a compound usually designated X but which may be a pteridine and which in turn transfers the electron to ferredoxin (a non-heme iron protein). The electron is then transferred from ferredoxin via the intermediate protein electron carrier *ferredoxin-NADP-reductase*, to NADP. Reduced NADP is as we have seen earlier, necessary for the operation of the carbon reduction cycle.

When a quantum of light is absorbed by the system II pigments, no matter which of the light gathering pigments (chlorophyll *b* or *a*) initially absorbs the light, the energy is eventually transferred to a special form of chlorophyll *a* termed chlorophyll P_{690} which can mediate the ejection of an electron which is accepted by a compound of unknown identity usually designated Y. Compound Y is sometimes known as Q since it is also the substance which causes quenching of the characteristic fluorescence of chlorophyll *a* in photosystem II; Y accepts energy from the pigment system which would otherwise be re-emitted as fluorescent light. From Y the electron passes along a series of compounds which include plastoquinone, plastocyanin and cytochromes *f* and *b*. The involvement of these substances has been demonstrated in experiments which have involved spectroscopic analysis of the photosynthetic systems—particularly their difference spectroscopy. As the electron moves through these energy transport components of gradually increasing redox potential some of the energy represented by the change in redox potential is coupled to the synthesis of ATP—another of the requirements of the carbon reduction cycle (see p. 124 for an outline of possible mechanisms whereby electron transport may be coupled to phosphorylation).

When the electron passes from system II to system I it can "fill the hole" left as the result of the ejection of an electron in the operation of system I as described above. What then fills the "hole" in system II? It is at this point that water is involved as an electron donor leading to the release of oxygen in a reaction as yet not fully understood, but which may involve a strong oxidant Z.

According to this reaction scheme the absorption of four pairs of quanta (8 in all) results in the transfer of four electrons from water to NADP, the concomitant synthesis of two molecules of ATP and the evolution of one molecule of oxygen. Qualitatively then the requirements of the carbon reduction cycle are fulfilled. However, only one molecule of ATP is produced per NADP reduced whereas the carbon reduction cycle in its present form requires 3 molecules of ATP per 2 molecules of NADP reduced. *Non-cyclic photophosphorylation*, as the above mechanism of ATP synthesis is termed, cannot quantitatively satisfy the needs of the carbon reduction cycle. However, by the addition of suitable co-factors isolated chloroplast preparations can be induced to synthesise ATP without the transfer of electrons from water to NADP. The electron path is short-circuited and the electron ejected from chlorophyll P_{700} in system I returns eventually to chlorophyll P_{700}. The return route, shown by a dotted line in Fig. 5.10, involves the synthesis of ATP. In this process, usually termed *cyclic photophosphorylation*, the only measurable product is ATP, oxygen is not evolved and NADP is not reduced. Cyclic photophosphorylation may balance the ATP–NADP stoichiometry for the operation of the Calvin cycle and could also supply ATP for other purposes such as the synthesis of polysaccharides.

Reference was made in the opening paragraphs of this chapter to the possibility that the earliest forms of life were unicellular heterotrophs living in an atmosphere devoid of oxygen and probably rich in hydrogen. Such organisms would of necessity have generated their essential ATP by the inefficient process of fermentation. The first effect of the biological invention of chlorophyll molecules would then presumably have been to enable such organisms to effect a cyclic photophosphorylation, as can still be observed in the anaerobic bacterium, *Chromatium*. This organism requires light for its growth and if provided with carbon dioxide as its source of carbon requires also the presence of hydrogen gas. It is an anaerobic organism. It contains an enzyme (*hydrogenase*) which catalyses the reduction of carbon dioxide by hydrogen gas. To synthesise its essential cell constituents it also requires ATP and this is the reason for its light requirement. Light functions in the economy of this organism, when supplied hydrogen gas, solely to effect cyclic photophosphorylation of ADP in which no hydrogen or electron donor is involved.

The next evolutionary step could then have been the development of a non-cyclic photophosphorylation perhaps utilising as electron donors hydrogen sulphide or thiosulphate and thereby making, with solar energy, a new reductant for CO_2 assimilation. The final step in this hypothetical sequence of biochemical evolution would be the development of a system with the higher oxidising potential which permitted the use of the ubiquitous substance, water, as an electron donor. This, in turn, by releasing molecular oxygen, made possible the evolution of aerobic respiration as an extension of the more limited and less energy-yielding reactions of fermentation.

LOCATION OF THE PHOTOSYNTHETIC APPARATUS

The isolation of chloroplasts capable of performing all of the reactions normally regarded as photosynthetic including the fixation of carbon dioxide, the evolution of oxygen and the synthesis of sugars and polysaccharides unequivocally demonstrates that the site of photosynthesis is the chloroplast. Experiments involving the fractionation of chloroplasts have further shown that the "dark reactions" associated with carbon dioxide fixation are located in the stroma of the chloroplast and the "light reactions", electron transport and photophosphorylation, takes place in the lamellar systems.

Relevant to this discussion are the experiments of Emerson and associates in which it was found that the maximum yield of oxygen from one intense short flash of light was one molecule for approximately 2500 molecules of chlorophyll. The interpretation of this apparent co-operation of 2500 chlorophyll molecules in the evolution of one molecule of oxygen led to the concept of a photosynthetic unit. In line with this it has been shown that the least abundant components in the electron transport system are not present at a ratio of less than one molecule for every 2500 chlorophyll molecules.

When electron micrographs showed particles in the chloroplast granal membranes which were about the right size to contain 2500 chlorophyll molecules and the necessary associated compounds it seemed that the photosynthetic unit had been structurally identified. However, further work has indicated that these *quantasomes*, as they are sometimes termed, probably do not contain chlorophyll and may not, therefore, represent the intact photosynthetic unit. Nevertheless, the concept of the photosynthetic unit remains and it may yet be shown that the numbers discussed above represent structural rather than statistical relationships.

Despite the very rapid recent growth of our knowledge of the phenomenon of photosynthesis much remains to be learned, particularly in relation to the photochemistry of photosynthesis; the mechanism of conversion of light to chemical energy. It is also becoming increasingly clear that the "dark reactions" of photosynthesis may vary more from species to species than has hitherto been realised and that we need more knowledge of the relationships between the metabolism of the chloroplasts and that of the remainder of the cell.

THE SYNTHESIS OF SUCROSE AND POLYSACCHARIDES

The simple carbohydrates and monosaccharides produced by photosynthetic reactions serve as precursors of more complex molecules produced from them by a variety of biosynthetic pathways. Carbohydrates such as sucrose and the more complex polysaccharide starch are usually considered as direct products of photosynthesis and are synthesised in reactions involving nucleotide sugars as intermediates.

For example in sucrose synthesis, glucose-6-phosphate produced by the action of *G-6-P isomerase* on fructose-6-P, an intermediate in the carbon reduction cycle, is converted to glucose-1-phosphate by the enzyme *phosphoglucomutase*. This compound can then combine with the nucleotide, uridine triphosphate (UTP) to produce uridine diphosphate glucose (UDPG) and pyrophosphate. The enzyme involved is *uridine diphosphoglucose pyrophosphorylase*:

$$\text{UTP} + \text{glucose-1-(P)} \rightleftharpoons \text{UDPG} + \text{pyrophosphate} \qquad (53)$$

This UDPG is an interesting and versatile compound:

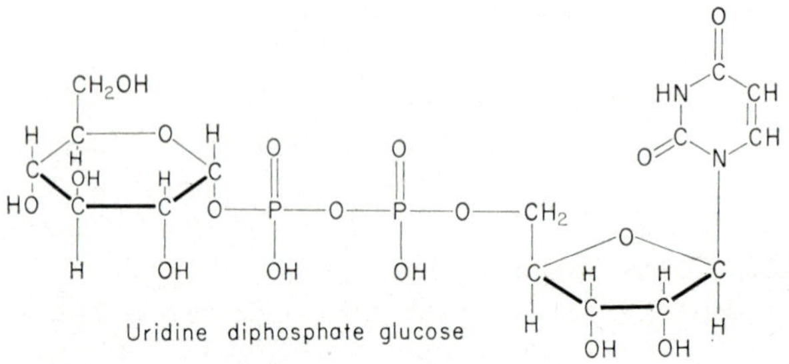

Uridine diphosphate glucose

Nucleotide sugar intermediates are also involved in the metabolism of starch, cellulose and a variety of sugars such as galactose, arabinose, and xylose. These latter sugars are precursors of the galactosan and pentosan cell wall polysaccharides. Further both UDPG and UDP galactose can suffer oxidation to the corresponding uronic acid derivatives which are the building blocks of the hemicelluloses and pectins of the cell wall.

As indicated above, UDPG is involved in the synthesis of the widely distributed and very important disaccharide, **sucrose** (cane sugar):

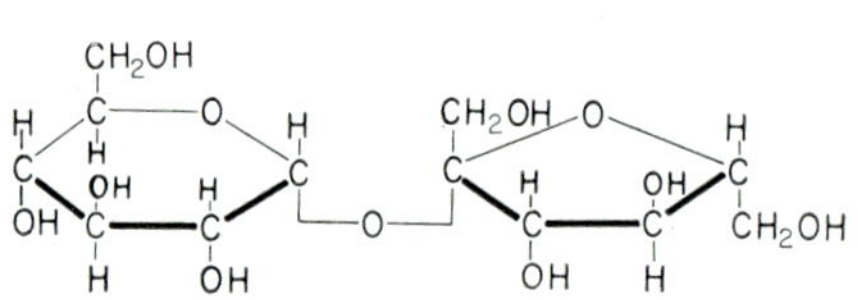

Sucrose

Higher plants have been shown to contain two separate sucrose-synthesising systems, each involving UDPG as an intermediate: *sucrose synthetase* which catalyses the transfer of glucose from UDPG to fructose,

$$\text{UDPG} + \text{fructose} \rightleftharpoons \text{sucrose} + \text{UDP} \qquad (54)$$

and *sucrose phosphate synthetase* which catalyses a reaction in which the accepting molecule is fructose-6-phosphate; free sucrose being subsequently released by dephosphorylation of the sucrose phosphate produced:

$$\text{UDPG} + \text{fructose-6-P} \rightleftharpoons \text{sucrose-P} + \text{UDP} \quad (55)$$

$$\text{sucrose-P} \rightleftharpoons \text{sucrose} + P_i \quad (56)$$

Sucrose-phosphate synthetase resides exclusively within the chloroplast and may be concerned with the initial synthesis of sucrose whereas sucrose synthetase is more abundant in non-photosynthetic tissues and may be primarily involved in the metabolism of translocated sucrose rather than its direct synthesis in association with photosynthetic reactions.

The reversibility of the reactions described above, particularly the sucrose synthetase reaction, is in marked contrast to the reaction catalysed by the enzyme *invertase*. The hydrolysis of sucrose by invertase to glucose and fructose proceeds with a decrease in free energy of 6·5 kcal ($\Delta G' = -6{\cdot}5$ kcal) and is, therefore, in aqueous solution (water being one of the reactants) virtually irreversible.

Thus, as might be expected of a compound of such importance as sucrose, its metabolism is complex and at least three distinct enzyme systems are involved in its synthesis and breakdown. Furthermore, isozymes of invertase have been described which reside in different parts of the cell and which may be associated with the control of sucrose transport and utilization. It is clear that we must await further information before the inter-relationships between the several pathways of sucrose metabolism can be fully clarified.

The polysaccharide, **starch**, is outstanding as a storage carbohydrate in plant cells. It occurs as grains 1–150 μ in diameter formed in the chloroplasts of photosynthetic cells or in the amyloplasts of colourless cells. Each storage starch grain shows a hilum or centre of initiation around which are deposited concentric layers of starch. Each layer represents the amount of starch deposited during 24 hours and the layers are often quite distinct because the starch deposited during the day is dense and more highly refractive than that deposited at night.

If starch is hydrolysed with acid it yields only glucose. When treated with the enzyme β-amylase it yields a dextrin (a polysaccharide unit giving a purplish colour with iodine in contrast to the blue-black reaction of starch) and the disaccharide, maltose. The disaccharide, maltose, consists of two glucopyranose units joined through a 1:4 α-linkage and the chemical evidence shows that starch molecules are built of chains of glucose units in which these 1:4 linkages are involved (Fig. 5.11).

When starch is heated in water to 60–80°C the starch grains swell and yield two components. One component (*amylose*) is water-soluble, is entirely degraded to maltose by the β-amylase and its molecules consist of unbranched spiral chains of glucose units (300–1000 units) linked by 1:4 α-linkages. The second component (*amylopectin*) is not soluble, yields both maltose and dextrin on treatment with β-amylase and has a *branched* chain structure due to the occurrence not only of 1:4 but also of 1·6 linkages between the glucose units (Fig. 5.11).

As mentioned above the biosynthesis of starch involves the participation of nucleoside diphosphate sugars as intermediates; in this case adenosine diphosphate derivatives are involved. Adenosine diphosphate-glucose (ADP-glucose) is synthesised in a reaction analogous with reaction eqn. 53 (p. 160) (the enzyme being *adenosine diphosphoglucose pyrophosphorylase*) and is involved in the transfer of glucose residues to pre-existing polysaccharide molecules (primers), the simplest of which is a 1:4α-polyglucan:

$$\text{ADP-glucose} + 1{:}4\alpha\text{-glucan} \rightarrow 1{:}4\alpha\text{-glucosyl glucan} + \text{ADP} \quad (57)$$

The glucose molecules are added on to the existing chain in such a way that they are joined in the 1:4α configuration and repetition of the process results in the production of the linear polymer amylose (Fig. 5.11). The synthesis of branched molecules of amylopection is achieved by the operation of another enzyme (*branching enzyme*) which joins linear polymers in the 1:6α configuration (see p. 108 for information on the phosphorylase reaction which may also be involved in a minor way in the synthesis and degradation of 1:4α-polyglucans).

Starch functions in both the short term (in leaves) and long term (in seeds and many storage organs, etc.) for the storage of energy and

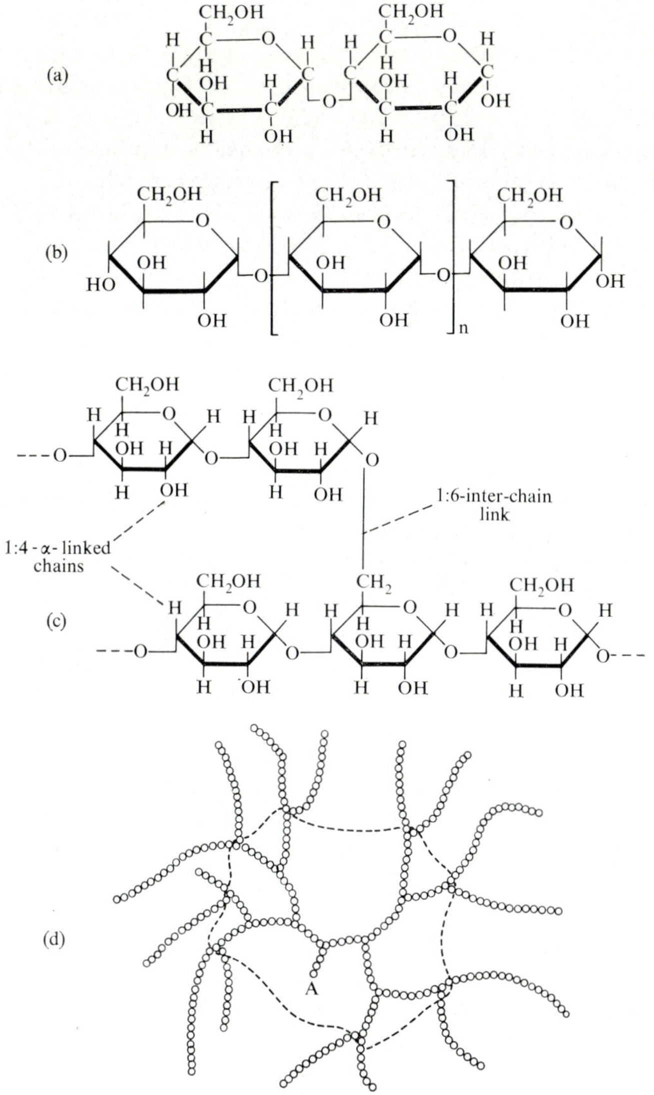

FIG. 5.11. The structure of starch: (a) maltose; (b) straight chain of amylose; (c) branching by 1:6 linkages as in amylopectin; (d) amylopectin. The dotted line shows inside it the dextrin-type molecule formed from amylopectin by the action of β-amylase.

carbon. Its synthesis and consumption (see p. 107) are under sophisticated control, exerted in the case of the biosynthetic pathway at the level of adenosine diphosphoglucose pyrophosphorylase, an effective point of control since ADP-glucose is the sole glycosyl precursor for starch biosynthesis in leaves. Adenosine diphosphoglucose pyrophophorylase is allosterically activated by some of the intermediates of photosynthesis (particularly 3-phosphoglycerate) and inhibited by orthophosphate. During photosynthesis an increase in the level of 3-phosphoglycerate occurs. This results in the activation of the pyrophosphorylase and therefore in a potentially increased rate of ADP-glucose production which may be utilized in starch synthesis. In darkness, however, the level in leaves of orthophosphate increases, the pyrophosphorylase is inhibited and ADP-glucose production and therefore starch synthesis is reduced or prevented.

Cellulose, the structural polysaccharide of the cell wall (see Chapter 2, p. 50 *et seq.*) is, like starch, a glucosan but here the glucose units are in 1:4β-linkage:

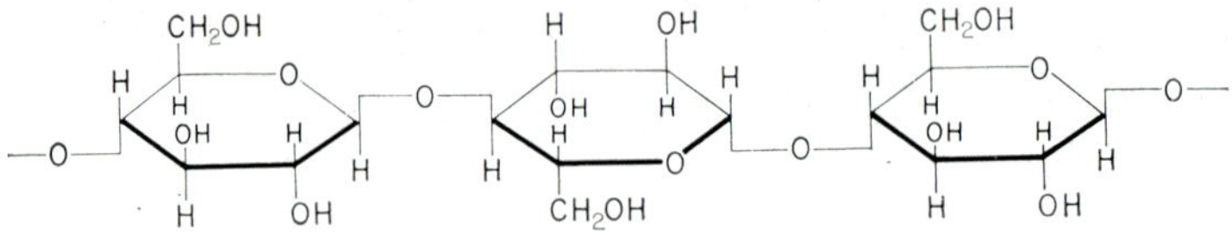

Cellulose

Unlike starch, it is completely resistant to hydrolysis by dilute acids and its hydrolysis, without carbonisation, to glucose is very difficult and involves the use of strong sulphuric acid at low temperature. During cell growth large amounts of cellulose are rapidly synthesised. Cellulose synthesis is, therefore, a very important aspect of the biochemistry of growth. Whereas, however, much is known of the mechanism of starch synthesis, of glycogen synthesis and of the synthesis of bacterial polysaccharides we do not fully understand the mechanism of cellulose synthesis. It is known that cellulose biosynthesis involves guanosinediphosphate-glucose and possibly a glucolipid intermediate acceptor. Information is accumulating as to how these biochemical reactions are involved in the construction of the highly ordered structures of the cell wall. It seems likely that the enzymes of cellulose synthesis operate at or outside the plasmalemma and may be contained in lipid bound organelles. The microtubules of the cytoplasm may participate in determining the orientation of the cellular microfibrils.

THE SYNTHESIS OF FATTY ACIDS

Early steps in the elucidation of the mechanism of fatty acid biosynthesis involved the isolation of a system involving two enzyme complexes which in the presence of ATP, Mn^{++}, biotin, reduced NADP and carbon dioxide, would synthesize fatty acids from acetyl-CoA. One puzzling feature was that although carbon dioxide was essential to the system all the carbon atoms of the fatty acid were derived from the acetyl-CoA. By using only one of the two enzyme complexes along with ATP, Mn^{++}, biotin and carbon dioxide, it was shown that the carbon dioxide was involved in the formation of an intermediate compound which in the presence of the second enzyme complex and reduced NADP gave fatty acid and released carbon dioxide. This intermediate compound was later shown to be the co-enzyme A derivative of malonic acid, malonyl-CoA, its formation being detailed in equation 58:

$$\mathrm{ATP+HCO_3^-+CH_3{-}CO\text{-}S\text{-}CoA} \xrightarrow[\text{biotin}]{\mathrm{Mn^{++}}} \mathrm{ADP+P_i+}\underbrace{\mathrm{COOH{-}CH_2{-}CO\text{-}S\text{-}CoA}}_{\text{malonyl-CoA}} \quad (58)$$

Subsequent reactions in fatty acid biosynthesis in plants (and *E. coli*) have been clearly shown to involve an acyl carrier protein (ACP) which serves the function of carrying the growing fatty acid chain and is also involved in the transfer of additional carbon atoms to the chain. Firstly the malonyl derivative of ACP is formed by the reaction between malonyl-CoA and reduced ACP (ACP-SH) thus:

$$\text{malonyl-S-CoA} + \text{ACP-SH} \rightarrow \text{malonyl-S-ACP} + \text{CoA-SH} \quad (59)$$

The malonyl-S-ACP then condenses with a primer fatty acid molecule which must also be attached to a molecule of ACP. The simplest primer is acetyl-S-ACP and the reaction between this and malonyl-S-ACP is:

$$\text{malonyl-S-ACP} + \text{acetyl-S-ACP} \rightarrow \text{acetoacetyl-S-ACP} + CO_2 + \text{ACP-SH} \quad (60)$$

It is at this stage that carbon dioxide is evolved. The acetoacetyl-S-ACP is then reduced, dehydrated and then reduced again to form butyryl-S-ACP:

$$\text{acetoacetyl-S-ACP} + \text{NADPH} + H^+ \rightarrow \alpha\text{-hydroxybutyryl-S-ACP} + \text{NADP}^+ \quad (61)$$

$$\alpha\text{-hydroxybutyryl-S-ACP} \rightarrow \text{crotonyl-S-ACP} + H_2O \quad (62)$$

$$\text{crotonyl-S-ACP} + \text{NADPH} + H^+ \rightarrow \text{butyryl-S-ACP} + \text{NADP}^+ \quad (63)$$

Butyryl-S-ACP may then condense with malonyl-S-ACP in a reaction analagous with eqn. 60 above when repetition of reactions 61, 62 and 63 will result in the production of propionyl-S-ACP. Thus each complete repetition of the system described above results in the addition of two carbon atoms to the primer fatty acid until eventually a long chain fatty acid may be built up.

The biosynthesis of fatty acids is superficially similar to the β-oxidation pathway of oxidation (p. 109) in that acetyl-CoA is involved and the reaction proceeds in steps each of which transfers two carbon atoms. However, the reactions involved in the synthesis of malonyl-CoA by carboxylation of acetyl-CoA and those utilizing ACP in the biosynthetic pathway have no counterparts in fatty acid degradation. Furthermore, as is usual in biosynthetic reactions the synthetic pathway uses NADP and not NAD.

As well as the saturated fatty acids so far described, plants also synthesise unsaturated fatty acids apparently by dehydrogenation of the ACP derivatives of saturated acids in reactions involving NADP and molecular oxygen:

$$\text{stearoyl—S—ACP} \xrightarrow[\text{NADP}]{O_2} \text{oleyl—S—ACP} \quad (64)$$

Although fatty acids are rarely found free in nature they are ubiquitous as constituents of various classes of lipids. In combination with long chain monohydroxy alcohols they form the plant waxes—widely distributed and important in water relations and resistance to pathogens. In combination with the trihydroxyl alcohol glycerol they form fats and oils which are biologically important as storage compounds in many seeds (e.g. peanut) and commercially important as a raw material in the manufacture of margarine and chocolate. More universally important, however, are the lipids (notably phospholipids which contain fatty acids, phosphatidic acid and often a nitrogenous base such as choline) which in association with proteins form the lipoproteins which are involved in the structure of almost all biological membranes including those of the cytoplasm, nucleus, chloroplast and mitochondrion.

THE ASSIMILATION OF NITROGEN AND SULPHUR

Plants usually obtain their nitrogen by absorption of nitrate or ammonium ions from the soil (symbiotic associations between higher plants and nitrogen fixing bacteria are of course exceptions to this). Ammonium ions may be utilized directly in the synthesis of amino acids (see p. 169), but nitrate must first be reduced to ammonia. This is accomplished in two stages; the reduction of nitrate to nitrite followed by the reduction of nitrite to ammonia. The first step—nitrate reduction—is catalysed by the flavo-protein enzyme complex *nitrate reductase* (Fig. 5.12) which contains molybdenum and FAD (flavin adenine dinucleotide) as a prosthetic group. Reduced FMN

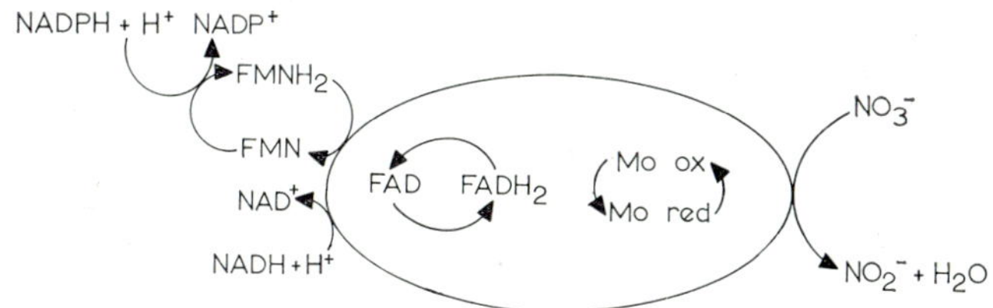

Fig. 5.12. Nitrate reductase and its electron donors. Key: FAD, Flavin adenine dinucleotide; FMN, Flavin adenine mononucleotide; Mo, Molybdenum.

(flavin adenine mono-nucleotide) and NAD act as electron donors to the complex. An *NADP reductase* enzyme, which catalyses the transfer of electrons from reduced NADP to FMN, is also closely associated with the enzyme complex so that reduced NADP can also act as electron donor.

Thus nitrate reductase shows considerable versatility in its electron donors—which may be related to the fact that the enzyme is widely distributed throughout the plant and may utilise whichever reductant is available—presumably reduced NADP in green tissue in the light and reduced NAD in non-green tissue and green tissue in darkness. (Although the enzyme does not appear to reside within the chloroplast photosynthetically generated reducing power is available outside the chloroplast in the light.)

The next step—the reduction of nitrite to ammonia—again may take place throughout the plant, although most is known about the process as it occurs in green tissues. *Nitrite reductase* in photosynthetic plant parts is located within the chloroplasts and the reaction it catalyses results in the transfer of 6 electrons from the photosynthetic electron transport system via ferredoxin (eqn. 65) to each molecule of nitrite.

$$\text{Reduced ferredoxin} \rightarrow \text{Oxidised ferredoxin} \quad \xrightarrow{6e^-} \quad NO_2^- \rightarrow NH_3 \tag{65}$$

The details of the reaction are not clear although the fact that hydroxylamine is reduced by nitrite reductase has led to the suggestion that the reaction proceeds in two electron steps and that hyponitrite and hydroxylamine are intermediates. However, these compounds have not been detected in nitrite reducing systems and if they are involved they apparently never leave the enzyme surface. An alternative view is that all six electrons are transferred at one enzyme surface—a most unusual reaction (see p. 174 for discussion of sulphite reduction).

The utilisation of nitrate in plants then involves absorption by the roots followed by reduction to ammonia which may be achieved in any part of the plant although measurement of the amounts of enzymes in the various plant parts suggests that in some plants at least (e.g. maize) the majority of nitrate reduction takes place in the leaves. Nitrate is reduced to nitrite in the leaf cytoplasm using electrons either from the respiratory or photosynthetic systems and the nitrite produced is reduced to ammonia within the chloroplasts in a reaction utilising photosynthetically generated reducing power. Nitrite reduction does also take place in non-photosynthetic tissue but the identity of the electron donor involved has not been established.

As mentioned above plants may utilise nitrate or ammonium ions as nitrogen source—the nitrate being converted to ammonia during assimilation. Associated with this variation in the otherwise monotonous plant diet of carbon dioxide, water and inorganic salts is one of the few examples of inducible enzymes which have been found in higher plants. The synthesis of nitrate reductase enzyme system is inhibited by ammonia in whose presence it is redundant and stimulated by nitrate for whose assimilation it is essential.

Feeding experiments with ammonium sulphate in which the nitrogen (at. wt. 14) is enriched with the heavy isotope of nitrogen (at. wt. 15, usually written ^{15}N) show that the nitrogen of ammonium is most actively incorporated into newly synthesised L-glutamic acid, and to a lesser extent into L-aspartic acid. This incorporation into glutamic acid is catalysed by the univerally distributed L-*glutamic acid dehydrogenase* (see Chapter 4, p. 114) which controls the reaction:

$$\text{oxoglutaric acid} + NH_3 + NADH + H^+ \rightleftharpoons \text{L-glutamic acid} + NAD^+ + H_2O \quad (66)$$

The existence of a corresponding enzyme promoting a reductive amination of oxalacetic acid to L-aspartic acid is uncertain.

Both oxoglutaric acid and oxaloacetic acid are intermediates formed in the respiratory breakdown of sugars. Respiratory energy in the form of reduced molecules of coenzyme is involved in both nitrate reduction and the primary synthesis of amino acid molecules by the reactions discussed above.

Glutamic acid is the primary product of ammonium assimilation and occupies a central position in amino acid metabolism. It functions as a precursor for the synthesis of other amino acids through the action of a group of enzymes termed *transaminases* (amino-transferases). These enzymes have as their prosthetic group pyridoxal phosphate, a derivative of vitamin B_6 (pyridoxine). The pyridoxal phosphate accepts the amino group of the amino acid to become pyridoxamine phosphate and then transfers the amino group to other organic acids which are thereby converted to the corresponding amino acids. The following reaction illustrates the role of pyridoxal phosphate:

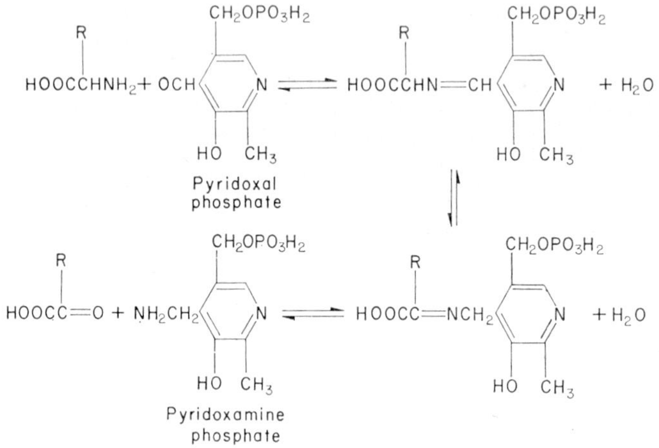

$CH_2OPO_3H_2$ R $CH_2OPO_3H_2$ R

$HOOCCHNH_2 + OCH$ N $\rightleftharpoons$ $HOOCCHN{=}CH$ N $+ H_2O$

HO CH_3 HO CH_3

Pyridoxal phosphate

(67)

$CH_2OPO_3H_2$ R $CH_2OPO_3H_2$ R

$HOOCC{=}O + NH_2CH_2$ N $\rightleftharpoons$ $HOOCC{=}NCH_2$ N $+ H_2O$

HO CH_3 HO CH_3

Pyridoxamine phosphate

Transamination from glutamic acid to a variety of keto acids accounts for the synthesis of a number of the amino acids. For example, transamination involving pyruvic acid yields alanine:

$$\text{pyruvic acid} + \text{glutamic acid} \rightleftharpoons \text{alanine} + \text{oxoglutaric acid} \quad (68)$$

Further metabolism of amino acids produced by transamination results in the formation of biogenetically related families of amino acids. Thus the pyruvate family comprises alanine, leucine and valine and similar families are based upon aspartic and glutamic acids and serine.

The biosynthesis of the carbon skeletons from which the aromatic acid phenylalanine is derived involves an important reaction sequence which results in the biogenesis of an aromatic ring (p 171).

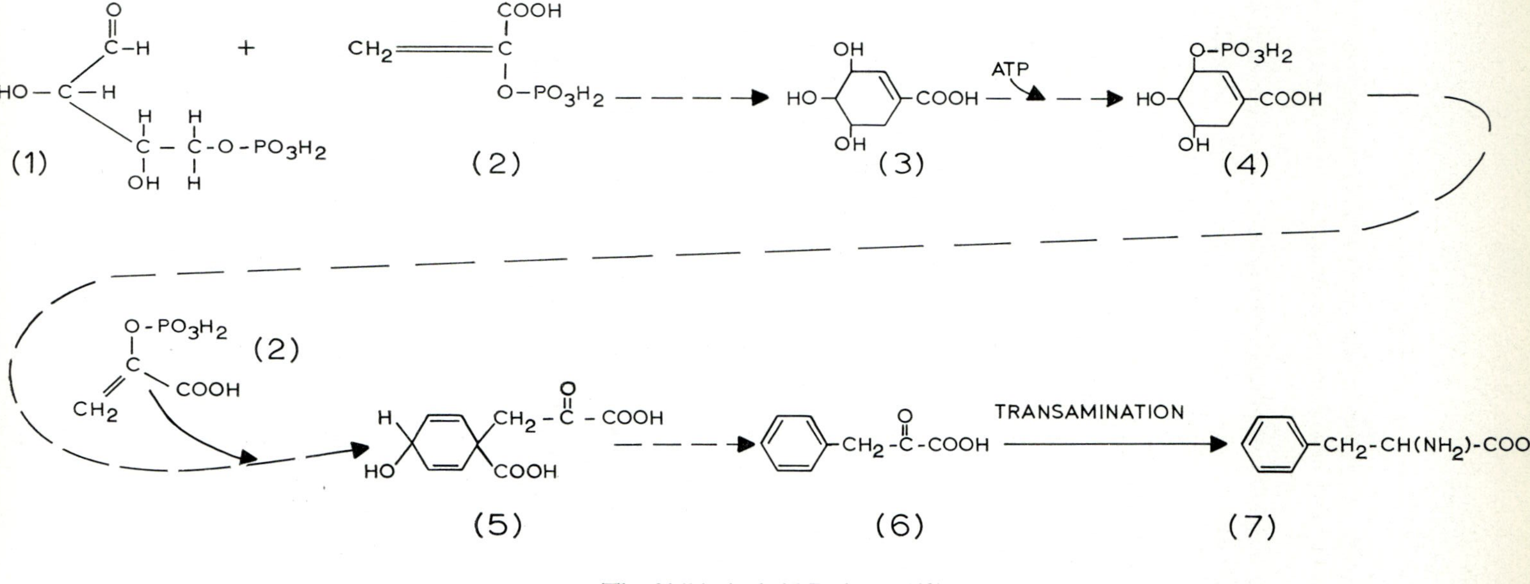

The Shikimic Acid Pathway (69)

This route, often called the *shikimic acid pathway* involves the condensation of phosphoenolpyruvate (2) and a 4-carbon sugar erythrose-4-phosphate (1) which is derived from the pentose phosphate pathway. The product of this reaction is converted to shikimic acid (3). Phosphorylation of shikimic acid to yield 5-phosphoshikimic acid (4) is followed by the addition of another molecule of phosphoenol pyruvate (2) which results in the synthesis of prephenic acid (5). Aromatization of the prephenic acid can give rise to phenylpyruvic acid (6) which upon transamination becomes phenylalanine. The carbon skeletons of the other aromatic amino acids, tryptophane and tyrosine are also synthesised via the shikimic acid pathway as is lignin and many of the aromatic secondary products described in Chapter 6.

The hydrolysis of proteins yields not only a mixture of amino acids but nearly always some ammonia. Part at least of this ammonia comes from amide residues (residues of glutamine and asparagine) present in the protein molecule. The free amides glutamine and asparagine often accumulate in cells, particularly when reserve proteins are being degraded and amino acids oxidised or following feeding with ammonium salts. These amides are not only to be regarded as stores of amide-nitrogen which can be readily used as a source of ammonia for amino acid synthesis but as units which are to be directly utilised in the synthesis of certain proteins. Study of the mechanism of glutamine synthesis suggests a further role of these amides. The synthesis of glutamine takes place under the influence of the enzyme, *glutamine synthetase*, and in the presence of magnesium ions by the equation:

$$\text{L-glutamic acid} + NH_3 + \text{ATP} \rightleftharpoons \text{L-glutamine} + \text{ADP} + \text{phosphate} \tag{70}$$

The synthesis of the amide group therefore involves the consumption of ATP. However, the reaction is reversible to a measurable extent so that energy stored in the amide group of accumulated glutamine could be re-utilised to synthesise ATP from ADP plus inorganic phosphate. The relatively high energy content of the amide group of glutamine may also explain why this compound functions as the donor for two of the nitrogen atoms incorporated into the purine ring during its synthesis.

Plants obtain their sulphur as sulphate ions (valency +6) from the soil and the assimilation of sulphate involves its reduction to the level of sulphide (valency −2). Early studies on sulphate utilisation using homogenates of mammalian liver indicated that formation of phenyl sulphates depended on the presence of ATP and involved the intermediate formation of "active sulphate", an adenosine phosphatosulphate analagous to "active acetate" (see p. 98). The activated sulphate group can be transferred to a variety of acceptors such as the phenols mentioned above or can be reduced as occurs in those organisms which use sulphate as their source of sulphur.

The reactions leading to the formation of "active sulphate" were elucidated in enzyme fractionation studies on baker's yeast. It is a two-step process leading first to the formation of adenosine-5′-phosphosulphate (APS) and then to the "active sulphate", 3′-adenosine-5′-phosphosulphate (PAPS) thus:

$$\text{ATP} + \text{SO}_4 \underset{\text{ATP - sulphurylase}}{\rightleftharpoons} \text{APS} + \text{PPi} \quad (71)$$

$$\text{PPi} + \text{H}_2\text{O} \underset{\text{Pyrophosphatase}}{\rightleftharpoons} 2\text{Pi} \quad (72)$$

$$\text{APS} + \text{ATP} \underset{\text{APS - kinase}}{\rightleftharpoons} \text{PAPS} + \text{ADP} \quad (73)$$

Reaction (71) proceeds with a gain in free energy (is highly enderonic) so that the equilibrium is hard over in the direction of ATP + SO_4. However this reaction proceeds to the right because it is "pulled" by the exergonic hydrolysis of pyrophosphate (PP_i) and by the high affinity of *APS kinase* for its substrate and the exergonic value of reaction (73). In consequence the overall reaction sequence proceeds with a slight decrease in free energy.

In studies on higher plants the presence of APS kinase has not been demonstrated and the suggestion has been made that in plants the sulphate activating system consists only of *ATP sulphurylase* and *pyrophosphatase*.

The reduction of sulphate to sulphide involves a change from valency +6 to —2 and if it is effected by steps each involving the transfer of two electrons can be represented thus:

$$\underset{\text{sulphate}}{SO_4''(+6)} \xrightarrow{2e^-} \underset{\text{sulphite}}{SO_3''(+4)} \xrightarrow{2e^-} ?\,(+2) \xrightarrow{2e^-} ?\,(0) \xrightarrow{2e^-} \underset{\text{sulphide}}{S''(-2)} \qquad (74)$$

The mechanisms involved are not fully worked out. It appears that all the intermediates remain bound to protein carriers. Evidence has been obtained that the initial reduction step (*sulphate reductase*) involves two reactions, a low molecular weight hydrogen transporting protein, and reduced FAD as the electron donor. The details of the further reduction steps are even more obscure. It has been suggested that the whole 6-electron oxidoreduction is catalysed by a single enzyme (*sulphite reductase*) and that reduced NADP is the electron donor.

The sulphide formed is then involved in the synthesis of reduced sulphur metabolites. The first step in the synthesis of the sulphur-containing amino acids is probably

$$\underset{\text{serine}}{CH_2OH.CH(NH_2)COOH} + H_2S \underset{}{\xrightleftharpoons[]{\text{serine sulphydrase}}} \underset{\text{cysteine}}{CH_2.SH.CH(NH_2)COOH} + H_2O \qquad (75)$$

Cysteine has been shown to be an effective source of sulphur for higher plant cells.

The sulphur-containing amino acids, cysteine, cystine and methionine make a special contribution to protein structure (disulphide bridges) and are often implicated in the mechanism of enzyme action. Sulphur is also a constituent of vitamins (thiamin and biotin), of co-enzyme A and of many odour-producing secondary plant products such as those responsible for the characteristic smells of onion, garlic and cabbage.

The mechanisms involved in the assimilation of nitrate and sulphate both involve stepwise reductions prior to incorporation of the element into important primary metabolites; their reduction reactions are endergonic and require energy made available from respiration or photosynthesis. The most central biological roles of these elements is as constituents of the amino acids involved in protein synthesis.

THE SYNTHESIS OF PROTEINS

The elucidation of mechanisms mediating the flow of information from the genes and how this is translated into the synthesis of cellular proteins from amino acids has been one of the major achievements of 20th-century biology.

A number of early observations pointed the way towards our current concepts of protein synthesis. Cells that are particularly active in protein synthesis are rich in cytoplasmic ribonucleic acid (RNA) and ribonucleoprotein (RNA-protein). Further, in a number of studies with intact cells it was found that their rate of protein synthesis varied directly as their cytoplasmic nucleic acid content and that their synthesis of protein was strongly inhibited by *ribonuclease* (which degrades RNA) and also by chloramphenicol. More recently, limited protein synthesis has been achieved in cell-free preparations obtained from bacterial and higher plant cells and reinforced by the addition of appropriate mixtures of the 20 protein amino acids and a suitable energy source such as ATP. Such *in-vitro* protein synthesis, most readily followed by studying the incorporation of ^{14}C-labelled amino acids into protein, is prevented by adding ribonuclease or chloramphenicol.

By disrupting cells in such a way as to liberate nuclei, mitochondria, chloroplasts and cytoplasmic ribosomes and separating these from one another by differential centrifugation it was shown that all these "particles" could incorporate amino acids into protein. Isolation of ribosomes (p. 38) from the chloroplasts and mitochondria confirmed that the ribosomes are the centres of protein synthesis. Studies in the electron microscope of cells active in protein synthesis show the prominence in such cells of endoplasmic reticulum studded with ribosomes (Fig. 2.9) and of ribosomes organised into clusters (polyribosomes).

Studies with cell-free systems capable of synthesising protein have shown that the synthesis of proteins is a complex process and it will be convenient first to outline the overall mechanisms and then to consider in more detail the several stages involved.

The first step in the process is the *transcription* of information relating to a particular protein from the deoxyribosenucleic acid (DNA) to a species of ribosenucleic acid, which because it serves to carry this information to the ribosomes is referred to as messenger or m-RNA. In association with m-RNA, the ribosomes assemble the amino acids, recognised by their combination with an adaptor molecule (transfer or t-RNA), and then polymerise them in accordance with information contained in the m-RNA strand. Thus the information stored in the DNA is *translated* into protein structure.

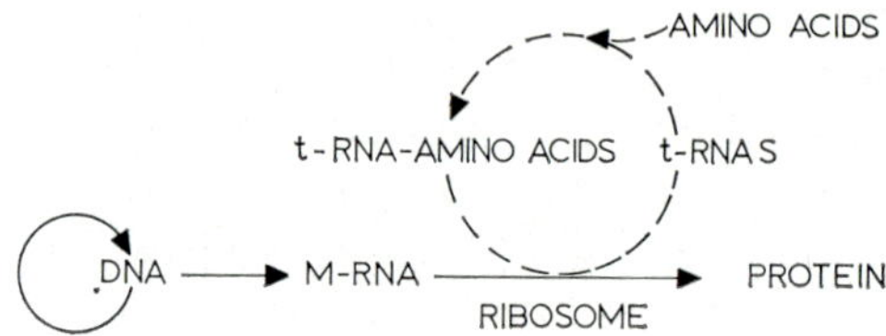

FIG. 5.13. The flow of information from DNA to proteins and the participation of t-RNA in protein synthesis. Solid arrows represent flow of information; dotted arrows the involvement of t-RNA and amino acids.

DNA is a linear heteropolymer containing four kinds of purine and pyrimidine bases, adenine (A), guanine (G), thymine (T) and cytosine (C) joined to a backbone of deoxyribose phosphate. As a result of hydrogen bonding between complementary bases A and T and G and C the helical DNA molecules usually associate in pairs as a double helix (Fig. 5.14). Genetic information is stored in the DNA molecule in a code which depends upon the sequence of the bases A, G, C and T along the molecule.

An essential prerequisite of a storage system for holding genetic information in a living organism is that precise duplication of the information must be possible. In DNA this is achieved by a mechanism depending upon the complementarity of A and G and C and T. Duplication is preceded by partial separation of the strands followed by alignment of free deoxyribonucleotides with their complementary bases in the separated DNA strands (Fig. 9.2, p. 267). An enzyme *DNA polymerase* then joins the nucleotides together. Thus two identical complete double helices of DNA are produced; each pair containing one new and one pre-existing strand (hence the application

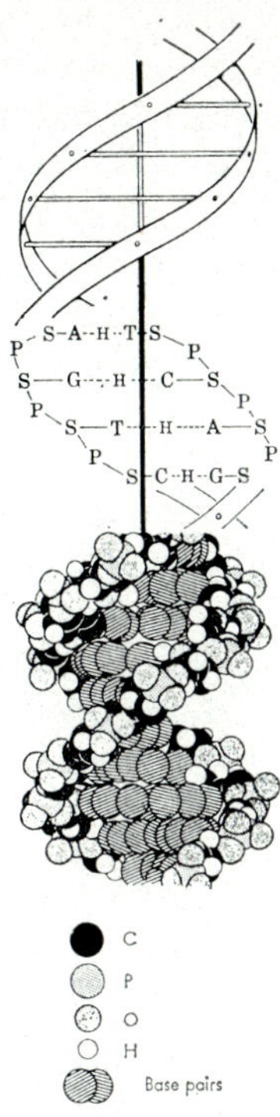

FIG. 5.14. The double helical structure of deoxyribonucleic acid (DNA). S, deoxyribose; C, cytosine; A, adenine; G, guanine; T, thymine; . . . H . . ., hydrogen bonding. (After C. Swanson, in *The Cell*, Prentice-Hall, New Jersey, 1960.)

of the term semi-conservative to DNA synthesis). The precision with which DNA replication occurs (or with which mistakes are rectified) is indicated by the infrequency of natural mutations. Genetic mutations may be visualized as changes in the nucleotide sequence of DNA—involving substitution or deletion of nucleotides.

As mentioned above, the genetic information is coded into DNA as the sequence of bases along the molecule. A great deal is known about this code. Each protein amino acid is represented by a triplet of bases or *codon*. This four-letter language permitting 64 different three-letter words gives ample scope for the specification of the twenty amino acids and other essential information such as when to start and when to stop protein synthesis—the "punctuation" of the code. In fact the code is "degenerate" in that more than one codon may specify a particular amino acid. A *structural gene* may be considered as a series of codons along a DNA molecule which specify a particular protein. A *cistron* is a length of DNA coding for a protein containing more than one type of polypeptide chain. The activities of these genes are influenced by *regulator genes* which code for proteins which exert a control upon the transcription process and *operator genes* which exert a control on a group of structural genes and hence upon the synthesis of a number of related proteins such as the enzymes of a particular metabolic pathway.

The major part of the DNA of a plant cell is located in the nucleus where it is usually condensed with proteins known as histones in the structures we recognise as chromosomes. Minor fractions of the cell's DNA (minor in quantity though not necessarily in importance) are located within the plastids and mitochondria and apparently specify some of the proteins synthesised in these organelles. The DNA of the organelles is different from the nuclear DNA in that it is not condensed with histones, is composed of double-stranded loops and is of a different density. All these features are common to prokaryotic† DNA and evidence such as this has been involved in the resuscitation of the "ancient symbiont" theory which suggests that the plastids of plant cells are derived from photosynthetic prokaryotic cells resembling present-day blue-green algae which invaded and formed a symbiotic relationship with heterotrophic eukaryotic† cells.

† Cells which contain membrane bound organelles (plastids and/or mitochondria) and an organised nucleus with chromosomes and spindle apparatus and which are characteristic of higher organisms including higher plants are termed eukaryotic cells (from the Greek, meaning good nucleus). In contrast the cells of bacteria and blue-green algae do not exhibit these features and are termed prokaryotic cells (from the Greek, meaning before nucleus).

The genetic information contained in the nucleus is believed to be carried to the cytoplasmic ribosomes as single-stranded m-RNA molecules into which the information is transcribed. Transcription is accomplished by a process which is in many ways similar to that of DNA duplication. The DNA strands separate and the free ribonucleotides of A, G, C and uracil (U) (instead of thymine) arrange themselves opposite their complementary bases in one DNA strand. *RNA polymerase* joins the bases together and a single strand of RNA (since only one DNA strand is transcribed) containing the genetic information of the DNA is produced, e.g. the DNA codon TTT being equivalent to the RNA codon AAA. Although this description has been applied to the synthesis of m-RNA the synthesis of t-RNA and ribosomal RNA is achieved by the same basic mechanism.

In translating the message contained in m-RNA into protein, specific amino acids are recognised, assembled and joined together in the sequence specified by the m-RNA. The amino acids do not, however, enter directly into the system but are first activated in reactions involving ATP and amino acid activating enzymes (*AAA enzymes*); a separate specific enzyme for each of the 20 individual protein amino acids:

$$\text{amino acid} + \text{ATP} + \text{AAAenzyme} \rightarrow \text{AAAenzyme} - \text{aminoacyladenylate} + \text{PP}_i \quad (76)$$

The amino acid adenylate enzyme complex combines with a t-RNA molecule which is specific for the particular amino acid (Fig. 5.15)—indeed there are more than one species of t-RNA for some of the amino acids.

$$\text{AAAenzyme} - \text{aminoacyladenylate} + \text{t-RNA} \rightarrow \text{aminoacyl t-RNA} + \text{AMP} + \text{AAAenzyme} \quad (77)$$

It is the aminoacyl-t-RNAs which enter the protein-synthesising system proper (ribosome–m–RNA complex). Each t-RNA species is recognised by virtue of a sequence of bases in a particular part (the coding site) of the t-RNA molecule. The sequence of bases in the coding site has been shown in some of the most intensively studied t-RNAs to be the complement (*anticodon*) to the m-RNA codon specifying the amino acid with which the t-RNA combines.

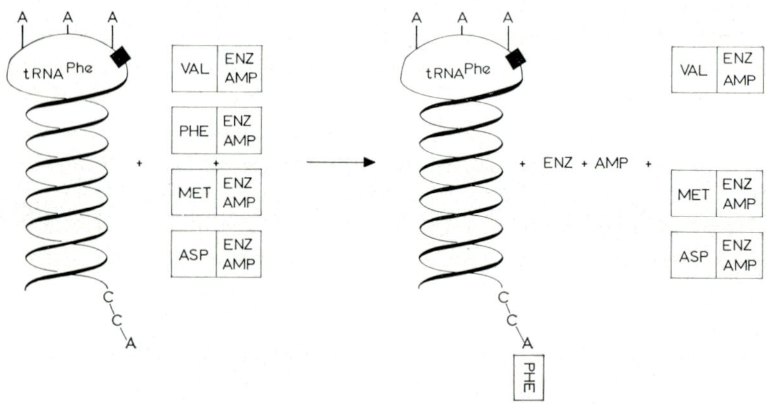

FIG. 5.15. The "charging" of t-RNA. The diagram shows the selection of phenylalanine by t-RNA^{PHE} (phenylalanine t-RNA) from a group of activated amino acids. ENZ, amino acid activating enzyme; PHE, phenlyalanine; VAL, valine; MET, methionine; ASP, aspartic acid. The diagram of t-RNA^{PHE} illustrates the anticodon loop and a double stranded helical portion of the molecule. The full "clover leaf" structure is not represented. The black square adjacent to the anticodon illustrates the position occupied by isopenteny-aminopurine in some plant t-RNAs (see p. 299).

t-RNA acts as an "adapter" which can on the one hand recognise and combine with a particular activated amino acid and which on the other hand is itself recognised by and can be utilised by the ribosome–m–RNA complex to link on the protein being synthesised.

t-RNAs play an essential role in protein synthesis and it has been further suggested that some control of protein synthesis may be exerted at the t-RNA level (p. 298). All t-RNAs terminate with the base sequence CCA and in a "charged" t-RNA an amino acid is attached to the terminal A residue of this sequence. The complete nucleotide sequences of several t-RNAs are known; their secondary structures are complex and have been likened to the shape of a clover leaf. Parts of the molecule are double-stranded and others are single-stranded loops. One of these loops carries the anticodon and for the sake of simplicity only this loop is represented in the diagrams of t-RNAs in Figs. 5.15 and 5.16.

Detailed analysis of t-RNA molecules (similar results have been obtained with other classes of RNA and with DNA) has shown that although A, G, C, U (or T) are the major nucleotides a variety of other nucleotides are present. Some such as inosine (I) and pseudouridylic acid (Ψ) may be scattered within the molecules whereas others are limited to specific sites. Isopentenylaminopurine (IPA) appears to be located at a specific position near to the anticodon in certain t-RNAs (Fig. 5.15) and may play a part in the control of t-RNA activity through affecting the conformation of the anticodon loop or even the whole secondary structure of the molecule. Much remains to be learned about the potentially important functions of these "uncommon" nucleotides and considerable research effort is presently being directed to this end (see p. 298).

The ribosomes of the protein synthesising system are small (*ca.* 150 Å diameter) organelles composed of protein and RNA. They may be found free in the cytoplasm, mitochondria and chloroplasts or attached to endoplasmic reticulum or organised into polyribosomes. Plant (and animal) cells have two types of ribosomes: one type, the 70S ribosomes, are found in chloroplasts and mitochondria and the other, 80S ribosomes, are found in the cytoplasm.

S, the Svedberg unit, is a measure of the rate of sedimentation of a particle in the ultracentrifuge. A high Svedberg number indicates that the particle sediments rapidly (sedimentation rate increases with particle size when shape and density are constant). The 70S ribosomes differ in a number of ways from the 80S ribosomes and are similar to those of micro-organisms—further evidence in favour of the "ancient symbiont" hypothesis.

Ribosomes are composed of two basic subunits—the 70S ribosome being composed of a 50S and a 30S subunit. These have been further fractionated into their RNA and protein constituents and in some cases have been reassembled to form active ribosomes. Omission of various components during reassembly allows investigation of the functions of the omitted component and it has been shown, for example, that one of the ribosomal proteins is involved in the

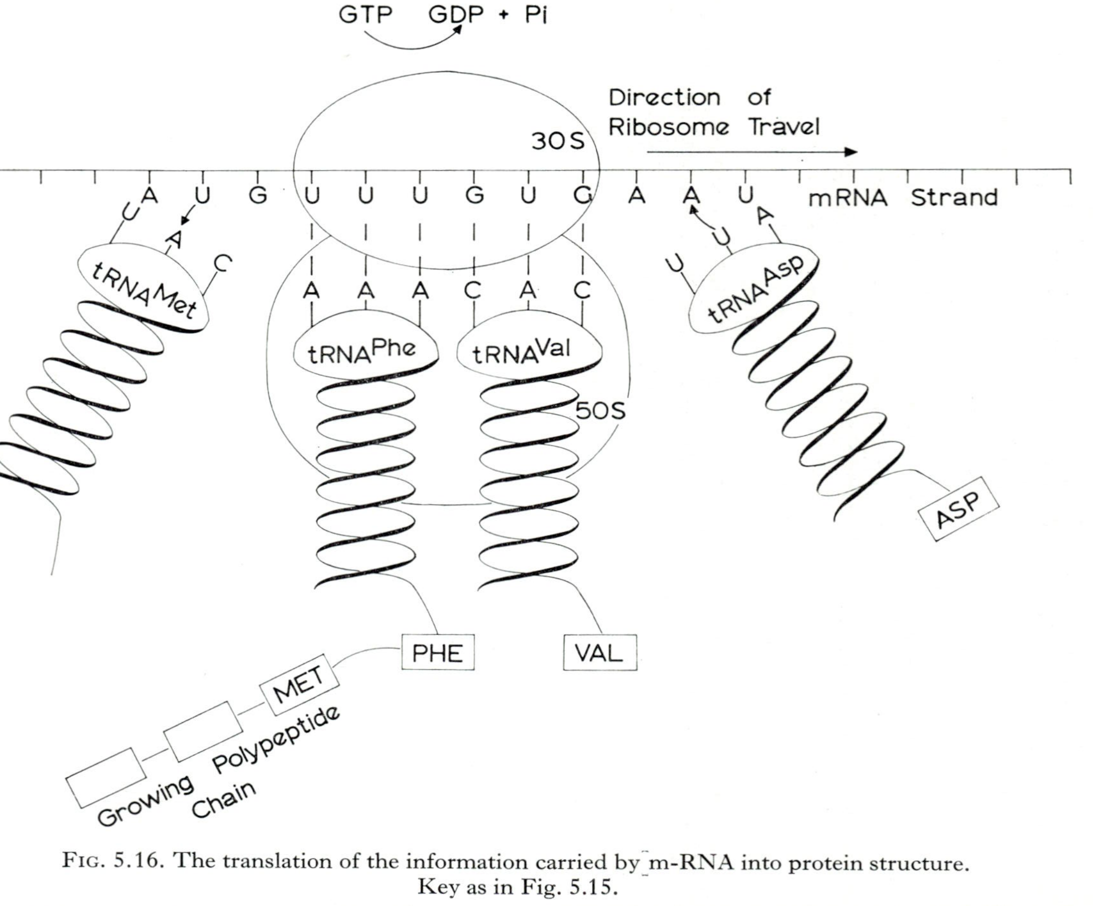

FIG. 5.16. The translation of the information carried by m-RNA into protein structure. Key as in Fig. 5.15.

initiation of protein synthesis and another affects the frequency of occurrence of translational errors.

Having described the individual components involved in translation we can now consider how these elements co-operate to synthesise a particular protein. A ribosome associates with a strand of m-RNA and moves along it (Fig. 5.16). At a given time there are two "charged" t-RNA molecules bound to the ribosome ($t\text{-RNA}^{Phe}$ and $t\text{-RNA}^{Val}$ in Fig. 5.16). As the ribosome moves along the m-RNA strand (translocation) a peptide bond is formed between the two amino acids attached to the bound t-RNAs (phe and val in Fig. 5.16) and simultaneously guanosine triphosphate (GTP), which is an essential co-enzyme in the reaction, is converted to guanosine diphosphate (DGP). The ribosome is now in a position to read the next codon in the m-RNA and by interaction between the m-RNA codon and the t-RNA anticodon a further charged t-RNA ($t\text{-RNA}^{Asp}$ in Fig. 5.16) becomes associated with the ribosome codon-recognition site. At the same time the t-RNA which carried what is now the penultimate amino acid in the polypeptide chain is released (this would be $t\text{-RNA}^{Phe}$ following translocation from the stage shown in Fig. 5.16). This procedure is repeated until the ribosome reaches the end of the m-RNA strand; the specified protein is then completed and the ribosome is released and may associate with another m-RNA strand. Often many ribosomes are associated with a single molecule of m-RNA in which case the whole is referred to as a polyribosome or polysome.

The effective life of m-RNA *in vivo* is important since it will determine the maximum number of protein molecules a particular molecule of m-RNA can produce. In bacteria the mean half-life of m-RNA is very short, of the order of 3–4 min. For eukaryotic cells our knowledge on this question is limited, but it seems that at least some species of m-RNA have mean half-lives which are measured in hours. Nucleotides released during the breakdown of m-RNA are presumably re-used in the synthesis of RNA.

Information is accumulating concerning the mechanism of initiation of protein synthesis. In bacterial, mitochondrial and chloroplast systems the initial amino acid in a polypeptide is always formylmethionine—the formyl residue being removed from the completed protein. Although the initiation mechanism in the cytoplasmic ribosomes of eukaryotic cells is not yet firmly established, it appears that the initiator amino acid is methionine which is usually removed from the completed protein. As well as the t-RNA^{Met} involved in the incorporation of methionine within a polypeptide chain another t-RNA^{Met} specifically associated with protein initiation has been discovered—an indication of the variety of roles played by specific t-RNAs and their involvement in the expression of the "punctuation" of the genetic code.

The reactions described above relate only to the synthesis of linear polypeptides; the complex structures characteristic of natural proteins presumably result automatically from the action of intra- and intermolecular forces following the release of the polypeptide chains.

Much of the foregoing discussion is based upon what has been learned from bacterial systems. However, rapid progress is being made in the technically more difficult study of protein synthesis in eukaryotic cells and so far there is reason to believe that the mechanism is likely to differ only in details from that of prokaryotic cells—a striking example of the fundamental unity of biological systems.

FURTHER READING

E. Rabinowitch and Govindgee. *Photosynthesis.* John Wiley & Sons Inc., 1969.

O. V. S. Heath. *The Physiological Aspects of Photosynthesis.* Heinemann Educational Books Ltd., 1969.

D. A. Walker. Three phases of chloroplast research. *Nature,* **226:** 1204–1208, 1970.

O. Kandler. Biosynthesis of poly- and oligosaccharides during photosynthesis in green plants. In *Harvesting the Sun,* pp. 131–152, edited by Anthony San Pietro, Frances A. Greer and Thomas J. Army. Academic Press, 1967.

A. T. James. Fatty acid biosynthesis in plants. In *Perspectives in Phytochemistry,* pp. 91–105, edited by J. B. Harborne and T. Swain. Academic Press, 1969.

R. S. Bandurski. The biological reduction of sulphate and nitrate. In *Plant Biochemistry,* pp. 467–490, edited by J. Bonner and J. E. Varner. Academic Press, 1965.

L. Fowden. Origins of the amino acids. In *Plant Biochemistry*, pp. 361–390, edited by J. Bonner and J. E. Varner. Academic Press, 1965.

M. Nomura. Ribosomes. *Scientific American*, **221**: 28–35, 1969.

C. A. Price, *Molecular Approaches to Plant Physiology*, chapter 5.1–5.3, The Dogma; Duplication and Translation. McGraw-Hill Book Company, 1970.

MORE ADVANCED READING

D. A. Walker and A. R. Crofts. Photosynthesis. *Ann. Rev. Biochem.*, **39**: 389–427, 1970.

M. Avron. Photoinduced electron transport in chloroplasts. *Current Topics in Bioenergetics*, **2**: 1–22, 1967.

J. Preiss and T. Kosuge. Regulation of enzyme activity in photosynthetic tissues. *Ann. Rev. Plant Physiol.*, **21**: 433–461, 1970.

L. Beevers and R. H. Hageman. Nitrate reduction in higher plants. *Ann. Rev. Plant Physiol.*, **20**: 495–522, 1970.

D. Boulter. Protein synthesis in plants. *Ann. Rev. Plant Physiol.*, **21**: 91–114, 1970.

CHAPTER 6

Secondary Plant Products

"The versatility of plants in effecting complex biosynthesis certainly outstrips man's ingenuity in suggesting rational ways in which these structures are elaborated or in providing reasons why they should exist at all."

L. Fowden, in *Endeavour*, vol. 21, pp. 135–42, 1962.

INTRODUCTION

THE chemical compounds whose metabolism has been discussed in Chapters 4 and 5 are central to the chemistry of life processes. They may be classed as essential or primary metabolites. However, in addition to these primary metabolites, many plants, particularly those of certain families and genera, synthesise compounds which are not in the mainstream of metabolism; other species not containing them are no less vigorous or successful. Such compounds are often grouped together as secondary plant products, although they are extremely numerous and chemically diverse (see Fig. 6.1). Despite the kaleidoscopic display of synthetic virtuosity evidenced by these compounds, most have their origin in a relatively few areas of primary metabolism.

Figure 6.1 shows that rubber and carotenoids are derived from acetyl-CoA via mevalonic acid and may both be classed as members of the isoprenoid group of compounds. Similarly shikimic acid (which is formed by a condensation of erthrose-4-phosphate and phosphoenolpyruvate; see p. 171) is a key intermediate in the

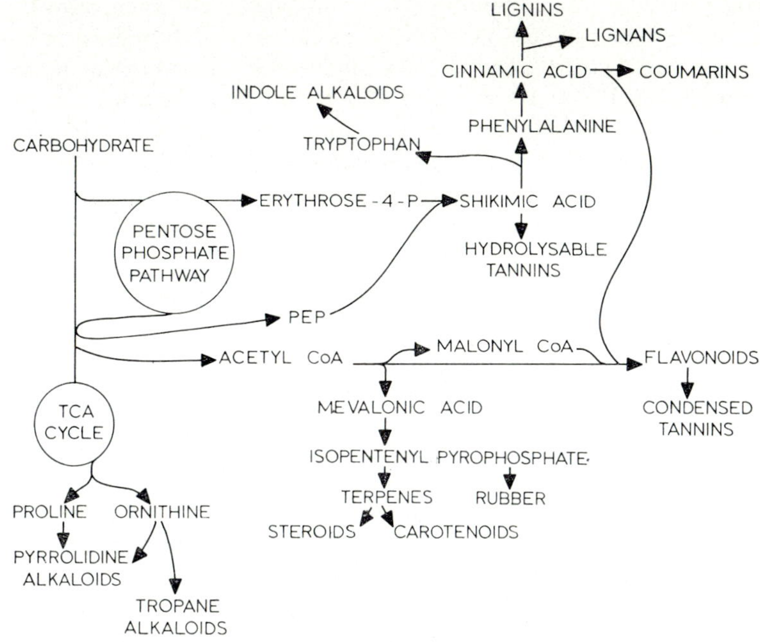

FIG. 6.1. Some relationships between primary and secondary metabolism. The probable origins are shown of the flavonoid, phenylpropanoid and isoprenoid group of compounds. The other groups discussed in the text, the non-protein amino acids and alkaloids, arise from a number of points in primary metabolism and only the origins of the tropane and pyrrolidine alkaloids from ornithine and proline and the indole alkaloids from tryptophan are shown.

synthesis of the phenlyalanine-related alkaloids, the phenylpropanoids, the flavonoids and the condensed tannins.

The above definition of what constitutes a secondary plant product although convenient may be difficult to apply in individual instances; how can we expect always to be able unequivocally to decide whether or not a compound is involved in primary metabolism when our knowledge even of the most central life processes is clearly far from complete? Perhaps, therefore, a better definition is that secondary plant products are compounds which have not *so far* been shown to be involved in primary metabolism; as far as their functions can at present be assessed they are accessory rather than central to the physiology of the plants in which they occur. This, however, should not be taken to imply that they are necessarily quantitatively unimportant in the biochemical economy of the plant. A 20 m length of *Hevea brasiliensis* trunk may contain some 6 kg of raw rubber and in guayule, *Parthenium argentatum*, the rubber may make up 20% of the plant's dry weight. The physiological significance, if any, of material produced on this scale is a matter of great interest.

This leads us to a consideration of the possible evolutionary origin of these secondary plant products. Any change in an organism conferring upon it an increased survival potential will come to be established in the population by natural selection; any deleterious change will by the same process be eliminated. Let us consider an intermediate (A) in primary metabolism which undergoes conversion to a second metabolite (B). Suppose, as a result of mutation, the opening up of a biosynthetic pathway converting A to X which is quite distinct from that involved in converting A to B. If this diversion of A to X is deleterious it will be selected against. If the presence of X confers a survival advantage, the mutant plants will become established and X will be classed as a secondary plant product if its function is obscure to us or if the function we assign to it is accessory (e.g. makes the plant less palatable to browsing animals) rather than part of a recognised essential metabolic pathway. If formation of X is neither advantageous nor disadvantageous it will not be actively selected for or against and may persist. Further mutations could then result in elaboration of the pathway $A \rightarrow X$ so that other compounds related to X would become natural products. Such compounds would become characteristic products of species or genera particularly if in the individuals containing these compounds there occurred further and now advantageous mutation. The "neutral" genes promoting the synthesis of the X family of compounds would survive by being associated with favourable genes.

Where we cannot identify a function for a secondary plant product it may be that it has no present significance but was important for survival at some earlier stage in the evolution of the species. A compound once effective against an important predator or parasite may no longer be so due to the development of immunity by the enemy; alternatively the predator may now be extinct.

The impressive range of secondary products to be found in plants contrasts with the paucity of such compounds in animals (the constitutents of the bile salts of vertebrates and the contents of the preening glands of birds are in some ways comparable). One factor here may be that animals excrete waste products of their metabolism to the exterior. This can only be achieved to a very limited extent by plants by the shedding of leaves, floral parts or roots. Hence plants may "seal" off such products in vacuoles, resin canals, tannin cells, heartwood and so on.

Secondary plant products are discontinuously distributed throughout the plant kingdom. Particular groups of secondary products may be characteristic of single or related genera. Species with identical or closely related secondary products may have arisen from a common ancestor. The presence or absence of a characteristic secondary product may therefore prove to be a useful taxonomic character when other characteristics make the placing of the plant between two related families or genera difficult. Classifications based upon other criteria may be vindicated by chemotaxonomic studies. The recent development of methods for the rapid elucidation of chemical structures and the associated rise of chemotaxomy have both stimulated the search for new secondary plant products; their number, already great, continues to rise rapidly. It is thus quite beyond the scope of this chapter to attempt any comprehensive survey of such compounds; references to some important texts in this field will be found under Further Reading. Some of the interesting problems of biosynthesis and function posed by secondary plant products will now be illustrated by a brief consideration of alkaloids, isoprenoids, nonprotein amino acids, flavonoids, and phenylpropanoids.

ALKALOIDS

The exact definition of what constitutes an alkaloid is difficult—partly because present knowledge of these compounds is a synthesis of information gained from botanical, pharmacological and chemical studies. However, most authorities would agree that a basic compound which contains its nitrogen as a constituent of a heterocyclic system, which is restricted in its distribution throughout the plant kingdom (say to certain genera) and which is active pharmacologically would certainly be regarded as an alkaloid. However, many compounds classed as alkaloids do not fulfil all of the above criteria, the alkaloid colchicine for example does not contain its nitrogen as part of a heterocyclic system.

The chemical structure of alkaloids varies from relatively simple, e.g. atropine (racemic hyoscyamine),

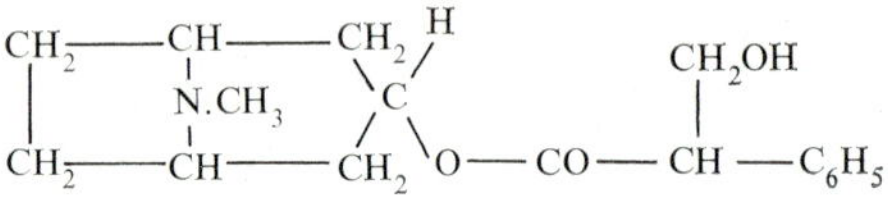

CH_2—CH—CH_2 H
N.CH_3 C CH_2OH
CH_2—CH—CH_2 O—CO—CH—C_6H_5

to complex, e.g. strychnine, the structure of which taxed the ingenuity of organic chemists for more than 100 years.

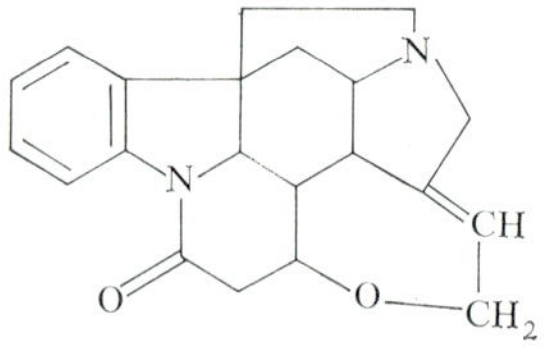

N
N
CH
O
O—CH_2

Notwithstanding the structural complexity of some of the alkaloids most can be related structurally and often biosynthetically to the amino acids. The indole nucleus of strychnine, for example,

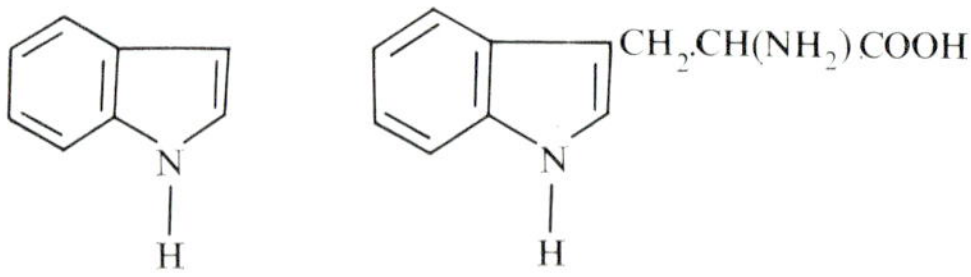

N
H
Indole

$CH_2.CH(NH_2)COOH$
N
H
Tryptophan

being derived from the aromatic amino acid tryptophan.

There are some twenty or so groups of alkaloids each group often being confined to related plants—several members of the group usually being found in the same species; a characteristic of a true alkaloid plant being that it is likely to contain a number of related alkaloids. Amongst the important groups are: the *pyridine-related* alkaloids which include nicotine which occurs in certain solanaceous plants, arecoline an alkaloid of the betel nut and coniine the toxin of hemlock (*Conium maculatum*) which was used to execute criminals in ancient Greece—Socrates having been the most distinguished victim; the *morphine* and *codeine* alkaloids which include heroin which is derived from the latex of the opium poppy; the *indole* alkaloids such as strychnine; and the *cinchona* alkaloids which include quinine, the antimalarial drug. Another large and pharmacologically important group are the condensed *pyrrolidine–piperidine* group (the tropane alkaloids) which will be discussed in more detail below.

The alkaloids mentioned above represent only a very small fraction of the large number (approximately 2000) which have been isolated from something like one-fifth of all vascular plant species. Alkaloids are also found in some *Lycopodium* species and in certain fungi—the ergot alkaloids synthesised by the parasite of rye, *Claviceps purpurea*, being the best known. Additions are still being made to the long list of plants which have been shown to contain alkaloids, a relatively recent discovery being that certain *Rauwolfia* species contain an alkaloid now named reserpine which is very useful in the treatment of hypertension.

The parts of the plant in which alkaloids are synthesised and accumulate are, as might be expected with such a heterogeneous group, very varied. They are not usually restricted to one location although high concentrations may accumulate at particular sites. For example, the seeds of *Strychnos nux-vomica* accumulate strychnine, the bark of the tree *Cinchona ledgeriana* may contain around 10% quinine sulphate by weight and the roots of the Mandrake (*Mandragora officinalis*) are a good source of tropane alkaloids.

Probably most is known about the sites of alkaloid biosynthesis in the Solanaceae although even in this family the picture is not entirely clear, the situation varying from species to species. Experiments in which the leaves and shoots of alkaloid-bearing Solanaceae such as *Atropa belladonna* were grafted onto rootstocks of non-alkaloid

Solanaceae such as tomato and vice-versa provide evidence that the roots are the site of alkaloid biosynthesis in these plants. Histochemical techniques utilising alkaloid precipitants coupled with microscopic examination of treated tissues have located accumulations of positively reacting materials in the meristems of both the roots and shoots of belladonna. These observations do not, however, unequivocally implicate the meristems in the initial synthesis of the alkaloids and experiments using tissue culture techniques have demonstrated that there is no simple relationship between cell division and alkaloid synthesis. Rapidly dividing cultures of belladonna cells do not synthesise alkaloids—only when root initials are induced to form does alkaloid biosynthesis commence. Thus the importance of roots in the synthesis is confirmed although the site of synthesis within these organs has not yet been rigidly established.

The significance of alkaloids in metabolism is obscure. Certainly rates at which they accumulate vary with the environmental and nutritional conditions under which the plants are grown and often also with the stage of development of the plant. Such observations do not necessarily indicate that the alkaloid is playing any part in general metabolism. However, although for the most part accumulation of alkaloids seems to be irreversible, there is evidence, in a few cases for the disappearance of alkaloids; alkaloids present in the seeds of *A. belladonna* disappear during germination. Such an involvement may perhaps indicate a function in the regulation of growth. Another suggestion has been that alkaloids may protect against predators. If this is so, the situation is often complex. Thus the berries of *A. belladonna* (deadly nightshade) are spread by birds which if they are to be effective as vectors must consume them. Some birds do this with impunity whereas two berries can be fatal to humans. Alkaloids are certainly not necessarily protective against pests; commercial crops of *A. belladonna* are sometimes ravaged by caterpillars! The diversity of compounds classed as alkaloids would suggest a diversity of functions if they have functions; in most cases such functions are at present obscure.

The tropane alkaloids, cited above, can be used to consider in more detail the relationships between the compounds of an alkaloid family and to illustrate how alkaloids can be synthesised.

The Tropane Alkaloids

The properties of these alkaloids have been known since ancient times; they have been employed in medicine and magic for thousands of years. The henbane (*Hyoscyamus niger*), which contains hyoscyamine, was used by the Babylonians to relieve toothache. The deadly nightshade (*Atropa belladonna*), still used in ophthamology, was used in past ages by women as a cosmetic; belladonna extract dilated the pupils of their eyes (hence the name belladonna). Other important alkaloidal species of the Solanaceae are *Datura* species (the thornapples) and *Mandragora officinalis* (mandrake). Tropane alkaloids occur in other families, notably in Erythroxylaceae; *Erythroxylon coca*, used by the South American Indians to help them withstand the hardship of their labours, is the natural source of cocaine.

The economic importance of the tropane alkaloids has resulted in them being much studied. They are esters of the base tropine (or a derivative of this base) with organic acids such as tropic, cinnamic or benzoic acids. Progress has been made in working out the biosynthesis of these alkaloids, although the associated enzymology is still obscure; some of the steps in biosynthesis may indeed be non-enzymic.

Leete and co-workers have shown that the amino acid ornithine (Fig. 6.2 (1)) is an intermediate. It gives rise to a keto acid (2) by transamination, and then by decarboxylation of the keto acid to the aldehyde, 4-amino-butanal (3). On methylation this gives 4-methylaminobutanal (4), a compound present in *Datura* roots. This undergoes cyclisation to Δ'-pyrroline (5), which combines with acetoacetate (6), and then following decarboxylation gives hygrine (8), a compound which can be detected in *E. coca* along with cocaine. Ring closure in hygrine gives tropinone (9), conversion of the $\rangle C=O$ group of tropinone to $\rangle C\begin{smallmatrix} H \\ OH \end{smallmatrix}$ gives tropine (10).

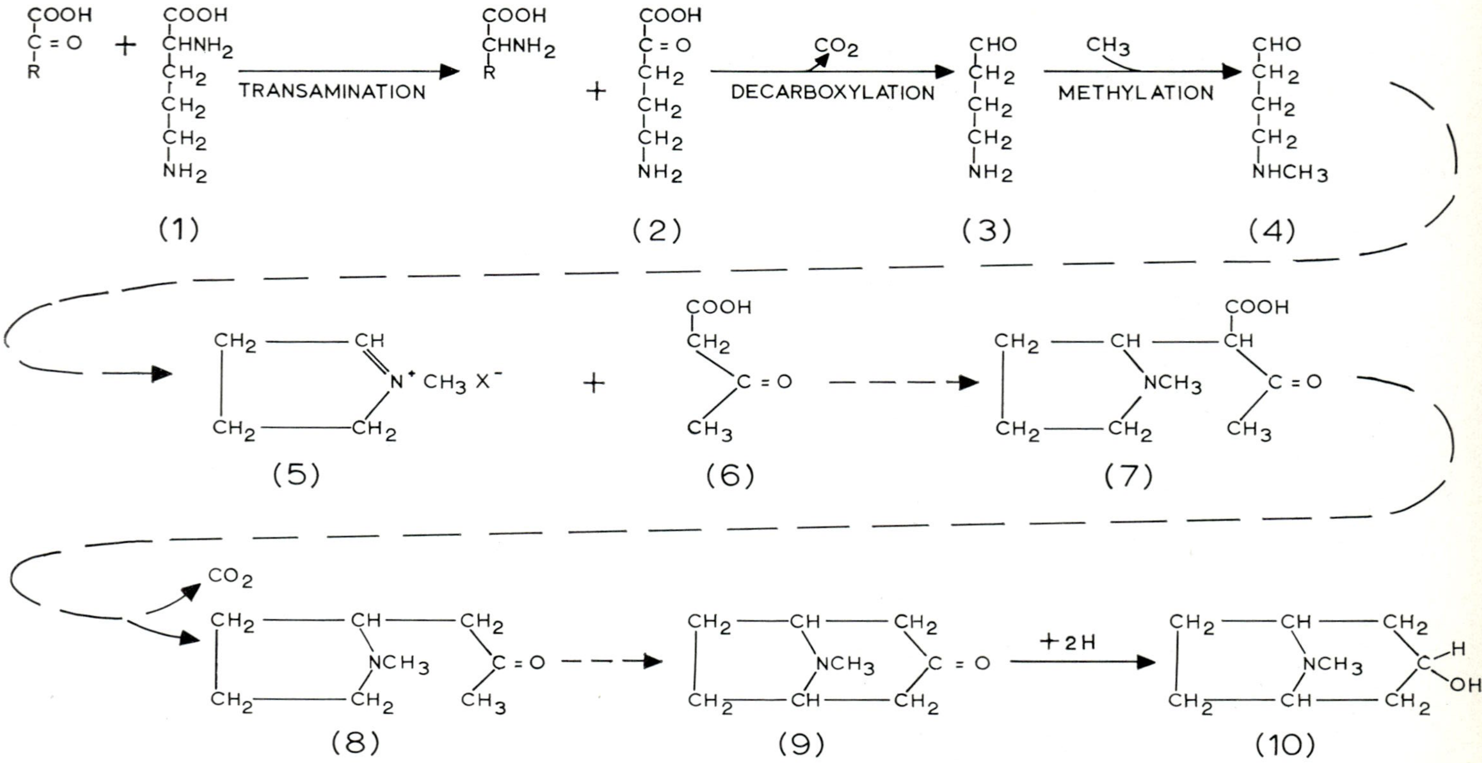

FIG. 6.2. The biosynthesis of the tropane skeleton from ornithine and acetoacetate. The broken arrows represent reactions which are as yet incompletely understood.

Esterification of tropine or a related base with an organic acid completes the synthesis of a tropane alkaloid. Tropic acid is a derivative of phenylalanine:

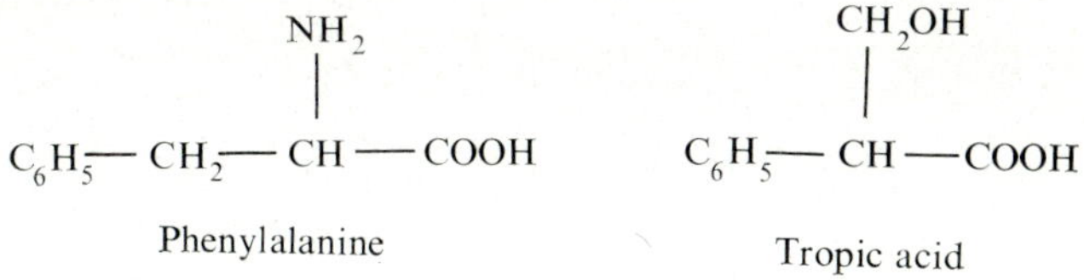

$$C_6H_5—CH_2—CH(NH_2)—COOH \qquad C_6H_5—CH(CH_2OH)—COOH$$

Phenylalanine Tropic acid

Combination of tropine with tropic acid gives hyoscyamine. Further elaboration of the alkaloid may then occur—thus hyoscine is synthesised from hyoscyamine by the introduction of an epoxy group.

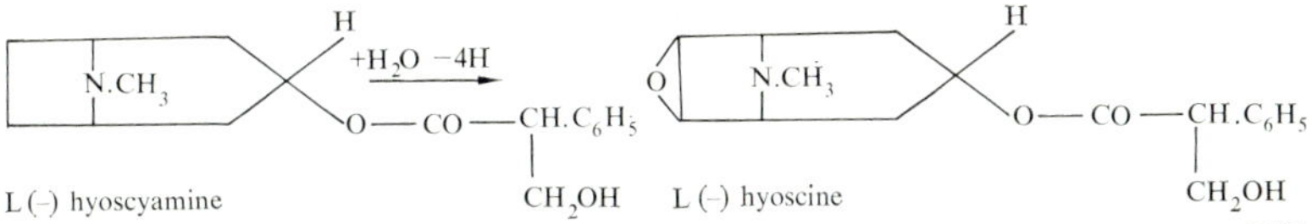

$$\text{L(−) hyoscyamine (N.CH}_3\text{; H; O—CO—CH(CH}_2\text{OH).C}_6\text{H}_5) \xrightarrow{+H_2O\ -4H} \text{L(−) hyoscine (O; N.CH}_3\text{; H; O—CO—CH(CH}_2\text{OH).C}_6\text{H}_5) \quad (78)$$

Reactions of this sort can result in the synthesis of a family of related alkaloids which are characteristic of each alkaloidal species.

The elucidation of the pathways outlined above have involved (1) the establishment of the chemical structures of the alkaloids and related compounds, (2) the postulation of reaction sequences based upon chemical experience and (3) demonstration of the occurrence of postulated intermediates. For example, if compound B is never found in the absence of compound A which sometimes occurs alone, the inference is that compound A is synthesised earlier in a biosynthetic pathway leading to B. The hypothetical pathways are tested by supplying suspected radioactive intermediates to the plant followed by isolation and often degradation of the products. The next stage in the confirmation of pathways—the study of individual reactions, enzymic or otherwise—is, in the case of the tropane alkaloids, far from complete.

ISOPRENOID COMPOUNDS

A diversity of secondary plant products including terpenes (monoterpenes, di- and triterpenes, sesquiterpenes), rubber, carotene and other carotenoids, and the plant sterols can be regarded as built up of isoprene units:

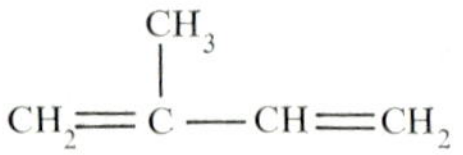

$$CH_2{=}C(CH_3){-}CH{=}CH_2$$

There is a common "biological isoprene unit" which is isopentenyl pyrophosphate (IPP) which, by means of an isomerase, is in equilibrium with its isomer dimethylallyl pyrophosphate. These two compounds can undergo condensation to give the C10 compound geranyl pyrophosphate which is the precursor of all monoterpenes.

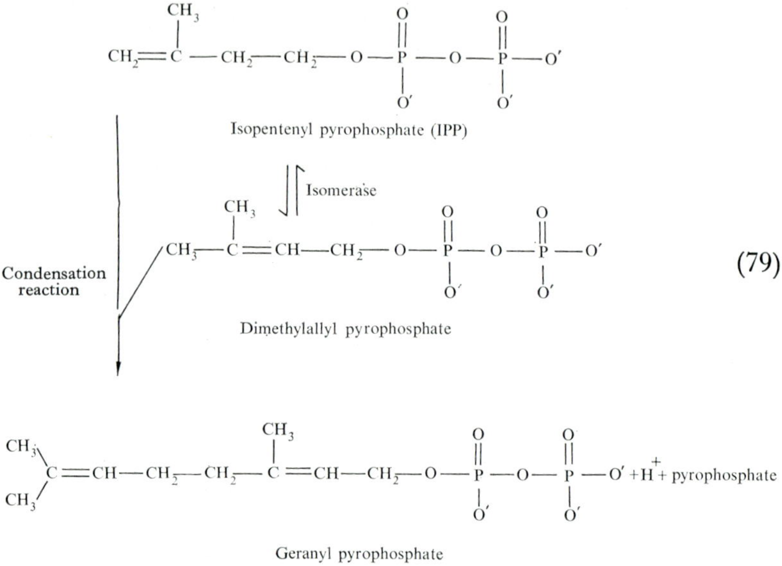

$$CH_2{=}C(CH_3){-}CH_2{-}CH_2{-}O{-}P(=O)(O'){-}O{-}P(=O)(O'){-}O'$$

Isopentenyl pyrophosphate (IPP)

Isomerase

$$CH_3{-}C(CH_3){=}CH{-}CH_2{-}O{-}P(=O)(O'){-}O{-}P(=O)(O'){-}O' \qquad (79)$$

Dimethylallyl pyrophosphate

Condensation reaction

$$(CH_3)_2C{=}CH{-}CH_2{-}CH_2{-}C(CH_3){=}CH{-}CH_2{-}O{-}P(=O)(O'){-}O{-}P(=O)(O'){-}O' + \overset{+}{H} + \text{pyrophosphate}$$

Geranyl pyrophosphate

Compare the formation of fructose diphosphate in glycolysis (Fig. 4.7):

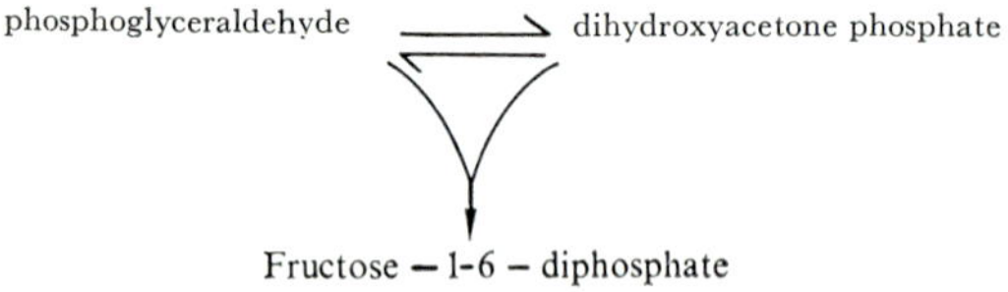

phosphoglyceraldehyde ⇌ dihydroxyacetone phosphate

↓

Fructose – 1-6 – diphosphate

Condensation of geranyl pyrophosphate with a further molecule of

IPP gives the C15 farnesyl pyrophosphate, the immediate precursor of the sesquiterpenes (C15), and by dimerisation gives rise to the triterpenes (C30) via squaline from which other triterpenes and the plant sterols are probably derived. Condensation of farnesyl pyrophosphate with a further molecule of IPP would give geranylgeranyl pyrophosphate (C20), the immediate precursor of the diterpenes and by dimerisation of the carotenoids (C40) (the precursor, tetrahydrophytoene, formed would be highly reduced and the steps between this theoretical precursor and such well-known carotenoids as the carotenes remain to be worked out).

Clearly great interest attaches to the biosynthesis of IPP. This is formed from acetyl-CoA, mevalonic acid (3,5-dihydroxy-3-methylpentanoic acid) (MVA) being a key intermediate.

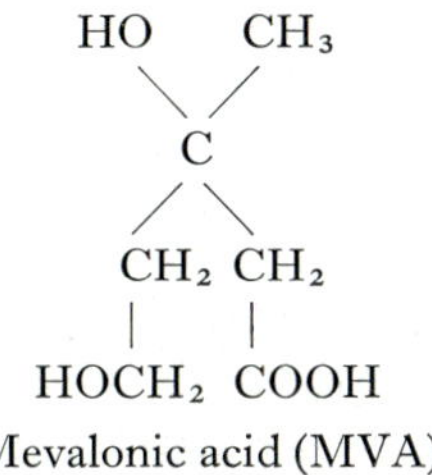

Mevalonic acid (MVA)

Two molecules of acetyl-CoA condense to form acetoacetyl-CoA. Condensation of this with a further molecule of acetyl-CoA gives β-hydroxy-β-methylglutaryl-CoA. This is then reduced and the resulting mevalonic acid released from CoA. Mevalonic acid is then converted by a two-step phosphorylation involving two molecules of ATP to mevalonic acid-5-pyrophosphate from which IPP arises by an oxidative decarboxylation in which ATP participates. There is evidence that this pathway of synthesis can occur both in chloroplasts or in the cytoplasm but that the key intermediate mevalonic acid does not readily pass across the plastid membrane. Thus during the early stages of seedling germination extraplastidic synthesis of sterols required for membrane formation occurs from endogenous food supplies and the mevalonate involved cannot penetrate the immature plastids to form unnecessary pigments. As soon as the seedling comes out of the ground and is illuminated, sterols are transferred to the plastids which rapidly become functional, fix CO_2 into mevalonate and actively synthesise the chloroplast terpenes.

Rubber

Rubber is a polyisoprene hydrocarbon, with the empirical formula $(C_5H_8)_n$ in which the isoprene units are in the *cis*-configuration.

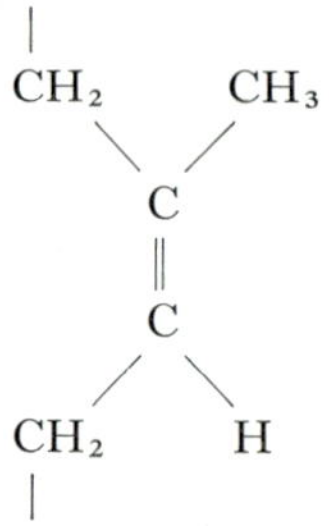

$$
\begin{array}{ccccc}
| & & & & \\
CH_2 & & & & CH_3 \\
 & \diagdown & & \diagup & \\
 & & C & & \\
 & & \| & & \\
 & & C & & \\
 & \diagup & & \diagdown & \\
CH_2 & & & & H \\
| & & & &
\end{array}
$$

Its molecular weight varies from 200,000 to 1,000,000. It occurs in the latex of rubber-bearing plants in the form of microscopic particles (0·01–50 μm diam.) suspended in a serum. Experiments with guayule (*Parthenium argentatum*) have shown that acetate, aceto-acetate and β-methyl crotonic acid ($CH_3.C(CH_3):CH.COOH$) can act as rubber precursors. Studies with *Hevea brasiliensis* and guayule have indicated the presence of enzymes converting ^{14}C-labelled acetate to β-hydroxy-β-methyl glutarate and the involvement of Co-A in these reactions. It remains for future work to elucidate the reactions in rubber-producing plants which effect the reductions of β-methyl crotonic acid and β-hydroxy-β-methyl glutarate to the C_5 intermediates which polymerise to give the rubber hydrocarbon.

The great variety of isoprenoid compounds found in plants appear to have a common origin from a C_5 unit, from which larger units are built up by condensation reactions. More intensive study of the synthesis of these compounds, often involving marked changes in the activity of the various pathways of synthesis during the ontogeny of the plant, could shed important light on the factors co-ordinating the relative activities of plant enzymes *in vivo*. This subject of metabolic regulation is further considered in Chapter 8.

NON-PROTEIN AMINO ACIDS

Plants synthesise all the amino acids required for the synthesis of their proteins (see Chapter 5, p. 169). In addition they synthesise at least 100 amino acids which are not incorporated into protein. Some of these amino acids occur only in a single species or small number of species. Many have unusual structures.

Some of the non-protein amino acids are intermediates in pathways of synthesis of the protein amino acids, thus homocysteine is a precursor of methionine, homoserine of threonine, and ornithine possibly of proline. Some are involved in the formation of other essential metabolites; β-alanine for instance is a constituent of co-enzyme A. Some of these non-protein amino acids can accumulate to high concentration in seeds and vegetative storage organs; examples are canavanine, homoarginine and azetidine-2-carboxylic acid. In such cases they probably represent utilisable stores of organic nitrogen. Some may be involved in nitrogen transport, thus citrulline can be detected in the xylem sap of many trees. In many instances, however, their role in metabolism is uncertain.

A number of the non-protein amino acids are homologues or simple substituted derivatives of the protein amino acids. They may therefore possibly arise by reactions analogous to those yielding the corresponding protein amino acids. This poses the further problem of whether new specific enzymes must be invoked to explain their biosynthesis or whether, alternatively, they may arise because of a limited specificity of enzymes involved in the synthesis of the protein amino acids. Thus the enzyme catalysing the last step in proline biosynthesis in some plants (the reduction of Δ′-pyrroline-5-carboxylic acid) will also reduce Δ′-piperidine-2-carboxylic acid to the non-protein amino acid, pipecolic acid. Pipecolic acid may be present in some and absent from other plants depending on the discrimination in each species of this enzyme. In other cases a single additional enzyme may convert an intermediate in the pathway involved in the synthesis of a protein amino acid, into a related non-protein amino acid. A single enzyme catalysing the formation of α,β-diaminopropionic acid from serine may lead on to the synthesis of the amino acids characteristic of the Mimosoideae (albizziine, mimosine and willardiine).

Some non-protein amino acids arise from the corresponding keto acids by transamination, their formation depending upon the presence of the keto acids. This applies to γ-methyleneglutamic acid, γ-hydroxy-methyleneglutamic acid and their amides. The amide synthetases involved are distinct from glutamine synthetase. These compounds appear to be actively metabolised; 90% of the non-protein nitrogen of the shoot apices of the maidenhair fern (*Adiantum pedatum*) is in the form of γ-hydroxy-methylene glutamic acid.

The study of these non-protein amino acids may reveal new pathways also implicated in the synthesis of protein amino acids. Although the enzyme responsible for the synthesis of L-glutamine from L-glutamic acid (glutamine synthetase, see p. 172) is well known and widely distributed, the evidence for a similar enzyme responsible for L-asparagine synthesis in higher plants is very much less satisfactory. *Ecballium* (squirting cucumber) and *Vicia sativa* contain γ-glutamyl-β-cyanoalanine. Studies with *Vicia sativa*, *Vicia faba* and *Ecballium elaterium* showed that they promote a reaction between cyanide and the amino acid serine giving rise to β-cyanoalanine.

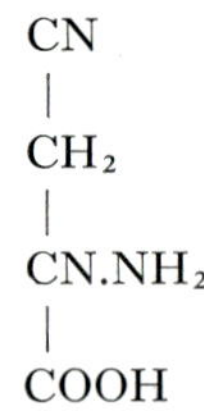

β-cyanoalanine

Vicia faba, *Ecballium elatorium* and *Lathryus odoratus* contain a *nitrilase* enzyme which hydrolyses β-cyanoalanine to give asparagine. *Vicia sativa* and the *Ecballium* species contain an enzyme promoting the transfer of a glutamic acid residue from the natural tripeptide, glutathione to the β-cyanoalanine giving γ-glutamyl-β-cyanoalanine. This pathway of asparagine synthesis via cyanide may prove to be of general significance despite our inability to detect free cyanide as a normal constituent of plant cells. *Ecballium* also contains two substituted asparagines, N^4-ethylasparagine and N^4-hydroxyethylasparagine which arise by interaction of asparagine respectively with ethylamine and ethanolamine. If *Ecballium* is fed with ^{14}C-labelled aspartic acid these substituted asparagines do not become ^{14}C-labelled. If, however, ^{14}C-cyanide is fed, ^{14}C-asparagine and the ^{14}C-labelled substituted asparagines are detected. This again points to asparagine synthesis from cyanide rather than from aspartic acid in this plant.

Some of the non-protein amino acids are very toxic to plants in which they do not occur naturally. Azetidine-2-carboxylic acid, which occurs in some liliaceous plants (*Convallaria majalis*, *Polygonatum multiflorum*) is an example. If azetidine-2-carboxylic acid is supplied to mung bean (in which this non-protein amino acid does not occur and to which it is very toxic) it becomes linked to the proline-t-RNA and hence incorporated into protein where proline should normally occur. By contrast proline-activating enzymes from the liliaceous plants do not activate azetidine-2-carboxylic acid so that this amino acid does not become incorporated into protein. The specificity of the proline-activating enzymes of the liliaceous plants protects them from the potential toxicity of azetidine-2-carboxylic acid. Further work, however, indicates that this interesting case does not provide a general explanation of the tolerance of plants to their own non-protein amino acids when they are toxic to other plants. The toxicities of canavanine (an analogue of arginine) and of mimosine (which can be regarded as an analogue of phenylalanine) are not to be satisfactorily explained by invoking the specificity of amino-acid activating enzymes.

FLAVONOIDS

This group of compounds is based upon the C_6-C_3-C_6 flavone nucleus:

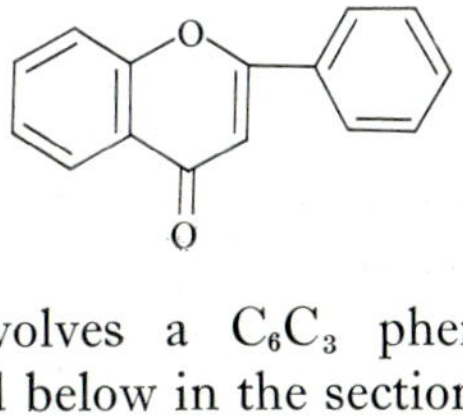

Their biosynthesis involves a C_6C_3 phenylpropanoid molecule synthesised as described below in the section devoted to these compounds. This then condenses with another aromatic ring which is synthesised by a quite different pathway which is also operative in fungi. Although the details of the reactions are not clear it is known that the aromatic ring arises by cyclisation of a carbon chain initiated by acetyl-CoA and extended by the addition of two malonyl-CoA units. Modification of the basic $C_6C_3C_6$ intermediate so formed, e.g. by hydroxylation and the introduction of methoxy groups forms the basis of the range of flavonoid compounds found in plants. The majority of flavonoids may be classified as anthocyanidins (e.g. peonidin shown below as the flavylium ion), flavonols (e.g. kaempferol) or flavones (e.g. apigenin).

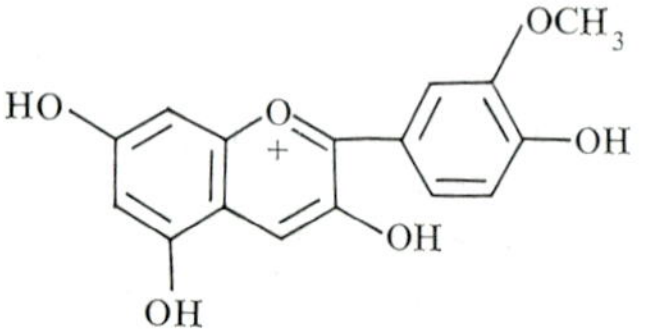

Peonidin

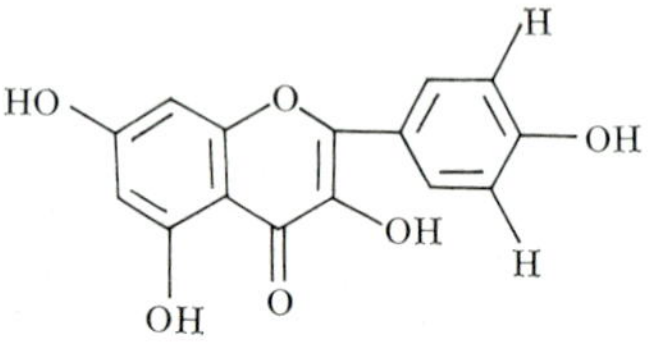

Kaempferol

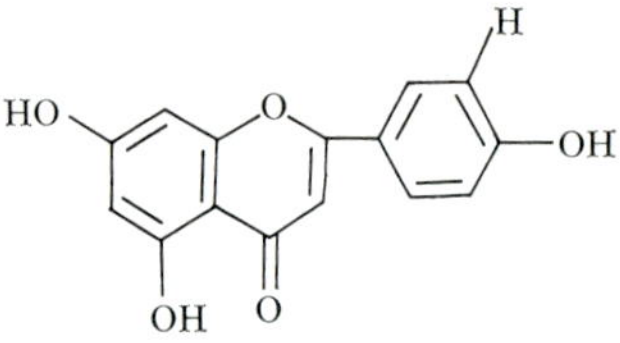

Apigenin

Free flavonoids are rarely found in plants. They occur more commonly as glycosides by combination with a range of mono- and oligosaccharides including glucose, rhamnose, xylose, galactose and gentiobiose. Such glycosides can occur in any part of the plant—flowers, leaves, shoots or roots. The major intracellular location is the vacuolar sap in which they are dissolved. The glycosides of the anthocyanidins, which are termed anthocyanins are responsible for most red and blue colours in plants whereas the flavonols, flavones and their glycosides tend to be yellowish in colour. Colourless flavonoids also occur. These pigments, particularly the anthocyanins are conspicuous and relatively easy to isolate and identify and have been used as markers in genetic and taxonomic studies.

There is little to be gained by multiplying examples of this extremely numerous and in some ways homogenous group of compounds. The possible variations can be imagined when it is realised that approximately 100 aglycones (sugar-free flavonoids) are known and that these may combine in several distinct ways with one or more of a range of sugars. Nevertheless, in most cases the chemical relationship to the basic $C_6C_3C_6$ flavone skeleton is clear.

The Biological Significance of Flavonoids

It seems certain that the flavonoid pigments which impart colours to flowers and fruits are associated with the attraction of agents of pollination and fruit dispersal. The significance of the pigments and colourless flavonoids found in stems, leaves and roots is less clear although some, e.g. pisatin in peas, have been shown to be phytoalexins—compounds which are produced in response to attack by pathogenic fungi and which may inhibit the growth of the invading organism. In a number of cases changes in the developmental stage or nutritional status of a plant have been found to be accompanied by changes in flavonoid content. For example, the morphological changes which follow the transfer of etiolated peas from darkness to light are associated with rapid qualitative and quantitative changes in flavonoid levels. The suggestion, made on the basis of these observations, that the flavonoids might in some way be involved in the control of morphological changes was strengthened by the finding that some flavonoids stimulate and others inhibit the enzyme destroying IAA (*IAA oxidase*), an enzyme whose activity is clearly potentially relevant to the control of growth.

There is as yet, however, no unequivocal evidence of a causal relationship between flavonoid levels and morphogenetic changes. Similar considerations apply to the finding that anthocyanin levels and carbohydrate metabolism appear to be closely linked. Tissues fed with sucrose solution and therefore having plentiful supplies of carbohydrate often have high levels of anthocyanins. However, although changes in levels of carbohydrate and anthocyanin may go hand in hand this need not of course imply any controlling function of anthocyanin level. It seems more likely that the accumulation of anthocyanin under conditions of plentiful carbohydrate supply may simply reflect a greater availability of material for conversion to flavonoids.

Thus although biological functions have been either established or postulated for a number of flavonoid compounds it seems certain that a considerable proportion of the vast array which occur in plants fit firmly into the category of secondary plant products whose function if any is obscure.

PHENYLPROPANOIDS

This group of compounds is based upon the C_6C_3 phenylpropane structure

C—C—C

and includes the simple phenylpropanoids, the coumarins, the lignans and lignin. Phenylalanine and tyrosine are of course members of this chemical group but since they are clearly primary metabolites they are relevant to the present chapter only insofar as they are involved in the biosynthetic pathways leading to the nitrogen free phenylpropanoids. Radiotracer work suggests that phenylalanine, which is synthesised via the shikimic acid pathway is converted by the enzyme *phenylalanine ammonia lyase* (PAL) to cinnamic acid, a compound believed to be a key intermediate in phenylpropanoid biosynthesis.

$$C_6H_5\text{—}CH_2\text{—}CH(NH_2)\text{—}COOH \xrightarrow[\text{PAL enzyme}]{-NH_3} C_6H_5\text{—}CH{=}CH\text{—}COOH \quad (80)$$

Phenylalanine → Cinnamic acid

PAL enzyme has been found to be widely distributed whereas the enzyme performing an analagous reaction with tyrosine as substrate, *tyrosine ammonia lyase* (TAL), has so far only been detected in grasses.

$$HO\text{—}C_6H_4\text{—}CH_2\text{—}CH(NH_2)\text{—}COOH \xrightarrow[\text{TAL enzyme}]{-NH_3} HO\text{—}C_6H_4\text{—}CH{=}CH\text{—}COOH \quad (81)$$

Tyrosine → p-coumaric acid

Simple Phenylpropanoids

The formation of other simple phenylpropanoid molecules results from the substitution of hydroxy and methoxy groups in the aromatic ring, e.g. sinapic and caffeic acids.

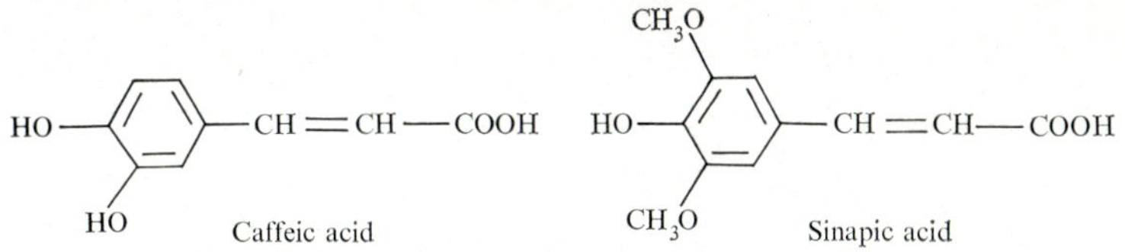

Caffeic acid

Sinapic acid

These acids are rarely found free and are usually in combined form as esters. Other variations on the basic theme include reduction of the side chain. Thus reduction of *p*-coumaric acid forms *p*-coumaryl alcohol which occurs as a glucoside in *Picea excelsa*. Large numbers of simple phenylpropanoid derivatives have been identified and doubtless many more await discovery.

Coumarins

Another group of related phenylpropanoids are the coumarins. These compounds are believed to result from the lactonisation of *cis-o*-hydroxycinnamic acid derivatives. Coumarin itself is widely distributed in plants and is the lactone of *cis-o*-hydroxycinnamic acid.

o-hydroxycinnamic acid —Lactonisation→ Coumarin $+H_2O$

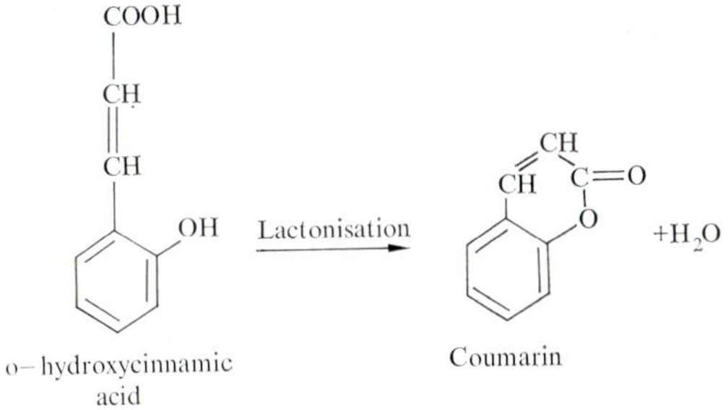

Dicoumarol, which is derived from two molecules of the coumarin type is the haemorrhagic factor in spoiled sweet clover hay and is used in medicine to prevent blood clotting.

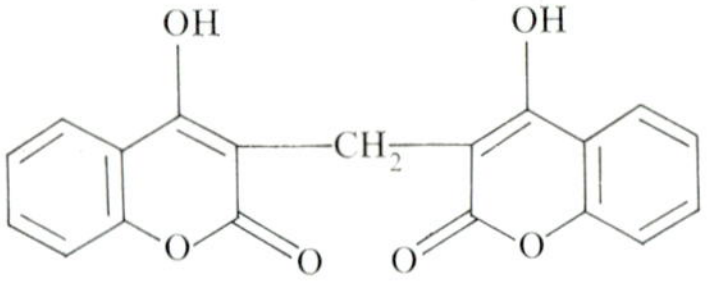

Dicoumarol

Warfarin, a haemorrhagic rat-poison, is also a coumarin derivative.

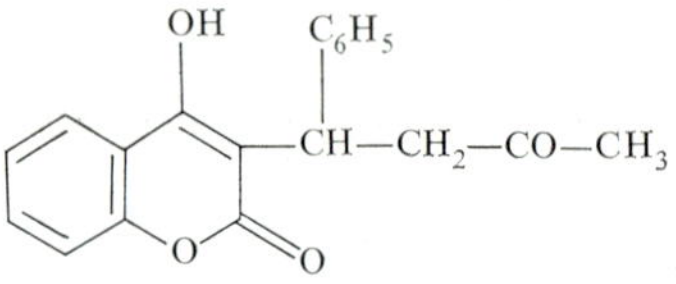

Warfarin

Lignans

The lignans consist of two phenylpropanoid molecules joined by carbon–carbon bonds between the middle carbon atoms of the side chain. Guaiaretic acid, a constituent of gum guaiacum, is a compound of this type.

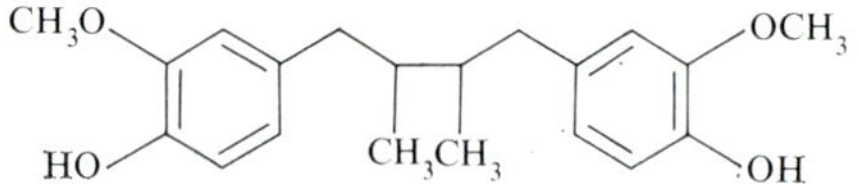

Guaiaretic acid

Lignin

Lignin is a complex thermoplastic polymer of phenylpropanoid units which encrusts and penetrates into the cellulose walls of certain tissues in vascular plants and contributes substantially to their mechanical strength. The secondary xylem vessels of angiosperms and tracheids of conifers are often heavily lignified. The detailed chemical structure of this compound remains a challenge to organic chemists despite recent progress. Much of the work, the contribution of Freudenberg being particularly significant, has involved the investigation of lignin of coniferous origin—partly because lignin is an embarrassingly plentiful waste product removed from wood pulp in the process of paper manufacture. The digestion of wood pulp using solvents such as an aqueous solution of sulphur dioxide and calcium bisulphite at high temperature results in the formation of soluble lignosulphonates for which no significant commercial use has yet been found. The waste is either burned or dumped—disposal becoming more and more difficult as awareness of environmental pollution increases.

One of the basic problems which has beset lignin research is that "standard" extraction methods involve quite drastic treatment which chemically modifies the natural lignin. Each extraction method therefore produces its own characteristic "lignin". Furthermore, as well as the extraction-induced differences there are quantitative and qualitative differences in the lignins of conifers and angiosperms and it seems that even interspecific differences of the same kind may occur. As might be expected this multiplicity of lignin types has not smoothed the course of lignin research. Nevertheless, although there are important gaps in our knowledge of the way in which the phenylpropanoid monomers combine and are deposited in or on the walls of cells the broad outlines of lignin chemistry and biosynthesis now seem fairly clear.

The modern phase of lignin research may be considered to have begun with Klason's work on the degradation products of lignin which led him to suggest in 1897 that coniferyl alcohol could be a lignin precursor and more recent tracer work has shown that this is indeed

HO—C_6H_3(—OCH_3)—CH═CH—CH_2OH Coniferyl alcohol

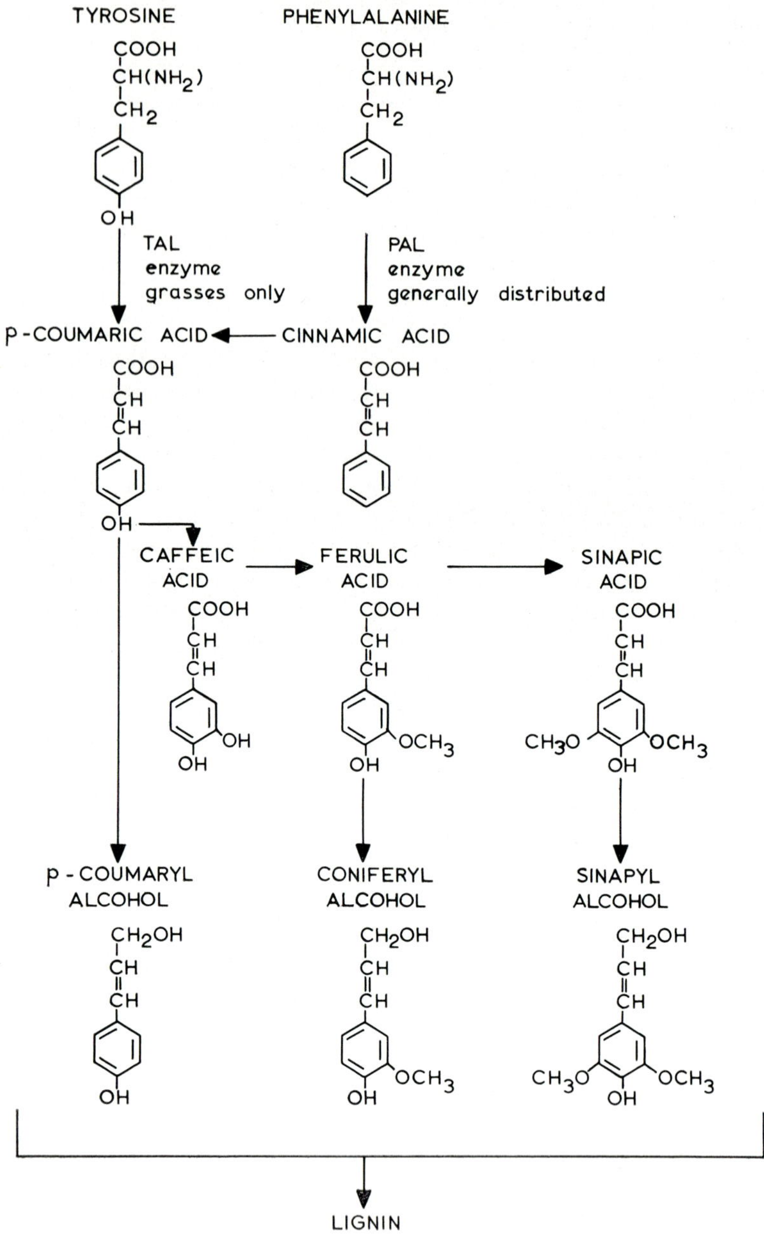

FIG. 6.3. The origins of the lignin monomers.
Key: PAL, phenylalanine ammonia lyase; TAL, tyrosine ammonia lyase.

the case. Radioactive incorporation studies have also shown that phenylalanine is also readily converted to lignin and the enzyme phenylalanine ammonia lyase, mentioned above in connection with simple phenylpropanoid biosynthesis has been implicated. Consistent with the observation noted earlier that grasses possess tyrosine ammonia lyase, they, unlike other vascular plants, readily incorporate carbon from tryosine into lignin (Fig. 6.3). The carbon skeletons of certain hydroxy and methoxy derivatives of *p*-coumaric acid such as caffeic, ferulic and sinapic acids can be incorporated into lignins and it seems therefore that the next stage in the biosynthetic pathway involves the introduction of additional hydroxy and methoxy groups. Reduction of the carboxyl groups of the acids so produced forms primary alcohols believed to be the immediate precursors of lignin. Higuchi and Brown confirmed the reduction of ferulic acid to coniferyl alcohol in wheat using a radioisotope technique which is designed to cause accumulation of radioactive intermediate compounds which are normally present at such low concentrations as to be undetectable by direct methods. The tissues are supplied with radioactive precursor and at the same time with the postulated intermediate in non-radioactive form. The non-radioactive intermediate will compete with any radioactive intermediate produced by the plant in reactions associated with the consumption of the intermediate. Thus the addition of non-radioactive intermediate should tend to encourage the accumulation of radioactive intermediate in the tissues.

The formation of lignin from the lignin monomers (predominantly though probably not exclusively from coniferyl alcohol in conifers and in addition from sinapyl alcohol in angiosperms) is believed to involve spontaneous polymerisation following enzymic oxidation of the phenolic hydroxy group. The action of the oxidising enzymes *laccase* or *peroxidase* results in the production of free radicals which couple spontaneously in a number of ways. Some of the large number of ways in which free radicals derived from coniferyl alcohol can combine are shown on p. 210.

These final non-enzymic reactions result in the formation of a very complex three-dimensional polymer the details of which may vary from species to species.

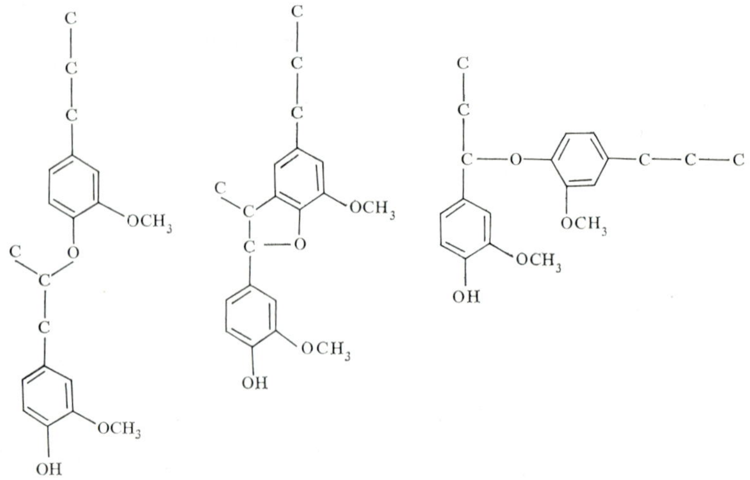

The glucoside coniferin has been implicated in the formation of spruce lignin.

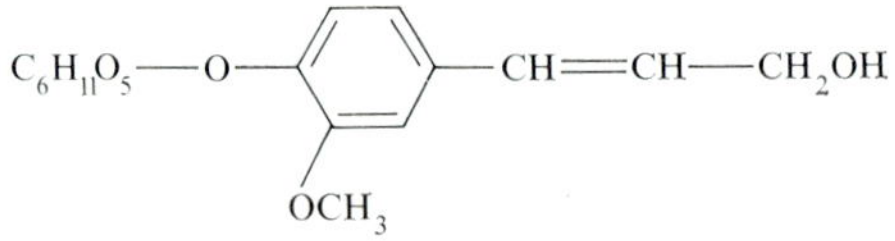

$C_6H_{11}O_5$—O—[ring]—CH=CH—CH_2OH (ring bearing OCH_3)

It is rapidly incorporated and it has been suggested that the glucose acts as a blocking group which is enzymically removed at the site of lignin deposition. Oxidation and polymerisation then taking place as described above. Although coniferin is not detectable in most lignin-producing species and may not in fact be generally involved in lignin synthesis it could be representative of a type of compound involved in the transport of lignin monomers from their site of synthesis to the site of lignin deposition.

It may seem inappropriate that lignin, which performs an important strengthening function, should be discussed in a chapter concerned with secondary plant products which by their very definition are off the mainstreams of metabolism. The reasons for this are twofold. Firstly, lignin monomers are clearly members of the phenylpropanoid group and their biosynthesis has common factors with other members of this class—most of which are true secondary plant products. Secondly, and perhaps more importantly, it has been suggested by Neish that lignin may be considered as having originated as a consequence of a mutation or mutations which resulted in formation of simple phenylpropanoids capable of spontaneous polymerization. The formation of lignin within the matrix of plant cell walls conferred greater structural rigidity and may indeed have been significant in the development of erect terrestrial plants.

We have here then the concept of the conversion of secondary to primary metabolites, an apparent evolutionary dead-end pathway to simple phenylpropanoids becoming a part of the important biosynthetic pathway leading to lignin. Who knows what uses evolution may yet make of other secondary plant products!

FURTHER READING

L. Fowden. The non-protein amino acids. *Endeavour*, **21**: 35–42, 1962.

G. A. Swan. *An Introduction to the Alkaloids*. Blackwell Scientific Publications, 1967.

A. C. Neish, Coumarins, Phenylpropanes and Lignin. Chapter 23 in *Plant Biochemistry*, edited by James Bonner and J. E. Varner, Academic Press, 1965.

MORE ADVANCED READING

L. Fowden, I. K. Smith and P. M. Dunhill. Some observations on the specificity of amino acid biosynthesis and incorporation into plant proteins. In *Recent Aspects of Nitrogen Metabolism in Plants*, edited by E. J. Hewitt and C. V. Cutting, Academic Press, London, 1968.

A. W. Galston. Flavonoids and photomorphogenesis in peas. Chapter 10 in *Perspectives in Phytochemistry*, edited by J. B. Harborne and T. Swain. Academic Press, London, 1969.

E. Leete. Alkaloid biosynthesis. *Annual Review of Plant Physiology*, **18**: 179–196. 1967.

W. Ruhland (Editor). *Encyclopedia of Plant Physiology*, vol. 10. *The Metabolism of Secondary Plant Products*. Springer-Verlag, Berlin, 1958.

CHAPTER 7

Absorption, Secretion and Translocation

"*During the . . . first decades of the present century there was a widespread tendency to think of living cells as if they were simply aqueous spaces isolated from their environment by selectively permeable, but inert, membranes. Today we realize that this 'collodion-bag concept' was a flagrant oversimplification. The principal defect of this mode of thought was, of course, that it totally neglected the active transport processes.*"

Runar Collander, in Cell membranes: their resistance to penetration and their capacity for transport, in *Plant Physiology*, vol. II, edited by F. C. Steward, Academic Press, New York, 1959.

"*What I do wish to stress is that solute movements in living cells are so intimately related to and dependent upon processes of metabolism that it should not be hoped that any real or hypothetical cell or membrane from which the factors of complex metabolism are absent, can go far in imitating the processes by which living cells absorb and accumulate solutes.*"

D. R. Hoagland, in *Lectures on the Inorganic Nutrition of Plants.* Chronica Botanica Co., Waltham, U.S.A., 1944.

INTRODUCTION

ACROSS the boundary between a living cell and its environment solutes continuously pass in both directions. When, during a given time interval, there is a net movement (*transport*) of a given solute into the cell we talk of *absorption* or *influx* of the solute and if this absorption leads to the establishment of a higher concentration inside than outside the cell we talk of solute *accumulation*. When the net movement is in the reverse direction we talk of solute *release* or *efflux* and of the resulting *depletion*. In some cases and at some times such movements across cell boundaries appear to be controlled in direction and rate by physical forces such as diffusion, particularly diffusion across cell membranes allowing free movement of some molecules or ions and preventing or restricting the movement of others (across membranes showing selective permeability). Such movements are often referred to as *passive*, in that they tend to decrease the energy potential of the system. In other cases the solute movement is opposite in direction and/or different in rate from that expected from physical considerations. The movement is then a process involving consumption of cellular energy and is said to be *active*, such movement is often referred to as secretion. When solutes are transferred from cell to cell through a tissue or over the longer distances between plant organs we talk of such solute transport as *translocation*.

This dictionary-like opening to the present chapter will not only facilitate our subsequent discussion but is necessary because there is considerable variation between authors in the use of the terms italicised above.

SALT REQUIREMENTS OF PLANTS

The development of the technique of "water culture" by Sachs, Knop and others at the end of the nineteenth century demonstrated that plants have requirements for the elements, nitrogen, phosphorus, sulphur, potassium, calcium, magnesium and iron and that the sources of these elements in plant nutrition are their soluble salts present in the soil or water in which plants grow. The development during the present century of refinements in the "water culture" technique and improved methods for purifying inorganic salts has shown that other elements, the so called micro-nutrient elements boron, copper, manganese, molybdenum, zinc and chloride are also essential plant nutrients although required by cells in very small amounts. If any of these *essential elements* are only available in soluble form in inadequate amounts deficiency symptoms develop as plant growth proceeds and if the deficiency is sufficiently severe death results. When available in excessive concentration these elements disturb normal growth and development, some of these elements being particularly toxic at high concentration. Thus boron, in the form of soluble borates is normally required at a concentration of not less than 0·05 parts per million but is markedly inhibitory to normal growth at a concentration of 1·0 parts per million.

The essentiality of most of the elements mentioned above could have been confidently predicted from the discussions we have developed in earlier chapters. Nitrogen and sulphur are constituent elements of proteins and many other cell constituents. The importance of phosphorus compounds such as nucleic acids, phospholipids and co-enzymes has been stressed several times. Magnesium is a constituent of the chlorophylls. Iron, copper, zinc, manganese and molybdenum are essential to the functioning of particular enzymes. In other cases, like that of calcium, their involvement in metabolism is less well understood and in still other cases, like potassium and boron, is almost completely obscure.

Analyses of the sap in the cell vacuoles and studies of the electrical conductivity of plant cells and cell fluids served during the first quarter of the present century to demonstrate that inorganic salts can be accumulated within plant cells to concentrations greatly in excess of their concentrations in the external environment and that this accumulation of salts was a selective process in that some elements were accumulated to a much greater extent than others. Such observations suggested that the uptake of inorganic salts or their ions could not be explained in terms solely of movement along an activity gradient by diffusion. E. Overton, as early as 1895, had suggested that "adenoid activity", by which he meant metabolic activity, might be involved in solute absorption and W. Pfeffer, in his famous book, *The Physiology of Plants* published in 1900, visualised the possibility that chemical reactions might be involved in the movement of solutes into living cells. Now it was necessary to examine such ideas experimentally.

Consequently from the early thirties onwards there occurred many studies of the quantitative aspects of salt uptake and of the factors which influence this process as it occurs in plant cells.

ION FLUXES AND MEMBRANE POTENTIALS

In the Introduction to this chapter it was stated that ion movements into and out of cells may be passive but in other cases they are active (secretory movements). How can we distinguish between these two kinds of ion movement? One approach is to examine the relationship between metabolism and ion movement. The other is to measure internal and external ion concentrations, membrane potentials and ion movements (influx and efflux) and to see how far the ion movements correspond in direction and rate with that to be expected in a system moving to thermodynamic equilibrium. It is this latter approach which is examined in this section.

The force motivating ion movement (the negative electrochemical potential, $-\bar{\mu}$) by diffusion is made up of two components, one depending on the concentration gradient and the other on the electrical energy per mole ion, $zF\Psi$ (where z = valency, F = Faraday, and Ψ = the electrical potential of the ion). When the electrochemical potential of an ion on two sides of a membrane is equal then the system is in thermodynamic equilibrium. Using the superscript i for inside, and superscript o for outside for the ion j we have

$$\bar{\mu}_j^i = \bar{\mu}_j^o \tag{82}$$

These electrochemical potentials are related to the standard chemical potential of the ion ($\bar{\mu}_j^*$) thus:

$$\bar{\mu}_j^i = \bar{\mu}_j^* + RT \ln C_j^i + z_j F\Psi^i \tag{83}$$

$$\bar{\mu}_j^o = \bar{\mu}_j^* + RT \ln C_j^o + z_j F\Psi^o \tag{84}$$

where R = gas constant expressed in joules per mol per degree,
T = absolute temperature,
C_j^i and C_j^o = the molar internal and external concentrations of the ion j.

Hence at any point corresponding to thermodynamic equilibrium, the difference in electrical potential on the two sides of the membrane (E_j^N, the so-called Nernst potential) is given by the equation

$$E_j^N = \Psi^i - \Psi^o = \frac{RT}{z_j F} \ln (C_j^o/C_j^i) \tag{85}$$

which at 20°C and converted to base 10 logarithms from natural logarithms becomes:

$$E_j^N = \frac{58}{z_j} \log (C_j^o/C_j^i) \text{ in millivolts} \tag{86}$$

Thus if $C_j^i = 100C_j^o$ for the cation j^+ then there is, at thermodynamic equilibrium, a potential across the membrane of —116 mV. Thus in a plant cell in which there is such a potential across the membrane between protoplast and external solution, the cation j^+ can move passively into the cell against the concentration gradient until its internal concentration is 100 × its external concentration. At this point the concentration gradient force is exactly balanced by the inwardly directed potential gradient. Ion accumulation can occur passively; can occur without work being done on the ion.

The net passive flux ($\mathcal{J}$) of the ion is the ratio of two fluxes $\mathcal{J}_{\text{in}}$ and $\mathcal{J}_{\text{out}}$ and the ratio of these two fluxes is related to the difference in electrochemical potential by the equation (the Ussing–Teorell equation)

$$RT\frac{\mathcal{J}_{\text{in}}}{\mathcal{J}_{\text{out}}} = \bar{\mu}_j^o - \bar{\mu}_j^i = -\Delta\bar{\mu}_j \tag{87}$$

Thus if we can determine $\Delta\bar{\mu}$ and the actual net flux of an ion we can see how far its movement does or does not correspond to that expected from passive ion flux; can see whether or not there is an active component in its uptake.

Now electrical potential differences arise across membranes when they permit different charged ions to pass through at different rates. What happens is that the permeating ions tend to carry electrical charges across the membrane at different rates. They cannot continue because the solutions on the two sides must remain at electrical neutrality. Hence an electrical potential differencea rises across the, membrane which slows down the fast moving ions and speeds up the slow moving ions to the same rate. Thus to take a very simple case, if a membrane separates a weak solution of NaCl from a more concentrated solution and if the mobility of Cl' is greater than that of Na^+, the Cl' tends to go ahead of Na^+. This sets up an electric field across the membrane which slows down the faster moving Cl′ and speeds up the slower moving Na^+.

The electrical potential difference across the membrane can be measured by placing appropriate electrodes on its two sides. To measure the potential across a cell membrane one electrode must be within the protoplast and preferably we should be able to ascertain whether it is in the vacuole or in the cytoplasm (since in one case it will be separated from the electrode in the external solution by tonoplast and plasmalemma and in the other only by the plasmalemma). To undertake such measurements very fine microelectrodes are needed; electrodes have been used, the glass tips of which have a diameter of *ca.* 1 micron (μm). Even with such electrodes large vacuolated cells rich in cytoplasm are needed, such large cells are also required to obtain adequate samples of vacuolar sap and of cytoplasm for accurate analysis of all their major ions, and for measurements of their ion fluxes. To meet these requirements experimenters have turned to the giant multinucleate cells of certain algae, notably species of *Nitella*, *Hydrodictyon* and *Chara*.

Radioactive ions are used to determine the actual ion fluxes. The influx (J_{in}) is obtained by measuring the initial rate of increase in radioactivity when an unlabelled cell is placed in a solution containing the radioactive ion. The efflux (J_{out}) is measured by allowing the cell to take up a radioactive ion and then studying how it loses radioactivity to an external flowing solution. If during such an experiment we plot the log of radioactivity within the cell against time we often obtain a graph of the form shown in Fig. 7.1. From such a graph the separate effluxes from the cell wall, the cytoplasm and the vacuole can be calculated from the successive rates of loss of radioactivity observed with time.

From equations (83) and (84) we have

$$\Delta\bar{\mu}_j = \bar{\mu}_j^i - \bar{\mu}_j^o = z_j F(\Psi^i - \Psi^o) - RT \ln (C_j^o/C_j^i) \qquad (88)$$

From the measured ion concentrations we have C_j^i and C_j^o and hence can calculate the Nernst potential (E_j^N) from equation (85):

$$E_j^N = \frac{RT}{z_j F} \ln (C_j^o/C_j^i) \qquad (89)$$

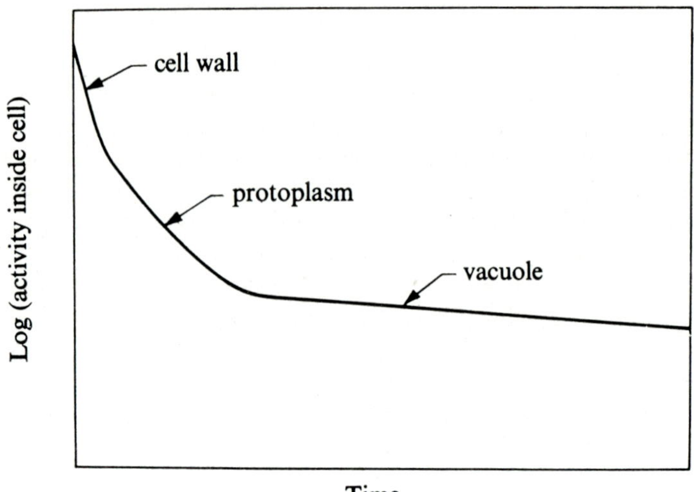

FIG. 7.1. Loss of radioactivity to an external flowing solution from a plant cell. From the slopes of the protoplasm and vacuole phases the effluxes across the plasmalemma and tonoplast can be calculated. (After J. Dainty, The Ionic Relations of Plants, Chapter 13, in *The Physiology of Plant Growth and Development*, edited by M. B. Wilkins, McGraw-Hill, London, 1969.)

The observed potential difference between the electrodes $(E) = \psi^i - \psi^o$,

so that
$$\Delta\bar{\mu}_j = z_j F(E - E_j^N) \tag{90}$$

and from the value for $\Delta\bar{\mu}_j$ the ratio J_{in}/J_{out} (equation 87) to be expected from passive flux can be calculated and compared with the measured flux actually taking place.

In an actual study on the cells of *Nitella translucens*, the E^N for Na^+ in the cytoplasm was —66 mV, the value for E (cytoplasm/external solution) was —138 mV. Thus although the ratio C^i/C^o was 14, the sodium was at a lower electrochemical potential in the cytoplasm than in the external solution and from this data the flux ratio J_{in}/J_{out} for Na^+ should be 12/1. However, instead of a strongly positive influx, the influx and efflux were almost equal. This indicates that the cell is operating an active mechanism for Na^+ exclusion. Similarly from the values for E^N, E and C^i/C^o for K^+ the calculated flux ratio is 1/5 for the cytoplasm–external solution boundary. However again, instead of a positive K^+ efflux there was virtually a nil net flux; the cell was actively transporting K^+ into the cell. The situation with regard to Cl′ departed even further from that corresponding to passive ion movement; when there was an actual nil net flux, the ratio J_{in}/J_{out} for passive flux was 1/10,000; Cl′ ions were entering the cell 10,000 times faster than they would by passive movement. Here then is evidence for active mechanisms of ion transport at the plasmalemma. However, from studies, with the same cells, of the electrical potential across the tonoplast it appears that while Na^+ is actively excluded from the vacuole, the ions K^+ and Cl′ are distributed between the cytoplasm and vacuole by passive movement.

FACTORS CONTROLLING ION ABSORPTION

Many studies of salt uptake have been undertaken with higher plant cells. Discs of storage tissue and excised seedling roots have been used in many of these investigations and permit us to distinguish between (i) an initial, rapid and usually short duration change in the salt content of living cells whenever they are transferred to a solution of changed salt content, (ii) a continuing uptake leading to an overall and often pronounced accumulation of inorganic ions.

F. C. Steward, in 1937, referred to the first of these processes as *induced* absorption and considered that its characteristics corresponded with physical processes and was therefore, by our definition, a "passive" relationship. The long term uptake, usually involving accumulation of both anions and cations, and which he termed *primary* absorption, is the process which has been shown in many recent studies to depend upon the metabolic activities of the living cells and which not only in this but in its other characteristics is clearly an "active" process. Further, if this distinction is valid we may expect rather different responses to environmental and nutritive variables according to the relative importance of induced (passive) or primary (active) absorption processes under the particular experimental conditions adopted by different workers. The virtue of the distinction drawn here is that it not only recognises the operation of both types of salt movement between the cell and its external environment but goes far to explain otherwise apparently contradictory results which in the past have been the subject of controversy between different laboratories.

The operation of two types of absorption is, for instance, illustrated by studies of the effects of temperature on the rate of salt uptake. In short term experiments (experiments of less than 2 hours' duration) or when absorption proceeds at temperatures at or immediately above 0°C the Q_{10} for the influence of temperature upon rate of salt uptake is about 1·2, indicative of a physical process. At higher temperatures and over more extended periods the Q_{10} is of the order of 2–3, indicative of the rate being dependent upon thermochemical reactions; reactions of metabolism. Similarly in short-term experiments and with tissues of low metabolic activity (such as recently isolated slices of the storage tissues of beet or potato) the rate of uptake of ions may be linearly related to their external concentration and the equilibrium distribution of particular cations and anions between the external solution and the cell approach that corresponding to a diffusion equilibrium (a Donnan equilibrium) such as is established across a membrane permeable to inorganic ions but impermeable to other larger (metabolite) ions. The uptake then shows the characteristics of a "passive" process. By contrast, when the metabolism of the same tissue is "activated" by a prior washing in aerated water (for 24 to 48 hr) then the tissue acquires the ability in the presence of oxygen, and at a rate which may be determined by the oxygen concentration, to accumulate progressively both cations and anions over a prolonged period. Further, under these conditions the rate of this "active" accumulation may be independent of the external concentration over a wide range. Further, if the "activated" tissue is placed under anaerobic conditions it may again be shown to carry out a short-lived uptake of ions corresponding with the "passive" uptake of the initial resting tissue.

Studies along these lines have emphasised the significance in the salt nutrition of plants of the metabolically actuated component of ion uptake. Our attention therefore is directed towards experiments which have contributed to our knowledge of this process and to the theories advanced regarding the mechanism whereby metabolism controls ion accumulation. Such an "active" process by definition involves the utilisation of metabolic energy and it is therefore not surprising that it is related to photosynthesis and/or respiration. To test for the involvement of photosynthesis in ion uptake by green cells it is only necessary to compare ion uptake in light and darkness. In the giant cells of *Nitella translucens* the K^+ efflux via the plasmalemma is the same in both light and dark, but the K^+ influx is $0{\cdot}9 \times 10^{-12}$ mol cm^{-2}s^{-1} in light and only $0{\cdot}3 \times 10^{-12}$ mol cm^{-2}s^{-1} in dark. The active Na^+ exclusion (efflux) is similarly reduced in the dark ($0{\cdot}55 \times 10^{-12}$ mol cm^{-2}s^{-1} in the light and $0{\cdot}10 \times 10^{-12}$ mol cm^{-2}s^{-1} in the dark). Recalling the evidence for a large active component in the Cl′ influx of these cells (p. 219), it is found that the Cl′ influx of $0{\cdot}85 \times 10^{-12}$ mol cm^{-2}s^{-1} in light falls to $0{\cdot}05 \times 10^{-12}$ mol cm^{-2}s^{-1} in dark. When working with non-green cells (storage tissue discs, roots) the energy for active uptake must come from respiration and when, in such tissues, respiration is inhibited by lack of oxygen or by using respiratory inhibitors there is a decrease in the "active" component of the ion uptake. The energy released in the oxidative reactions of respiration is conserved in a "useful" form by the simultaneous synthesis of high-energy phosphate bonds. The substance, dinitrophenol, at appropriate concentrations, "uncouples" the oxidative reactions of respiration from the synthesis of the energy-rich phosphates. Oxygen uptake is not inhibited, it may even be stimulated, but the released energy of the respiratory substrates is not conserved in a utilisable form. Such concentrations of dinitrophenol depress or completely inhibit "active" ion uptake.

The ions present in a solution affect the absorption of one another. In a complex solution such as is required to present all the essential nutrient ions, these interactions are of great complexity. However, quantitative studies of ion uptake from two-salt solutions have enabled interactions between pairs of cations and anions to be examined. Such work has shown that chemically related anions interfere with one another during absorption whereas more diverse ions do not. Thus, chloride uptake is reduced by bromide or iodide ions, sulphate uptake by selenate ions, phosphate uptake by arsenate ions. The alkali cations similarly "compete" with one another for absorption. Calcium ions compete with strontium but not with magnesium. Studies of the uptake of alkali cations from mixed solutions (with a common anion) illustrate ion interactions in a very interesting way. Firstly potassium ions are usually more rapidly absorbed than the other alkali cations (sodium, caesium, rubidium, lithium). This discriminative uptake is characteristic of "active" uptake and such selectivity is not marked under experimental conditions permitting only a physical or "passive" uptake. Secondly, the quantitative effects of one cation upon the uptake of a second cation can be treated kinetically to see if there is a competitive inhibition of uptake similar to the competitive inhibition of enzyme reactions previously discussed (p. 74). Thus, if we plot the reciprocal of the rate of K^+ uptake at several different external concentrations of K^+ against the reciprocal of the concentration of a second ion simultaneously present the straight lines will be parallel for a non-competitive interaction. If, however, there is a competitive interaction (resembling competitive inhibition) the straight lines will meet at a point corresponding to the reciprocal of the maximum rate (V) of K^+ uptake. Such studies (Fig. 7.2) have shown that K^+, Rb^+ and Cs^+ compete with one another whereas the inhibition of Rb^+ uptake by Li^+ is not competitive. Studied in this way cations and anions fall into a number of competitive groups. From the comparison of this competitive inhibition of uptake with the competitive inhibition of enzyme reactions it may be suggested that the "active" uptake of ions involves the formation of ion-complexes (carrier-ion complexes) comparable to substrate–enzyme complexes. This implies that the absorption of ions involves their chemical combination with substances (carriers) located at the barrier preventing or resisting free diffusion of ions between the cell and its external environment. Further, it may be postulated that ions within each competitive group combine with the same cellular constituent at the diffusion barrier. These observations however do not enable us to decide how many different kinds of carrier molecule are operational. Thus some recent work has indicated that active K^+ uptake can occur at at least two "sites", one with a high K^+ affinity and specificity and which operates at low external K^+ concentrations (system 1) and another which has less specificity and lower affinity and which operates at higher external K^+ concentrations (Na^+ is a competitive inhibitor of K^+ uptake at this second site) (system 2). In young barley roots the K_s (see p. 74) of system 1 for K^+ is 0·025 mM, the V_{max} is

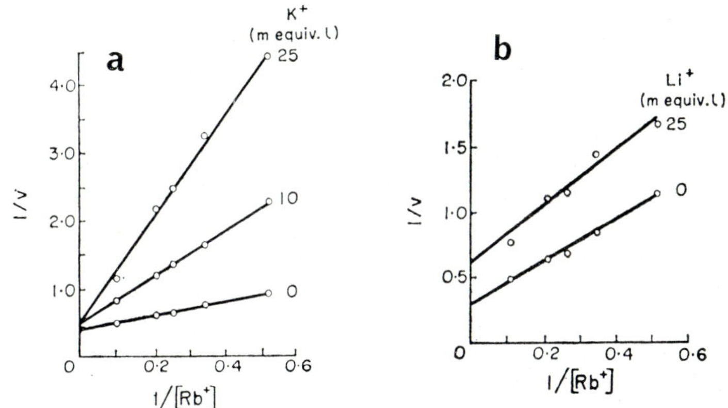

Fig. 7.2. Plots of the reciprocal of the concentration of rubidium ions $[Rb^+]$ against the reciprocal of the rate of uptake of these ions ($1/V$). Concentrations in m-equivalents per litre. Uptake rates in m-equivalents per gram of tissue per hour. (a) shows competitive inhibition of Rb^+ uptake by potassium ions, K^+. (b) shows non-competitive inhibition of Rb^+ uptake by lithium, Li^+. (After E. Epstein and C. E. Hagen, *Plant Physiology*, **27**: 457, 1952, from J. F. Sutcliffe, *Mineral Salts Absorption in Plants*, Pergamon Press, Oxford, 1962.)

12 μmole/g fresh weight^{-1} hr^{-1} and Na^+ is a very weak inhibitor of this uptake. By contrast for system 2, K_s for K^+ is 17 mM, and for Na^+ 0·8 mM. The affinity of system 1 for K^+ is approximately 700 times that of system 2 and favours K^+ uptake; system 2, by contrast, favours Na^+ over K^+ uptake by a factor of 20. Such studies therefore lead us on to enquire both as to the location and nature of the diffusion barrier and the nature of the chemical reactions which link metabolism with ion uptake.

THE SITES OF ION ACCUMULATION WITHIN CELLS

As previously indicated it is possible by using radioactive ions to trace their rate and extent of penetration from an external solution into the cell walls, into the cytoplasmic phase and separately into cytoplasmic structures, particularly mitochondria (separated by centrifugation) and vacuoles. By previous feeding with radioactive ions followed by immersion of the cells in "cold" solutions it is possible to study (by determination of the labelling of the external solution) how far the absorbed ions can exchange with the external solution and to see if different parts of the total content of any chosen ion, exchange at different rates with the external solution. It is also easily possible to study how far "absorbed" ions can be rapidly or only slowly washed out of tissues and cells. Experiments along these lines with higher plant and algal cells have, in general, indicated that, for each ionic species, part can be very rapidly exchanged or washed out, part (in mature vacuolated cells by far the greater part) is strongly retained within the cells, and a third part is retained with an intermediate tenacity. Thus, for instance, in the large highly vacuolated cells of the alga, *Nitella*, a small part (about 0·1%) of its potassium exchanges very rapidly (50% exchange is reached in 23 sec), a second part (1–2%) is exchanged to this extent in 5 hr, and the third, overwhelmingly larger part, is exchanged to this extent only in about 40 days.

The region of the cell into which ions penetrate rapidly by diffusion and from which they are readily removed by washing has been referred to as the *free space*. The regions into which, by contrast, ions penetrate slowly, in which they accumulate and from which they are only slowly exchanged is then referred to as *non-free space*. The real difficulty comes in trying to identify the cell locations corresponding to "free space" and "non-free space". It is, however, generally agreed that many ions enter the vacuoles slowly, that this entry is an "active" process and that the ions of the vacuolar sap account for most, if not all, of the slowly exchanged ions of fully expanded cells (hence the reference by many authors to vacuolar non-free space). It is also agreed that ions rapidly diffuse into and out of the cell walls so that they constitute a free space. The difficulty comes in trying to assess the exchangeability of ions in the cytoplasm. Studies with isolated mitochondria have shown not only their ability to carry out active ion accumulation but indicate the mitochondria as part of the non-free space of the cells. It is particularly the exchange status of ions in the hyaloplasm which is still the subject of controversy although as we have seen there is strong evidence that the plasmalemma is a barrier to solute movement (see p. 219) so that as a result of it and, perhaps, also of its associated endoplasmic reticulum, there is a cytoplasmic non-free space in which ions are retained although with less tenacity than following their accumulation into the vacuoles and mitochondria. The similarity of structure of all cytoplasmic membranes as revealed by electron microscopy also seems to fit in with this conclusion. The "active" uptake of ions therefore seems to involve their movement at a speed and in a direction not determined by the concentration gradient and across permeability barriers represented by the lipo-protein membranes of the cell. The two sites of K^+ uptake referred to above (systems 1 and 2) are apparently spatially separated; system 1 is located in the plasmalemma, system 2 in the tonoplast of root cortical cells.

THE LINKAGE OF METABOLISM WITH ION ABSORPTION

The lipoprotein membranes of the cell act as diffusion barriers; they have a high resistance to passive penetration by ions. How then can we explain the rapid active movement of ions across such membranes? One explanation is to postulate that the ions undergo reversible binding with some constituent of the membrane (this constituent is usually called the *carrier*). The ion is then visualised to pass across the thickness of the membrane, not as a free ion but as an ion-carrier complex (Fig. 7.3). This concept has the immediate attraction that it suggests an explanation of selectivity; selective absorption would reflect the abundance in the membrane and chemical affinities of the carrier molecules. Ions which compete with one another would be ions capable of combination with the same carrier but the affinities of the carrier for the separate ions of the group would differ. Selective uptake of potassium, as compared with other alkali metal cations, would reflect the high affinity of the carrier involved for potassium.

Studies on Na^+ efflux and K^+ influx in the cells of the alga *Hydrodictyon* suggest that they have a sodium–potassium pump similar to that in animal cell membranes. It is suggested that in such a pump there is a single carrier which transports Na^+ on its outward journey and K^+ on its inward journey. If this is so the Na^+ efflux should depend upon the external K^+ concentration; indeed if there is no external K^+ the carrier should not be able to make its return

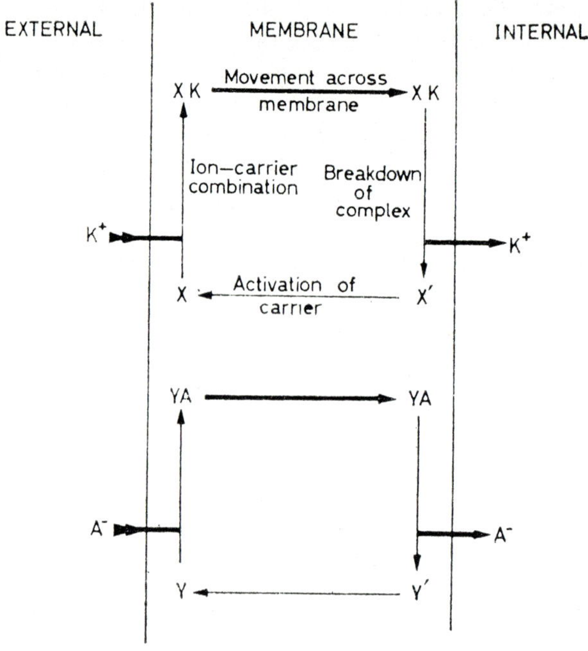

FIG. 7.3. The general concept of the operation of a carrier mechanism in ion uptake by cells. X and Y, carriers; X′ and Y′, precursors of the carriers; XK and YA, carrier-ion complex; K^+, cation; A′, anion. (From H. E. Street, The physiology of roots, in *Viewpoints in Biology*, vol. I, Butterworths, London, 1962.)

journey through the membrane and the Na^+ efflux should cease. Certainly it is found that the Na^+ efflux is dependent upon $[K_{outside}]$ and this relationship is almost identical with the influence of $[K_{outside}]$ on the active K^+ influx.

The hypothesis that active ion uptake operates through carrier systems requires us to consider how the operation of such systems could depend upon energy released by metabolism. Referring to Fig. 7.3, we see that energy could be used for regeneration of the carrier ($X' \longrightarrow X$), for promoting combination of the carrier with the ion ($X \longrightarrow XK$), for breakdown of this complex ($XK \longrightarrow X' + K^+$) or for transport of the carrier ion-complex (XK) and/or of the carrier precursor (X′) across the membrane. There are a disconcerting number of points at which the operation of the system could depend upon energy. Clearly, to be able to choose between these alternatives, to formulate the chemical reactions connecting metabolism with ion uptake and to really explain specificity of uptake we must know the exact chemical nature of the carrier molecules. This we do not know though modern biochemical techniques should lead in the near future to the isolation of such carrier molecules if indeed they are the basis of active uptake.

However, the mechanism of active ion uptake is being studied in a number of botanical laboratories and, therefore, as you might have expected, there is no lack of hypotheses regarding the chemical nature of carriers nor of how energy enters carrier systems. Such hypotheses merit the reader's attention because they illustrate the nature of original thought in plant physiology and challenge him (or her) to think how such hypotheses could be subjected to the test of experiment.

In Fig. 7.3 both a cation and an anion carrier are depicted; cation and anion uptake are visualised as involving *separate* carriers. However, not only is it possible to visualise amphoteric carriers (carriers capable of binding both cations and anions) but it could be that only the uptake of anions (or of cations) involves a carrier system. Thus, if anions are transported actively in this way, the cations might then move passively along the electrical gradient created by the accumulation of the negatively charged anions at the inner surface of the transport path. The Swedish botanist, Lundegårdh, was led to consider such a hypothesis from evidence he obtained that the outer surfaces of the absorbing cells of the root were negatively charged and might therefore be cation permeable but repel anions. Further, from studies of the rate of respiration during active salt uptake he concluded that the rate of oxygen uptake of the absorbing cells was the result of two component uptakes, one (the "ground respiration") unrelated to ion uptake and the second whose magnitude was determined by the rate of *anion* uptake (the "anion respiration"). He thus concluded that anion uptake was quantitatively geared to respiration and that cation uptake was a passive inward movement along an electrical gradient created by anion absorption. Then, in 1939, Lundegårdh found that the "anion respiration" of his roots was extremely sensitive to inhibition by cyanide and by carbon monoxide in the dark. From this he concluded that the anion carrier system was synonymous with the cytochrome system. At the exterior surface of the cell reduced cytochrome (p. 115) reacts with oxygen thus:

$$4Fe^{++}_{cyt} + 4H^{+} + O_2 \rightarrow 4Fe^{+++}_{cyt} + 2H_2O \tag{91}$$

or considering only the oxidation of the cytochrome:

$$Fe^{++}_{cyt} \rightarrow Fe^{+++}_{cyt} + (e') \tag{92}$$

or if the cytochrome cations are associated with anions then the release of an electron allows the cytochrome to pick up a further univalent anion thus:

$$\begin{Bmatrix} Fe^{++}_{cyt} \\ 2A^{-} \end{Bmatrix} + A^{-} \rightarrow \begin{Bmatrix} Fe^{+++}_{cyt} \\ 3A^{-} \end{Bmatrix} + (e') \tag{93}$$

Then the extra anion now associated with the oxidised cytochrome may be presumed to pass along the cytochrome electron-transport chain or move inwards by diffusion of the cytochrome-anion complex and be released at a point of lower oxidation potential, for instance, where cytochrome *b* reacts with flavoprotein. For this process to lead to anion uptake it is postulated that the release of the anion occurs in the inside of the diffusion barrier membrane. This hypothesis when first enunciated aroused considerable interest, not only because of Lundegårdh's personal reputation as a plant physiologist, but because it was clearly capable of experimental test.

The Lundegårdh hypothesis required that 4 monovalent anions and not more than 4 be absorbed per uptake of one O_2 molecule in "anion respiration" and that all anions stimulate respiration to the same extent per anion charge absorbed. Experiment showed that neither condition was fulfilled. It also became clear that one of the ions whose uptake most enhanced respiration was the cation NH_4^+ and that respiration was enhanced when ion uptake is restricted to cation uptake as occurs when a root is surrounded by a moist anion exchange resin. This hypothesis also precludes an active uptake of ions occurring under anaerobic conditions following upon a period of active aerobic respiration; it is a mechanism which would not provide for any storage of the ability to promote active ion uptake. In this it is contrary to a number of experimental observations. Further, only one carrier is postulated for all anions and it therefore fails to explain how there is no competition in uptake, for instance, between halide ions and sulphate ions, nor between sulphate and nitrate. It equally leaves unexplained competition between cations.

The Lundegårdh hypothesis formed the basis of the suggestion by Robertson that the anions being accumulated by plant cells were taken up by mitochondria (in which the electron transport chain is located) at the external surface of the cell (the plasmalemma) and then released from these organelles into the vacuole at the tonoplast. For this to happen mitochondria must circulate between plasmalemma and tonoplast and conditions must promote influx of anions at the exterior of the cytoplasm and efflux at the tonoplast–cytoplasm junction. Transport across the mitochondrial membrane had to be determined in direction by some factor like cytoplasmic pH or cytoplasmic oxidation-reduction potential. The high activity of mitochondria supplied with oxidisable substrates and of illuminated plastids in ion accumulation suggested the possibility that the separation of charge associated with electron transport could promote ion flow (see p. 126 for an outline of the possible mechanisms of charge separation and how this may be linked to phosphorylation). Cations would tend to move towards the negative side of the membrane and anions to the positive side.

These concepts do not preclude the operation of carrier species which facilitate the flow of ions through the membranes but they use charge separation (what Mitchell termed the "chemiosmotic" gradient) as the driving force for active ion transport. Further, a mechanism in which membrane transport of ions is explained in terms of the special metabolism of mitochondria or plastids does not seem to take account of the apparent importance of the tonoplast (and possibly also of the plasmalemma) in ion accumulation by plant cells.

This exercise of examining critically hypotheses regarding the mechanism of active ion uptake can now be extended to some suggestions regarding the chemical nature of carriers. The mechanisms postulated all suggest that the carrier molecules are amphoteric (hence capable of binding both cations and anions), and thereby evade a major criticism of the Lundegårdh hypothesis. In 1952, R. J. Goldacre reported rhythmic movements of root hair vacuoles and suggested that they arise from ordered contraction and unfolding of protein molecules orientated in the vacuole membranes. A similar folding and unfolding of protein molecules was considered to motivate protoplasmic streaming. Goldacre suggested that such contractile proteins within the membrane could when they were in the unfolded form bind ions by free valencies exposed at the membrane surface. Contraction would draw these ions through the membrane and the act of contraction could lead to liberation of the ions as the free valencies of the protein become satisfied amongst themselves in the folded molecules (Fig. 7.4). Unfolding of the protein molecule would reset the trap. Knowledge of the behaviour of myosin, the contractile protein of muscle, suggests how respiratory released energy could work such a mechanism. The extended form of myosin is energy-rich and may contract spontaneously, the unfolding of the myosin molecule is linked with the simultaneous degradation of ATP to ADP. The Goldacre hypothesis is in accord with the importance of proteins in membrane structure, it clearly defines two states of the carrier interconverted by utilisation of ATP which seems now the most universal source of the energy involved in cellular work and it explains transport across the thickness of the membrane by making this a forceful displacement of the ions. It is difficult to devise experimental tests of this hypothesis. However, it is opposed by one interesing experimental observation. If both the rhythmic movements in root hair vacuoles observed by Goldacre and protoplasmic streaming are dependent upon the folding and unfolding of protein molecules, as Goldacre himself contended, then any treatment which does not impair protoplasmic streaming would not be

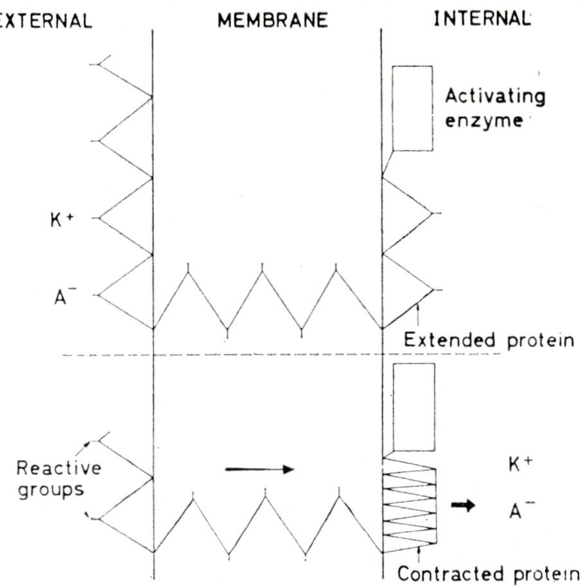

Fig. 7.4. Diagrammatic representation of the Goldacre concept of the functioning of a contractile protein as an amphoteric carrier of ions. *Above* the broken line, "trap set"; *below*, intake and release into the cell of ions as a consequence of contraction of the protein fibre. Energy could be fed into the system to extend the protein fibre and re-expose reactive groups external to the membrane (to set the ion trap). (From H. E. Street, as Fig. 7.3.)

expected to inhibit ion uptake. However, in work with cells of the red beet and of the leaf of *Elodea canadensis*, it has been shown that concentrations of chloramphenicol, which do not inhibit either oxygen uptake or protoplasmic streaming, do inhibit ion uptake and protein *synthesis*.

In 1956, Bennet-Clark suggested that the carrier could be a protein associated with the phosphatide, lecithin. This takes account of the fact that it is particularly lipo-proteins which are involved in the building of cell membranes and that certain enzymes seem to be located in cell membranes. It requires the presence in the cell membrane of the required number of distinct phosphatides to corre-

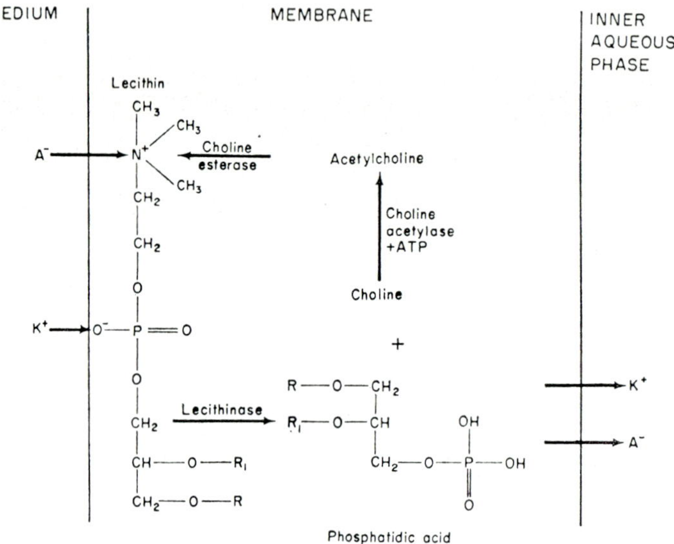

FIG. 7.5. Diagrammatic representation of the hypothesis of Bennet-Clark that a cyclic formation and breakdown of the phospholipid, lecithin, within the membrane would enable it to act as an amphoteric carrier. Energy as ATP is consumed in lecithin synthesis. (From Sutcliffe, as Fig. 7.2.)

spond with the number of known "competitive groups" of cations and anions. On this hypothesis the phosphate group in the phosphatide is regarded as the active centre binding cations and the basic choline group as the anion binding centre (Fig. 7.5). The liberation of the ions into the cell is regarded as being effected at the inner surface of the membrane by decomposition of the lecithin by the enzyme, *lecithinase*. The regeneration of the carrier from phosphatidic acid and choline is postulated as involving enzymes (*choline acetylase* and *choline esterase*) and ATP as a source of energy.

The concept of ATP as the source of energy for ion pumps, which has been developed above, is in accord with the observation that dinitrophenol and other reagents (CCCP = carbonylcyanide-m-chlorophenyl hydrazone, DCMU = dichlorophenyldimethylurea) which prevent ATP formation are inhibitors of active salt transport; all these strongly inhibit the sodium-potassium pump of giant algae cells.

The opening of stomata which is promoted by light requires oxygen and is inhibited by substances known to inhibit ATP formation. This opening involves an increase in solute concentration within the guard cells of the stomata leading to water uptake (increase in cell turgor). Many botanists have endeavoured to identify the process responsible for this increase in solute content within the guard cells. Recently a new analytical technique, that of electron micro-probe analysis has been brought to bear on this problem. By this technique it is possible to estimate the concentrations of elements in regions as small as 1–2 μ in diameter. Applying this to the study of stomatal movement in tobacco leaves evidence has been obtained that there is a rapid transport of K^+ ions from the epidermal cells to the guard cells prior to their opening and that this, if accompanied by a balancing anion uptake, is quantitatively sufficient to cause the required change in turgor. Guard cells therefore may contain a potassium pump similar to that of giant algal cells.

The reference above to the inhibition of salt absorption by chloramphenicol, an inhibitor of protein synthesis, can now serve to introduce one of the most interesting recent hypotheses. This arose from the observations that when the "dormant" cells of storage organs (potatoes, beets) are activated by washing in aerated distilled water then they develop not only a capacity for "primary" salt absorption but also for protein synthesis, and certain factors seem to affect these two processes in the same way. However, the concept that there was a close functional relationship between these two processes seemed to face the difficulty that active ion uptake can take place into cells in which there is no net protein synthesis. However, studies involving the use of the mass isotope of nitrogen (N^{15}) have shown that in such cells protein is being rapidly broken down and resynthesised (there is a rapid protein "turnover"). This has led Steward and his co-workers to suggest that peptides in the cell membranes bind both cations and anions and that such peptides are liberated and travel to the microsomes where the ions are liberated as their amino acids are incorporated into protein. Much along the same lines, Sutcliffe (1962) suggests that proteins are constantly entering into and being released from the structure of cell membranes. The released proteins carry ions to the ribonucleic acid templates of the endoplasmic reticulum and there, when protein and template combine "clusters" of ions are released. Such ions could attract to themselves water to form ion-rich vesicles and these could move in the cytoplasm and by bursting at the vacuole surface release ions to the central pool.

It has been suggested that ions could be engulfed by invagination of the plasmalemma leading to the release into the cytoplasm of vesicles enclosing external solution. A process of this kind, called pinocytosis, is seen in many animal cells and occurs when protoplasts are released from plant cells by enzyme dissolution of their cell walls. However, there is no convincing evidence that pinocytosis occurs in intact plant cells.

THE TRANSLOCATION OF SALTS

From the evidence that cellulose cell walls are readily permeable to water and to salt ions it would seem that salts could travel through living tissues by diffusion in the cell walls. However, various factors (such as illumination) which tend to divert salts from the cytoplasm into the central vacuole seem to reduce markedly the export of salts to surrounding cells. This suggests that salt movement from cell to cell in a living tissue may involve the protoplasmic continuity of cells via plasmodesmata. Further, if the hypothesis of minute mobile ion-rich vesicles is correct one might visualise that these can travel, not only across the cytoplasm to the central vacuole but through cytoplasmic connections into adjacent cells.

Considering the root, it would seem that salt movement across the cortex could proceed to the parenchyma surrounding the xylem conducting elements entirely within the cytoplasmic continuum (the symplast). Alternatively, movement as far as the endodermis could occur predominantly in the cell walls. At the endodermis the impermeability of the cell walls due to the Casparian strip would divert salts into the cytoplasm of the endodermal cells and living cells of the stele. This problem is not resolved, nor do we fully understand the mechanism whereby the living cells of the stele release salts into the xylem elements. It is, however, tempting to suggest that the efflux of salts from the living cells into the xylem is comparable to the release of ions into central vacuoles. Ions are selectively accumulated into vacuoles and there is some evidence pointing to selective secretion of ions into the young xylem vessels and tracheids. The alternative hypothesis postulates a passive leakage of ions into the xylem from the living cells of the stele. This hypothesis is based upon work with maize roots from which the cortex may be readily removed as an intact cylinder, leaving an apparently inviolate stele. The freshly isolated stele in contrast to the freshly isolated cortex rapidly releases its ions into a bathing solution.

The xylem sap differs in composition and is often more concentrated than the soil solution. This is easy to understand if there is selective secretion of ions into the xylem from the stelar parenchyma. However, selective uptake of ions mediated by the root cortical cells, for example preferential uptake of K^+ into the cytoplasm by a system 1 mechanism (see p. 223) with further enhancement of the $K^+:Na^+$ ratio in the symplast by a system 2 secretion into the vacuoles could result in selective supply of ions from the root to the shoot even if the release of ions into the xylem was itself passive and non-selective. From the region where ions enter the xylem they are transferred by a mass flow of sap maintained by transpiration or induced by the difference in osmotic potential between xylem sap and soil solution (root pressure).

Although most nutrient elements are translocated as free ions in the xylem, analysis of xylem sap always reveals the presence of organic compounds. These include not only soluble sugars but organic compounds of nitrogen and phosphorus. The evidence indicates that such organic compounds are of quantitative significance in the movement in the xylem of at least the elements nitrogen and phosphorus.

THE TRANSLOCATION OF ORGANIC COMPOUNDS

Extensive experiments involving ringing (removal of the tissue (bark) external to the xylem) demonstrated the importance of the "bark" in the longitudinal transport of organic substances, particularly of the sugars synthesised in the leaves. Knowledge of the anatomy of the bark dating back to Hartig's discovery of the sieve tubes in 1837, immediately focused attention upon these conducting elements of the phloem as the major pathway for the translocation of organic compounds. Subsequent work has confirmed that although organic substances occur in the xylem vessels and tracheids and hence are translocated in the transpiration stream, nevertheless, by far the most important channel for the translocation of carbohydrates, amino acids and other organic compounds is the phloem. Often functional phloem does not contain phloem fibres; in a number of instances companion cells are absent or discontinuous. The sieve-tubes are the only constant phloem elements to which a role in long-distance transport can be ascribed. The failure of eosin transport (in contrast to the transport of a number of other dyes, e.g. fluorescein) in phloem is to be correlated with eosin inducing closure of the pores of the sieve plates by the polysaccharide callose. Winter dormancy and the cessation of the longitudinal transport of sugar is also correlated with a callose sealing of the sieve plates.

We know the channel of translocation, but the mechanism of the translocation process in the sieve tubes has proved a most puzzling problem. The nature of this problem can be illustrated by considering the translocation of the sugars, produced in the leaves by photosynthesis, to the developing storage organs like fruits and tubers. This carbohydrate moves as sucrose. In 1953, Kennedy and Mittler showed that the willow aphid feeds by inserting its proboscis into a single sieve tube unit. The aphid can be anaesthetised and then cut free from its proboscis. The proboscis remains as a very fine tube leading down into the sieve tube and from this tube there exudes pure phloem sap. This contains about 10% of sucrose and usually traces of other oligosaccharides (raffinose and stachyose). This sucrose moves very rapidly in the sieve tubes. The phloem sap exuding from a single sieve tube (via an aphid stylet) can exceed 5 mm^5 hr^{-1} (implying filling of each sieve tube unit 10 times per second). The rapidity of sucrose transport can be illustrated by quoting the value calculated in 1922 by Dixon and Ball for the rate of movement of sucrose along the stolon during the development of a potato tuber. From the duration of the development of the tuber, from measuring its carbohydrate content, from the overall cross-sectional area of the sieve tubes in the stolon and by assuming that the sucrose moved as a 10% solution they concluded that if the sieve tubes were the route of transport their contents must travel at about 40 cm hr^{-1}. Other estimates of the velocity of sucrose movement fall within the range of 10–150 cm hr^{-1}. This movement is far faster than could occur by diffusion of sugar molecules; it was calculated, for instance, that the rate of movement of sucrose in the phloem of the cotton plant proceeds at a rate 40,000 times faster than would be expected by diffusion. This inevitably leads us to consider whether the phloem sap flows in the sieve tube as water through a pipe; whether a mass flow of liquid occurs in the sieve tubes.

To consider this question further it is necessary to review present views on the structure of sieve tubes.

THE STRUCTURE OF THE SIEVE TUBE

Sieve tubes are made up of longitudinal series of tubular sieve-tube units or elements connected together through perforated end walls known as sieve plates. The sieve tubes are usually 20–30 μ in diameter and the sieve plates occur 100–500 μ apart. The sieve pores are 0·1–5·0 μ in diameter and occupy up to 50% of the sieve plate surface. Although during the last 30 years numerous research papers have been published on the structure of sieve tubes as seen by light and electron microscopy we do not know what is the *in vivo* structure of functional sieve tubes. Because this is so all hypotheses of the mechanism of phloem transport are tentative and opposed by some apparently well authenticated observations. It seems to be generally agreed that the sieve tube contains an enucleated plasmolysable protoplast with a parietal layer rich in vesicles and in ER membranes and bounded by an external membrane (plasmalemma) continuous through the sieve tube pores. Many workers have reported that the lumen of the sieve tube units and the sieve plate pores contain plasmatic filaments visible in the electron microscope (they are up to 250 Å in diameter) and which appear to be tubular.

The crux of the problem of the relationship of form to function in sieve tubes relates to the disputed nature of the interconnections between the sieve tube units via the sieve plates and in a recent review it was concluded that at least four or five possible arrangements could not be excluded by the structural evidence (Fig. 7.7 a, b, c, d). These structure "models" can be considered in relation to hypotheses of the mechanism of solute transport in the sieve tubes.

MECHANISM OF TRANSPORT IN SIEVE TUBES

In 1930 the German botanist, E. Münch, put forward a *mass-flow hypothesis* based upon the concept that the sieve tube units were in open communication via the sieve plates (Fig. 7.7a). To illustrate his concept, Münch depicted an osmotic "model" in which a pressure-actuated mass flow of solution along a tube would occur (Fig. 7.6). In this model, spherical semi-permeable membranes enclose at A a solution of sugar and at C either a weaker solution of sugar or water and the two spheres are connected by a tube B. Now when the two spheres are immersed in water a greater hydrostatic pressure will develop by osmosis in A than in C and hence sugar solution will flow from A to C along B and water will be released through the pores of C into the surrounding water. This flow of sugar solution will continue until the concentrations of sugar in A and C are equal. To maintain the flow, sugar would have to be continuously added to A and withdrawn from C. Now, in relating this model to events in the plant, Münch postulated that A could represent the ending of a file of sieve tube units in the photosynthetic tissue of the leaf, where sugar synthesised in the mesophyll cells could be continuously secreted into the sieve tubes. Then C would represent say the ending of the sieve tube system in a developing storage organ where sugar is being continuously withdrawn from the sieve tubes into the surrounding cells for storage or metabolic utilisation. B would be the continuous sieve tube connection between A and C. Here, then, an "active" movement of sugar into and at another level out of the sieve tube leads to a maintained difference in concentration at different levels and this, in turn, leads to a mass flow of sap (sugar solution) in the sieve tube units from a region of high concentration to one where the sugar concentration is lower. The associated water movement would then presumably lead to release of water at the storage organ back into the xylem where it could enter the transpiration stream (D). Münch estimated that this solvent water liberated from the sieve tubes can contribute as much as 5% to the main transpiration stream.

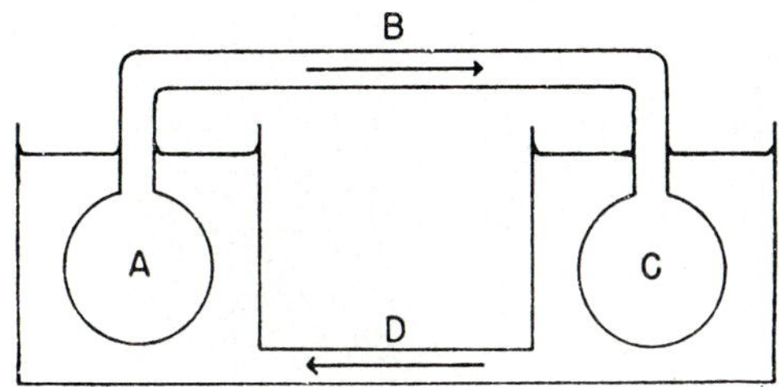

FIG. 7.6. Diagram of an osmotic model illustrating the principle of the Münch mass-flow hypothesis of translocation. (For discussion see text.)

This hypothesis visualises the sieve tube structure shown in Fig. 7.7a. It can be shown, as expected from the Münch hypothesis, that a gradient of hydrostatic pressure experimentally induced in the xylem of a cut twig causes an enhanced transport of sugar in the phloem and along the direction of falling pressure in the xylem. This enhanced flow occurs without dilution of the sieve tube sap. In many trees a cut into the phloem causes an immediate release of sugary sap. There always seems to be an immediate "expulsion" phase to this release of sap followed by a slower "bleeding". The initial expulsion is clearly suggestive of the rupturing of a system under considerable hydrostatic pressure.

From the osmotic potentials of phloem exudates a total pressure difference between source and sink of 15 atm is quite feasible. The Poiseuille equation which relates pressure drop per unit length, viscosity of fluid, radius of tube and velocity of flow can be used to calculate one of these when values for the others are known. Taking velocity of flow in the sieve tubes as 100 cm hr^{-1}, assuming the viscosity of the phloem sap to approximate to that of a 10% solution of sucrose, and using known values for sieve-tube diameter, sieve plate pore diameter and proportion of sieve plate occupied by open pores it can be calculated that the hydrostatic pressure gradient in

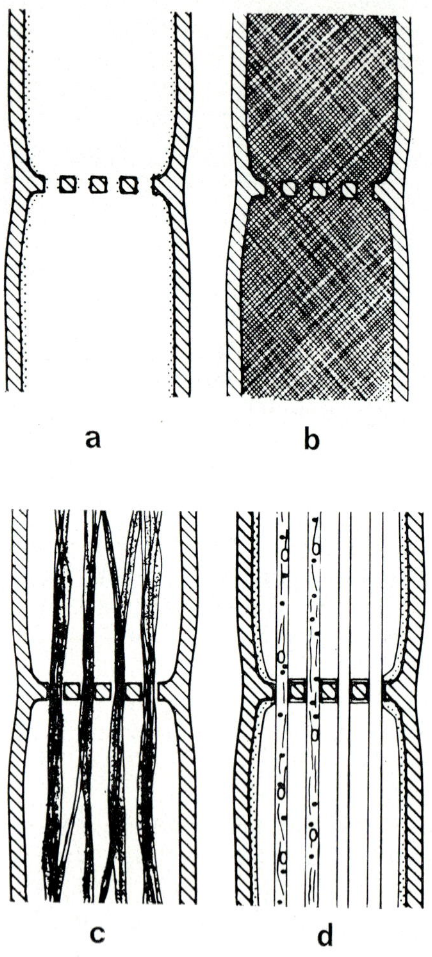

the sieve tubes would have to be about 0·6 atm m^{-1} to account for the postulated flow rate. Under these circumstances a pressure drop of 15 atm would permit transport by mass flow over some 25 m (tall trees therefore pose a problem for a mass-flow hypothesis). These calculations are however based upon the structure depicted in Fig. 7.7a. If, however, there exists in the sieve tubes either a uniform network of plasmatic filaments (Fig. 7.7b) or if strands of filaments pass through the sieve pores (Fig. 7.7c) impossibly higher hydrostatic pressure gradients would be demanded by the mass-flow hypothesis. The many electronmicrographs showing filaments, "slime" or "cytoplasm" in the sieve pores have to be dismissed as artefacts of fixation if the mass-flow hypothesis in the form enunciated by Münch is to be upheld.

To explain transport in sieve tubes having the structures depicted in Fig. 7.7 b, c or d, it is necessary to postulate a mass flow *generated within each sieve tube unit* by the expenditure of energy derived either solely from the respiratory activity of the sieve tube units or from this supplemented by a contribution from the associated companion cells and/or phloem parenchyma. Calculations have been made, assuming dimensions for the plasmatic filaments and their spacing based upon measurements from electronmicrographs, of the pressure gradients which would have to be established to achieve the known rates of transport in sieve tubes constructed as depicted in Fig. 7.7 b and c. Then the energy needed to establish such pressure gradients can be

---

FIG. 7.7 (*opposite*). Diagrammatic "models" of possible structure sieve tubes.

a. Pores of sieve plate unobstructed so that lumina of sieve tube unite in open communication (only model compatible with Münch hypothesis).

b. Lumina and pores filled with a uniform network of plasmatic filaments (a fixed lattice filling the lumen).

c. Bundles of plasmatic filaments passing through the sieve pores (the essential feature here being the high flow resistance of the sieve pores).

d. Transcellular strands (unobstructed tubes with fluid contents in motion) as envisaged by Thaine.

(After P. E. Weatherley and R. P. C. Johnson, *International Review of Cytology*, **24:** 149–192, 1968.)

expressed in terms of g glucose which would have to be respired per day per cm^3 of sieve tube and this compared with value recorded for phloem respiration. Such calculations indicate that the energy demand in a sieve tube containing throughout its lumen a fixed lattice of filaments (Fig. 7.7b) would be many times greater than even the highest values for phloem respiration make feasible. In the model Fig. 7.7c, however, the resistance to flow between the strands of filaments in the lumen would be small relative to the resistance in the sieve pores and the energy need to achieve the required rate of flow through the pores (assuming these to contain filaments 50–250 Å in diameter spaced at an average distance apart of 200 Å) is within the range calculated from the respiration measurements. The sieve tube model Fig. 7.7d is considered to be less demanding of energy and therefore in this sense even more feasible. It must, however, be emphasised that such calculations are not concerned with the mechanisms whereby the energy available for respiration might be geared to impel the sieve tube sap along the strands.

Model Fig. 7.7d is a diagrammatic representation of the structure of the sieve tube proposed by Thaine who has reported the observation by phase-contrast microscopy of active streaming in fine transcellular strands (1–7 μ diameter) running for long distances and passing unbroken through the sieve plates. Particles similar in size to mitochondria or small plastids are described as travelling along these transcellular strands and their movement is interpreted as that of particles being swept along in a flowing stream of liquid. These transcellular strands are considered by Thaine as indicating the nature of the flow occurring in sieve tubes. It is, however, extremely controversial whether the cross-sectional areas of these strands and their rate of flow could account for the rate and mass of movement of solutes in the sieve tubes. Because of the inadequate rate of the observed streaming it has been suggested that the movement of particles may reflect a more rapid rate of flow of the solution surrounding them. The mechanism causing the flow in the transcellular strands remains quite obscure, no contractions or regular vibrations have been observed. There is also the problem of their "turn round" at the ends of a particular sieve tube unit sequence. Thaine claims the strands to be membrane enclosed and if this is so the permeability of the membranes will be critical since on this hypothesis both a plasmalemma and the membrane of the transcellular strands separate source and sink from the moving solute column. The occurrence in sieve tubes of these transcellular strands, resembling the trans-vacuolar cytoplasmic strands of parenchymatous cells, has been strongly challenged by a number of workers.

Spanner has advanced the hypothesis that an activated electro-osmotic flow occurs through the sieve pores which contain roughly parallel orientated cytoplasmic filaments and that this induces a mass flow of sap within the sieve tubes. Electro-osmosis involves a movement of water or other nonionised hydrophilic molecules (such as sucrose) as hydration shells or by induced frictional flow resulting from a polarised flow of ions along a maintained electrical potential gradient and through a charged membrane. Spanner suggests that the electro-osmotic flow in sieve tubes is due to the maintenance of a high potassium ion (K^+) concentration on one side of the sieve plate and a lower concentration on the other side. Potassium ions, carrying phloem sap, then stream from the more positive to the less positive side through the negatively charged micro-channels between the cytoplasmic filaments in the sieve plate pores. To maintain this potassium ion gradient it is postulated that potassium ions on the downstream side of the sieve plate "leak" into the longitudinal walls of the sieve tube elements and on the upstream side are "pumped" back into the sieve tube unit by a potassium pump driven by ATP derived either entirely from the sieve tube unit or in part from the metabolically active adjacent companion cells (Fig. 7.8). This hypothesis is supported by evidence of electrical potentials across sieve plates and by the knowledge that electro-osmosis can be a very powerful force. There are however many aspects of this hypothesis for which there is no quantitative data: the actual differences in K^+ concentration occurring across sieve plates; the exact path of K^+ which maintains such differences if they exist; the quantitative relationship between the net potassium flux and sap movement.

This discussion has indicated that the companion cells of the phloem may be involved in supplying energy to the sieve tubes both for the initial transport of sucrose into and out of sieve tubes and for its transport within the tubes. They may also be involved in the polarised

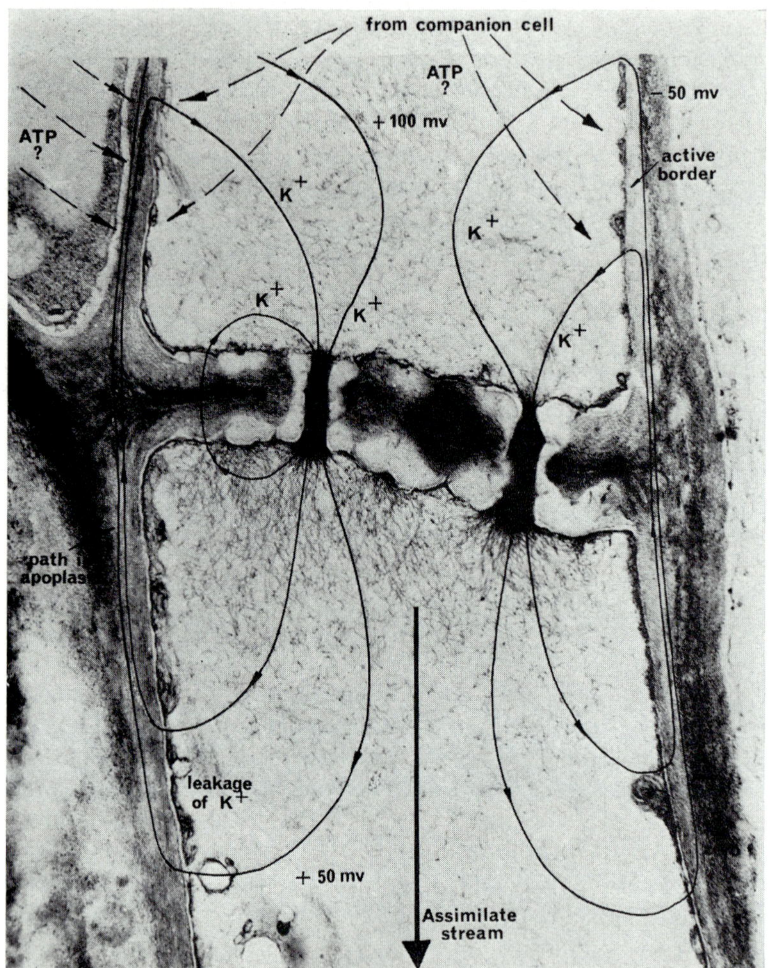

FIG. 7.8. Electron micrograph through a sieve plate of *Helianthus annus* superimposed by a diagrammatic representation of the K^+ ion movements demanded by the Spanner hypothesis. (From D. C. Spanner and R. L. Jones, *Planta*, **92:** 64–72, 1970.)

flow of K^+ ions at the levels of the sieve plates as postulated in the Spanner hypothesis. At other sites, particularly at the fine vein endings in leaves where sucrose is being imported from the photosynthetic cells into the phloem, in cotyledons where solutes are being exported to the growing embryo and at sites where sugar is being removed from the phloem for storage, the presence has been reported, in a range of plant species, of special "*transfer*" *cells* rich in cytoplasm and with extensive wall ingrowth associated with a greatly increased plasmalemma area.

The content of this chapter emphasises to the writers both the importance for the understanding of physiological processes of extending our knowledge of the fine structure of plant cells and the decisive role of "active" processes (processes consuming cellular energy) in the movement of solutes within and between cells.

FURTHER READING

J. Dainty. The ionic relations of plants, pp. 455–487 in *The Physiology of Plant Growth and Development*, edited by M. B. Wilkins, McGraw-Hill, London, 1969.

D. R. Hoagland. *Lectures on the Inorganic Nutrition of Plants*. Chronica Botanica, Co., Waltham, Mass., 1944.

H. E. Street. The physiology of roots, in *Viewpoints in Biology*, vol. I, pp. 1–49, edited by J. D. Carthy and C. L. Duddington. Butterworths, London, 1962.

J. F. Sutcliffe. *Mineral Salts Absorption in Plants*. Pergamon Press, Oxford, 1962.

P. E. Weatherley and R. P. C. Johnson. The form and function of the sieve tube: a problem in reconciliation. *International Review of Cytology*, **24**: 149–192, 1968.

MORE ADVANCED READING

G. E. Briggs, A. B. Hope and R. N. Robertson. *Electrolytes and Plant Cells*. Blackwells, Oxford, 1961.

M. J. Canny. The rate of translocation. *Biological Reviews*, **35**: 507–532, 1960.

R. N. Robertson. *Protons, Electrons, Phosphorylation and Active Transport*. Cambridge University Press, Cambridge, 1968.

D. C. Spanner. The translocation of sugar in sieve tubes. *Journal of Experimental Botany*, **9**: 332–342, 1958.

C. P. Swanson. Translocation of organic solutes, in *Plant Physiology*, vol. II, pp. 481–551, edited by F. C. Steward. Academic Press, New York, 1969.

R. Thaine. The protoplasmic streaming theory of phloem transport. *Journal of Experimental Botany*, **15**: 470–484, 1964.

CHAPTER 8

The Regulation of Metabolism

"Most studies of the cell at a molecular level have had to be primarily qualitative, identifying its micro- and macro-molecular components and working out the metabolic interrelations that are so conveniently symbolised by arrows. But now that many biosynthetic pathways have become more or less completely known, it has become possible not only to describe flow rates but to analyse in detail the mechanisms that control them."

Bernard D. Davis, "Opening Address" in *Cellular Regulatory Mechanisms.* Cold Spring Harbor Symposia on Quantitative Biology, vol. 26. Biological Laboratory, Cold Spring Harbor, New York, 1961.

INTRODUCTION

WE have discussed the importance and underlying chemical reactions of various aspects of plant physiology. It is clear that these physiological processes are interrelated, that they are each facets of an integrated metabolism. However, to understand the metabolic activities of living cells we must inevitably first consider these facets in isolation; as separate multi-enzyme systems. This then serves as a basis from which we can move on to consider the interdependence of physiological processes. Finally, we must endeavour to uncover the control mechanisms by which equilibria ("steady states") are maintained and changed to new equilibria (as cells grow and change their chemical activities during cell differentiation). It is in the study of such control mechanisms that we shall probably expose the causes of senescence and death.

It is a tenet of genetics that the metabolic activities and potentialities of cells depend upon and are limited by their inheritance. This implies that the factors controlling metabolism must be determined in their number and intensity by the inherited genes. In line with this concept is the extensive experimental evidence that the genes control the synthesis of the essential proteins of the cell including those which function as enzymes. Further, in view of the evidence that the nuclear genes are composed of DNA–protein then one way in which they could exert this function would be by determining the structure of a "messenger RNA" which, moving out from the nucleus could, in turn, confer upon the ribosomes the ability to synthesise a particular enzymic or structural protein (see Chapter 5, p. 175). Any change of the structure of the genic DNA (*a mutation*) could then be reflected either in a change in the stability or specific activity of an enzymic protein or, if sufficiently profound, lead to the synthesis of biologically inactive protein. Such a change could manifest itself as the deletion of an enzyme having, if it occupied a sufficiently central position in metabolism, a lethal effect, or, alternatively, inducing an absolute requirement for an external supply of some metabolite whose biosynthesis involves the lost enzyme. Mutants of this kind in which a particular enzyme is missing can be readily induced in certain micro-organisms (like the fungus *Neurospora*) by mutagenic agents like X-rays and the study of such biochemical mutants has not only established the role of the genes in controlling enzyme synthesis but also enabled a number of biosynthetic pathways in metabolism to be worked out. Work with micro-organisms has also shown that genes also control factors involved in solute absorption by cells. The factors concerned here are also probably "enzyme-like" and have hence been termed *permeases* (they could also be described as specific solute "carriers"—see Chapter 7, p. 222).

Here then we have as a starting point to a discussion of the regulation of metabolism the evidence that enzyme synthesis is controlled by the genes. This control is exercised through the agency of "messenger RNA" molecules synthesised under the direct control of the genes within the nucleus and then released into and exerting their effects within the cytoplasm, the site of synthesis of most, if not all, enzymes. This is, however, a qualitative concept. To understand its quantitative aspects we have to try to answer such questions as: What controls whether certain genes function and what controls the activity with which they function in enzyme synthesis? Once enzymes have been synthesised are there mechanisms which control their activity? If self-regulatory systems lead to the establishment of "steady states" what factors disturb such balances to cause cells to change both qualitatively and quantitatively their metabolic activities?

The regulation of the activity of a metabolic pathway is achieved through the control of certain "key" enzymic steps. Such key reactions have been termed "pacemaker" or "cross-over" points of metabolism. In general such steps are regulated by changes in either the *activity* or cellular *concentration* of the enzymes involved. There is widespread evidence for the operation of both types of regulation in living organisms, and they are distinguishable by the rapidity at which they are effected. Changes in enzyme concentration take much longer to be effected than changes in enzymic activity. In view of this these mechanisms have been termed "coarse" and "fine" control respectively.

"Coarse" control involves then, regulation by induction and repression of enzyme synthesis. "Fine" control involves regulation by activation and inhibition of enzyme activity. Other factors, such as availability of co-factors, will naturally also affect the rate of enzyme reactions (see p. 69).

INDUCTION AND REPRESSION OF ENZYME SYNTHESIS ("COARSE CONTROL")

In certain micro-organisms it has been possible to demonstrate the activity of certain enzymes under all tested environmental and nutritive conditions. Such enzymes have been described as *constitutive*. However, a number of these micro-organisms have been shown to be capable of developing the activity of additional enzymes if supplied with certain potential metabolites or molecules similar to such metabolites. Thus yeasts are capable of actively metabolising glucose by the activity of constitutive enzymes, but constitutive enzymes are not available which will metabolise other sugars such as galactose or arabinose. However, if these sugars are supplied the yeast quickly develops a capacity for metabolising them and this *adaptation* can take place without cell growth or division. The "adapted" cells show newly acquired enzymic activity. Such new enzymes whose activity builds up in response to the "inducing" molecules are termed *adaptive* enzymes. In such cases it is very difficult to prove that the enzymes concerned are absolutely absent before induction but we do know that the induction involves synthesis of new enzymic protein (and not an unmasking of the activity of preformed enzyme molecules).

In other cases we encounter not induction but repression of enzyme *synthesis* by a metabolite in whose synthesis the enzyme is concerned (the same may also be inhibited in its *activity* by other components of the reaction sequence in which it it involved, see p. 252). The enzyme whose activity is lost or markedly reduced is not necessarily that which completes the synthesis of the *repressor* molecule (i.e. not that catalysing the last reaction of the biosynthetic pathway). For instance, uracil can suppress, in certain strains of the bacterium, *Escherichia coli*, the activity of the enzyme *aspartate carbamyl-transferase* which promotes the interaction between aspartic acid and carbamyl phosphate, a reaction which is the first step in the reaction sequence involved in pyrimidine biosynthesis. Further, the experimental evidence indicates that the uracil acts as a repressor by preventing synthesis of the enzyme.

Here, then, we have examples of the regulation of metabolism by a promotion or suppression of *enzyme synthesis*. What is not yet fully understood is how "inducers" promote and "repressors" inhibit the synthesis of new enzyme protein. When two "inducer" molecules are supplied simultaneously and when the rate of protein synthesis is limited by nitrogen supply then it is possible to demonstrate competition between the systems synthesising the adaptive enzymes; in so far as one enzyme is synthesised the amount of the second enzyme formed is correspondingly reduced. Secondly, induction of an enzyme can sometimes be effected by a compound (chemically similar to the substrate) but not acted upon by the enzyme whereas other molecules which can act as substrates for the enzyme may be ineffective as inducers. Thus methyl-β-D-thiogalactoside is a powerful inducer of *β-galactosidase* in *E. coli* although inactive as a substrate whereas phenyl-β-D-thiogalactoside which can act as a substrate does not induce synthesis of the enzyme. Thirdly, enzyme induction can be a very rapid process, a very marked rise in enzyme content occurring within hours or even minutes. This, for instance, is the case with one of the very few enzymes known, in the case of higher plants, to be adaptive, the enzyme, *nitrate reductase* (Chapter 5, p. 167). These observations have been regarded as indicating that the sites for the synthesis of adaptive enzymes are preformed in the cells but require activation, in which case inducers may function by releasing the enzyme protein from the RNA templates and thereby permitting further protein synthesis. Recent and very intensive study of enzyme induction and repression in bacteria has, however, led to the postulation of an alternative concept. This is that the chromosomes carry not only "structural" genes or gene groups (*operons*) which synthesise the molecules of messenger RNA and hence regulate protein synthesis but other genes ("regulator" genes) whose function is to synthesise a *repressor* which, either directly or after cytoplasmic modification, suppresses the functioning of the *operon*. Within this framework the induction of enzymes is interpreted as the consequence of the inactivation of the repressor substance by the inducer molecule, and repression of enzymes as involving cytoplasmic activation of the repressor substance.

Enzyme synthesis depends upon the availability of the necessary amino acids and the necessary specific transfer RNA (t-RNA) molecules. When an amino acid deficiency limits protein synthesis, m-RNA synthesis may proceed normally but synthesis of t-RNA and ribosome RNA be blocked. This seems to be a consequence of the presence of unchanged t-RNA (t-RNA not combined with amino acid). Mutants are known in which this block fails to operate and t-RNA and ribosome RNA continue to be synthesised despite the failure of protein synthesis.

There are some cases where micro-organisms adapt only very slowly to a metabolite, where the organism can be gradually *trained* over a number of cell generations either to metabolise a compound or overcome its inhibitory effect on growth and metabolism. For instance, the bacterium *Bacillus lactis aerogenes* can be slowly trained to utilise glycerol and if fully trained can retain, at least for a time, the ability to utilise this source of carbon when grown in its absence and supplied with glucose. The alga *Chlorella vulgaris* can slowly develop resistance to the "anti-metabolite" selenomethionine and the resistance is due to enhanced activity of the enzymes reducing sulphate to the normal metabolite, methionine. Thus, resistance to selenomethionine once developed is retained in its absence. However, sulphur starvation or methionine feeding lead to a decrease in the activity of the sulphate-assimilating enzymes and, in consequence, loss of resistance to the "antimetabolite". By growing a particular strain of *Chlorella vulgaris* in darkness we have slowly trained it to utilise galactose as an effective carbon source and have shown that the "trained" cells have enhanced activity of the enzyme *galactokinase* (the enzyme which promotes the formation of galactose-6-phosphate by transfer of phosphate from ATP). In this case the activity of the galactokinase quickly decreases when galactose is replaced by the readily utilisable sugar, glucose.

The fact that several generations of cells have to be traversed in this phenomenon of training raises the possibility that here we are selecting mutant cells, cells which have acquired by gene change an ability to synthesise a new enzyme. However, in the *Chlorella* studies described above the enzymes whose activity is enhanced are already present in low activity in the "normal" cells and the experimental evidence strongly supports the view that *all* the cells can be trained and that there is no selection of some mutant arising during culture of the organism. In these cases we are probably witnessing an increase in the number of enzyme-synthesising sites; we are eliciting a *greater* flow of the necessary messenger-RNA from the nucleus to activate more ribosomal centres of synthesis. If this proves to be so then it will provide evidence that substances in the cytoplasm can not only initiate but enhance the functional capacity of genes.

INHIBITION AND ACTIVATION OF ENZYMES ("FINE" CONTROL)

The activity of any metabolic pathway will depend upon the availability of the precursor molecules and the activities of the separate enzymic steps involved. Thus, reaction pathways using common metabolites as starting materials will be in competition and each pathway will be limited in its rate by the pace of the slowest step in the reaction chain. Again the reaction pathway may be very sensitive to the concentrations of the reaction products. This may be due to the position of equilibrium in certain key reactions. Such a situation is exemplified by the dehydrogenase reactions of respiration. Unless the co-enzyme (NAD) is kept almost completely in the oxidised form (by the reactions of terminal oxidation) the respiratory intermediates instead of being oxidised will remain reduced.

Co-factor levels and the rate of turn-over of co-factors may also affect the activities of certain enzymic steps. Present evidence suggests that in a number of tissues the activities of *glucose-6-phosphate dehydrogenase* and *6-phosphogluconate dehydrogenase* may be closely related to the rate of oxidation of the NAPH formed.

The enzymes in a biosynthetic or other metabolic reaction chain may be strongly inhibited by intermediates or by the essential end-product itself. Oxaloacetic acid inhibits its own synthesis from malic acid by the *malate dehydrogenase*, it also strongly inhibits the enzyme, *succinate dehydrogenase*. Thus, oxaloacetic acid, an intermediate in the Krebs cycle, is an inhibitor of two of the enzymes of the cycle. Oxaloacetic acid is, during the active operation of the Krebs cycle, removed by reaction with acetyl-CoA to give citric acid. If, however, the supply of acetyl-CoA drops, then oxaloacetate begins to accumulate and its further formation is checked by the inhibitory action of the oxaloacetate on Krebs cycle enzymes. A similar situation can be seen in the synthesis of the amino acid isoleucine from aspartic acid, in which the amino acid, threonine, is an intermediate. One of the steps which precedes threonine formation is the phosphorylation of homoserine to homoserine phosphate by *homoserine kinase*. Threonine strongly inhibits this enzyme. In turn threonine is deaminated to α-ketobutyric acid by *threonine dehydratase* and this enzyme is strongly inhibited by the end-product of the biosynthesis, isoleucine. This type of control has been defined by Krebs (1957) as a *negative feed-back*, whereby a product of a reaction sequence inhibits one of the earlier reactions of the sequence, thereby slowing down the overall rate until the product is removed. Such *negative feed-back* mechanisms may operate in many of the reaction sequences of metabolism.

Pasteur in his studies on yeast noted that when this organism is growing under anaerobic conditions it multiplies slowly, evolves a large amount of carbon dioxide and consumes a large quantity of sugar which it ferments to ethanol. Pasteur noted, however, that when oxygen is supplied less carbon dioxide is evolved and less sugar consumed although cell growth and division now proceed more rapidly. It is this action of oxygen in suppressing the fermentation of sugar and initiating its aerobic respiration which is usually referred to as the *Pasteur effect*. In recent years a number of hypotheses have been advanced to explain the mechanism of this effect. It is now considered to involve the inhibitory effect of high cellular levels of ATP on specific kinases (see p. 129) involved in the EMP pathway of respiration. In particular the "irreversible" conversion of fructose-6-phosphate to fructose-1-6-diphosphate by the enzyme *phosphofructokinase* appears to be a key reaction. Such inhibition of activity of phosphofructokinase by high levels of ATP has been demonstrated in cells from a variety of organisms. Under conditions of anaerobiosis there is, despite the high rate of sugar utilisation, only a low level of ATP. When conditions are changed from anaerobiosis and fermentation to aerobiosis and oxidative phosphorylation there is a marked rise in the rate of synthesis and in the cellular content of ATP, which reaches the level where it is partially inhibiting the phosphofructokinase activity with the result that sugar is metabolised less rapidly.

Studies of the mechanism of end-product inhibition have revealed that the configuration and activity of certain enzymes can be modified by combination with substances ("*effectors*") unrelated to their substrates. The "effector" binds to a site(s) separate from that involved in substrate–enzyme combination and in so doing alters the configuration of the enzyme, e.g. by altering the degree of aggregation of the sub-units of the enzyme molecule. Such enzymes are described as *allosteric* enzymes. Aspartic acid can be the precursor of pyrimidine bases and the first reaction involves the allosteric enzyme *aspartate carbamyl-transferase*. This enzyme is inhibited by one of the end-products, cytidine triphosphate (CTP). Clearly there is considerable dissimilarity between the structure of the substrate and the effector.

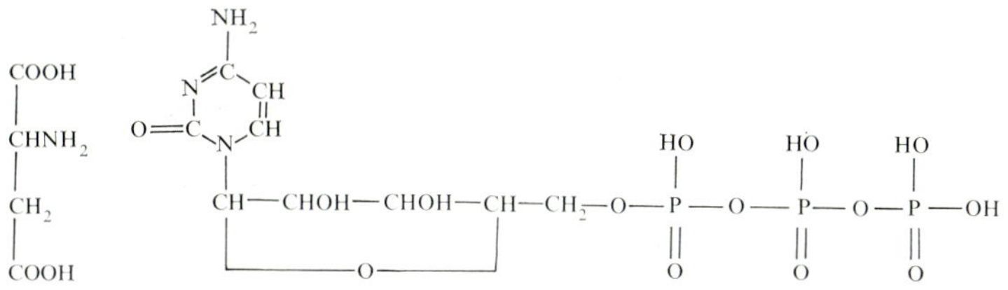

COOH
|
CHNH$_2$
|
CH$_2$
|
COOH

Aspartic acid

NH$_2$
|
N≠C—CH
| ||
O=C—N—CH
|
CH—CHOH—CHOH—CH—CH$_2$—O—P(=O)(HO)—O—P(=O)(HO)—O—P(=O)(HO)—OH
|__________O__________|

Cytidine triphosphate (CTP)

Treatment of aspartate carbamyl-transferase with heat, with Hg^{++} ions or with urea abolish the end-product sensitivity without destroying its catalytic activity. These treatments cause the enzyme (M. wt. 300,000) to break up into sub-units; four of these are catalytic units (M. wt. 48,000) and are insensitive to end-product inhibition, the other four are regulatory units (M. wt. 28,000) which are catalytically inactive but bind CTP. When the sub-units are linked together combination of CTP at sites on the regulatory units alters the configuration of the enzyme unfavourably at the sites concerned with substrate change. While CTP causes allosteric inhibition of activity, ATP will activate the enzyme by competing with CTP. This phenomenon of the allosteric regulation of aspartate carbamyl-transferase activity is to be contrasted with the repression of synthesis of this same enzyme by uracil (see p. 250).

Reference has previously been made to allosteric enzymes in Chapter 3 (p. 75) and in the discussion of starch synthesis in Chapter 5 (p. 164).

By contrast, the Calvin cycle in photosynthesis illustrates a *positive feed-back* mechanism. The concentration of Calvin cycle intermediates persisting in photosynthetic cells in the dark is very low. When photosynthesis commences the availability of ribulose diphosphate (RuDP) may limit the rate of CO_2 assimilation. However, the formation of phosphoglyceric acid (PGA) immediately raises the level of other intermediates in the cycle, including that of RuDP, and this in turn speeds up the rate of CO_2 assimilation. Similarly, when cells are starved of respiratory substrate the ATP of the cells may be markedly depleted. The initial utilisation of sugar may then be limited in rate by the rate of the phosphorylation reactions which involve ATP and lead to the formation of fructose diphosphate. However, once sugar again begins to enter the respiratory pathway oxidative phosphorylation proceeds and ATP is synthesised at a far greater rate than is required for the initial phosphorylation reactions of respiration. The level of ATP therefore rises and the initial reactions proceed at a faster rate.

It is known that certain cell constituents are involved in the reversible activation and inactivation of enzymes. Thus, the enzyme *phosphoglucomutase* which catalyses the reaction

$$\underset{\text{(G-1-P)}}{\text{glucose-1-phosphate}} \rightleftharpoons \underset{\text{(G-6-P)}}{\text{glucose-6-phosphate}} \qquad (94)$$

requires for activation a primer (catalytic) amount of glucose-1-6-diphosphate and the active catalyst is an enzyme phosphate in which a phosphate group is combined with a β-hydroxyl group of one of the constituent amino-acids, serine. The enzyme phosphate can transfer its phosphate to either G-1-P or G-6-P to give the diphosphate and when the diphosphate reacts with inactive enzyme a mixture of the monophosphates of glucose results. A catalytic amount of the glucose diphosphate is needed for the enzyme to interconvert massive amounts of the monophosphates. Some enzymes (particularly many of the enzymes involved in cellular respiration and proteolytic enzymes) are only active when certain sulphur-containing groups (thiol groups) in their molecules are in the reduced or sulphydryl form (—SH). One natural compound which seems to be of significance in maintaining the activity of such —SH enzymes is the tripeptide, glutathione (γ-glutamylcysteinylglycine), which reduces disulphide groups being itself simultaneously oxidised. It is interesting that the glutathione content of cells rises just before their division and is usually high in actively growing cells. The glutathione content of seeds, particularly of the embryo tissues, rises during germination and this may be an important factor involved in the activation of enzymes occurring during the early stages of germination (Chapter 1).

Whether there are natural inhibitors of enzymes distinct from the intermediates and end-products of reaction sequences, to which reference was made in discussing negative feed-back, is still a matter of controversy. Certain powerful inhibitors of growth and metabolism have been demonstrated to occur in dormant seeds, buds and storage organs and evidence has been obtained that the amounts of such inhibitors decrease during the natural breaking of dormancy. It has, however, proved difficult to identify these chemically although recent work on the dormancy of sycamore buds and the control of abscission of the fruit of cotton has led to the isolation and chemical identification of one such natural growth inhibitor, *abscisic acid.* At present abscisic acid appears from most studies to be a repressor of enzyme synthesis rather than an inhibitor of enzyme activity, although an allosteric inhibition of invertase activity by abscisic acid has been reported.

Reference has previously been made (Chapter 5, p. 172) to the enzyme, *glutamine synthetase*, which promotes the synthesis of the amide, glutamine. We find that some plant roots when grown in excised culture (see Chapter 9) contain a powerful inhibitor of this enzyme and that the content of this inhibitor is markedly increased when the roots are supplied with glutamine as their sole source of nitrogen. Here the inhibitor is quite distinct from the end-product of the reaction (glutamine) but its content in the cells rises in response to glutamine accumulation. This may indicate that where end-products or intermediates inhibit enzymes they may do so, not directly but by giving rise to or promoting the synthesis of inhibitor molecules.

It is a characteristic feature of metabolism that many reactions are coupled together through a common co-enzyme or energy carrier molecule. One obvious example of such coupling is the linkage between electron transport and ATP synthesis in the process of terminal oxidation (Chapter 4). It has been demonstrated that certain mitochondria cease oxygen uptake unless supplied with phosphate and ADP, and as soon as all the ADP has been converted to ATP the electron transport system stops. This explains how energy-consuming processes like salt accumulation, protein synthesis and cell wall growth by consuming ATP enhance respiration. Respiration rate can also be enhanced by dinitrophenol (DNP) which acts by uncoupling electron transport from phosphorylation. Electron transport proceeds at an enhanced rate and the energy release is lost as heat. Students of animal physiology have long known that the first effect of DNP on animals was to raise their temperature to an abnormal level. What is of particular interest is that while the coupling between electron transport and phosphorylation is very tight in recently isolated mitochondria, this coupling loosens as the mitochondria "age", and that mitochondria from senescent cells also show a lower production of ATP per electron transported than do those from young actively metabolising cells. Further, an "uncoupler" can be isolated from "aged" mitochondria. It is therefore possible that breakdown in coupling can occur in cells during differentiation or senescence as a result either of the formation of uncoupling substances or of failure to maintain certain spatial relationships at the catalytic surfaces within organelles such as mitochondria.

CELL STRUCTURE AND THE CONTROL OF METABOLISM

Any consideration of coupling and uncoupling mechanisms naturally leads on to a consideration of the importance of cell structure and the control of metabolism. Studies of the activities and enzyme contents of nuclei, plastids, mitochondria and microsomes show clearly that these structures, separated from one another and from the hyaloplasm and endoplasmic reticulum by cellular membranes, are the centres of particular aspects of metabolism. This means that the flow between these centres and the concentrations within them of metabolites, co-enzymes and energy carriers are regulated by the process of membrane transport. It is also clear from our knowledge of the fine structure and chemistry of these cytoplasmic "particles" that within them there is a further organisation and segregation of enzymes. Thus, in the chloroplasts, the photochemical reactions of the grana lamellae are separated from "dark" reactions occurring in the stroma. In the mitochondria the enzymes promoting oxidative phosphorylation and occurring in the cristae are separate from other mitochondrial enzymes. Further, from the metabolic activities of disrupted chloroplasts and mitochondria, we have evidence that within these structures reaction chains are not only chains in the sense that one reaction produces the substrates for the next reaction of the chain but that the enzymes involved are spatially arranged at catalytic surfaces to form a reaction chain or production line in space. In consequence overall concentrations of metabolites in cells give no idea of effective concentrations at catalytic surfaces, surfaces where intermediates may undergo successive chemical changes without ever entering into aqueous solution.

Apart from the mechanisms of "coarse" and "fine" control discussed above there occur in plants and animals rhythms of metabolic activity and extrinsic control mechanisms involving, in both plant and animals, hormones and, in animals only, nervous system control. The role of plant hormones in the control of growth and differentiation in plants is discussed in the next chapter. The present chapter concludes with a brief description of circadian rhythms of metabolic activity in plants and of a particular example of a timing reaction.

RHYTHMS IN CELLULAR METABOLISM

Circadian Rhythms

Rhythmical changes in leaf position, growth rate and metabolic activity have been known in plants for a very long time. Most of the rhythms have a natural period close to but not precisely of 24 hr (periods range from 21 to 28 hr). They are therefore usually referred to as circadian rhythms (*circa*, about; *diem*, day). These rhythms are endogenous in so far as they will persist (at least for a time; in plants often for 1–2 weeks) in a uniform environment and they can be initiated by a single stimulus (then perodicity is inherent and not acquired from cyclical environmental variations). The existence of these endogenous rhythms suggests the operation of a system capable of sustained oscillations, of a biological clock regulating physiological activity.

The best known case of a rhythm in metabolic activity is that of the dark fixation of carbon dioxide in the leaves of Crassulacean plants like *Bryophyllum*. Dark CO_2 fixation in *Bryophyllum fedtschenkoi* has a natural rhythm of 22·4 hr at 26°C. The period of this rhythm can be shortened by subjecting the plant to shorter light and dark periods than occur naturally; however as soon as the plant is returned to continuous darkness the rhythm reverts to its natural period with a single peak of dark CO_2 fixation activity once every 22·4 hr. The rhythm can be completely suppressed by continuous bright light or high temperature but reappears on return to darkness at a suitable temperature, the first peak of CO_2 fixation occurring at a definite predictable time after return to the compatible conditions. The rhythm is also inhibited by low temperatures or anaerobic conditions. Within a temperature range the period of the rhythm can be shortened by rise in temperature but the process has a low Q_{10} of *ca.* 0·8–1·2 and temperature compensation occurs; the period at first lengthened by lowering temperatures slowly reverts to its normal period at the lower temperature.

The action spectra of light as a stimulus initiating circadian rhythms have, in most cases, not been carefully determined; the most active region of the spectrum for initiating the *Bryophyllum* rhythm is 600–700 mμ. The rhythm is shown by leaf discs (with or without the epidermis attached) and by tissue cultures initiated from mesophyll cells; the rhythm is therefore a property of the cells involved in the fixation reaction and does not depend upon the cells being within a functional leaf. Throughout the cycle the cells retain a high activity of the critical enzyme of the fixation process, *phosphoenolpyruvate carboxylase* (see p. 101). In this instance, therefore, the rhythm cannot be explained in terms of variation in the activity of an enzyme; although in certain other metabolic rhythms there are synchronous changes in the activity of particular enzymes. Infiltration of *Bryophyllum* leaves with phosphoenolpyruvic acid fails to enhance fixation at times in the cycle when it is zero or very low suggesting that there is no rhythmic variation in substrate availability. One possible mechanism now being investigated is that the fixation reaction is subject to end-product inhibition and that the timing reaction before the next burst of fixation is the removal of the end-product from the site of its formation to a new intracellular location.

The Timing Reaction in Photoperiodism

Certain plants only flower when subject to an appropriate day length. Both light and dark reactions are involved in this phenomenon of response to day length (photoperiodism). One critical aspect of the induction to flower is the period of darkness and photoperiodic plants can be classified according as to whether a period of darkness *in excess of* a certain length is necessary for them to flower (short day plants) or is inhibitory to their flowering (long day plants). In many species the appropriate photoperiodic treatment need only be given for a few days to induce flowering, even though the plants are subsequently maintained under unfavourable photoperiods. For some plants, e.g. *Xanthium*, *Lolium*, *Pharbitis*, only a single day of appropriate length is necessary although prolonging the sequence of favourable photoperiods will hasten flowering and increase the number of flowers produced.

Interruption of the dark period by a short period (minutes) of low intensity light will prevent flowering of short day plants or promote flowering of long day plants (long day plants will flower in continuous light). The light which is effective in this light break effect is red light (peak activity at 665 mμ) and its effect can be reversed by an immediately following exposure to far red light (peak activity at 725 mμ). These observations led to the discovery and subsequent isolation of a photoreceptor pigment termed *phytochrome*, now known also to be implicated in other developmental responses to light. This pigment exists in two forms (designated P_r and P_{fr}) which are interconvertible by absorption of the appropriate radiation

$$P_r \underset{725\ \text{m}\mu}{\overset{665\ \text{m}\mu}{\rightleftharpoons}} P_{fr} \qquad (95)$$

Further in the cells there occurs a dark reversion of P_{fr} to P_r.

From these observations it can be postulated that in short day plants the dark period must be long enough for there to be little or no P_{fr} for a critical length of time; for long day plants that they must not be exposed for too long to a lack of P_{fr} (dominance of P_r). This concept implies that during darkness a timing reaction starts from the point when P_r has become the dominant form of phytochrome in the cells. Unfortunately we do not know how phytochrome functions biochemically.

Although there is some experimental evidence that phytochrome controls a timing reaction, zero time being counted from the establishment of a dominance of P_r, there are other observations which indicate that this is not the whole story. Many of the additional observations can be accommodated by adding the further hypothesis that the responses to light and darkness of photoperiodic plants are influenced by an endogenous circadian rhythm in which a phase of responsiveness to light alternates with a phase of responsiveness to darkness. Thus when certain short day plants have been placed in a cycle in excess of 24 hr and hence submitted to a very long night, the effectiveness of light breaks in suppressing flowering has been found to vary rhythmically during the prolonged period of darkness.

FURTHER READING

G. N. Cohen. *The Regulation of Cell Metabolism.* Holt, Rinehart & Winston, 1969.

M. Dixon. *Multi-enzyme Systems.* Cambridge University Press, 1951.

H. E. Umbarger. Intracellular regulatory mechanisms. *Science,* **145:** 674–679, 1964.

E. O'F. Walsh. *Introduction to Biochemistry.* English University Press Limited, 1968.

MORE ADVANCED READING

Cold Spring Harbor Symposia on Quantitative Biology. *Cellular Regulatory Mechanisms,* vol. 26. The Biological Laboratory, Cold Spring Harbor, New York, 1961.

H. L. Kornberg. The co-ordination of metabolic routes. *Symp. Soc. gen. Microbiol.* **15:** 8–31, 1965.

E. R. Stadtman. Allosteric regulation of enzyme activity. *Advances in Enzymology,* **28:** 41, 1966.

M. B. Wilkins (Ed.). *The Physiology of Plant Growth and Development.* McGraw-Hill, London, 1969.

G. E. W. Wolstenholme and C. M. O'Connor (Editors). *The Regulation of Cell Metabolism.* CIBA Foundation Symposium, Churchill, London, 1959.

CHAPTER 9

Growth and Differentiation

"*I am a firm believer that without speculation there is no good or original observation.*"

Charles Darwin in a letter to Alfred Russell Wallace.

"*Morphogenesis, the science that treats of the cause and origin of form, is a field where many different disciplines meet. Morphology is obviously concerned in it, and physiology, embryology and genetics. Biochemistry and biophysics have an important place here as well. There is no field in biology which touches so many problems. This, I believe, is because we are dealing, in morphogenesis, with the basic phenomenon in biology—how protoplasm builds organized living systems.*" Edmund W. Sinnott, in *Growth and Differentiation in Plants.* Iowa State College Press, Iowa, 1953.

INTRODUCTION

It would be going beyond the scope of the present text to develop any general consideration of plant growth and development. Further, many of the phenomena of growth and development are not, in the present state of knowledge, interpretable in metabolic terms. One of the immediate and major tasks facing plant physiologists is the extension of our knowledge of the regulation of metabolism and, particularly, of energy-flow in the cell and in the multicellular organism so as to be able to interpret in these terms the phenomena of cell growth and differentiation and organ growth and development. Within the limits of this chapter little more can be attempted than to indicate in a very general way the kind of problems which immediately arise when one begins to probe in this direction.

MERISTEMS AND CELL DIVISION

The growth of plant organs is the outcome of cell division, enlargement of the new cells and their differentiation into the different kinds of tissue cells. These processes of cell division and growth are localised in *meristems*. The extreme apices of roots are occupied by primary meristems and in the older parts of the root system secondary meristems (cambia) give rise to additional vascular tissues and to protective layers of cork cells. Growth and morphology are the outcome of the activities of these meristems, and further the activity of each meristem influences the activity of other meristems (particularly those near to it) giving rise to *growth correlations*. For instance, while the main root apical meristem is active it retards the activity of the more recently initiated lateral root tip meristems, a phenomenon usually referred to as *apical dominance*.

By preparing longitudinal sections of the root it can be seen that its extreme tip is occupied by a cap of cells which protect the permanently dividing tissue or *promeristem* from which the root cap cells and all the new cells which give rise to the primary tissues of the growing root have their ultimate origin. Some of the tissue cells arise directly from new cells initiated in the promeristem, others after these cells have undergone a limited number of further divisions. In some species it seems that the root promeristem is a group of dividing cells lying on the surface of a hemisphere. The cells on the proximal surface of the hemisphere (that surface furthest from the extreme tip) give rise to the cells from which are formed the stele, the cortex and the piliferous layer and sometimes also the outer cells of the root cap. The distal (apical) flat face of the hemisphere is covered by a plate of promeristem cells which give rise to the central cells of the root cap. The hemisphere itself is composed of a group of cells (500–1000 cells) which rarely divide and these together are usually therefore referred to as the *quiescent centre* of the meristem (Fig. 9.1). During root development there may be quite large changes in the number of dividing cells and in the number of quiescent cells in the promeristem: thus the quiescent centre is absent or represented by a very few cells in young lateral roots and embryonic roots.

Despite changes in promeristem size and in the cell division activity of its constituent cells, the whole apical meristem of the root is always an organised structure with a pattern in the arrangement of the cells and with cell division and cell enlargement proceeding in such a way that this pattern is not destroyed. The maintenance of pattern is the outcome of a control of the planes along which division walls are laid down in cell division and of a balance between rates of

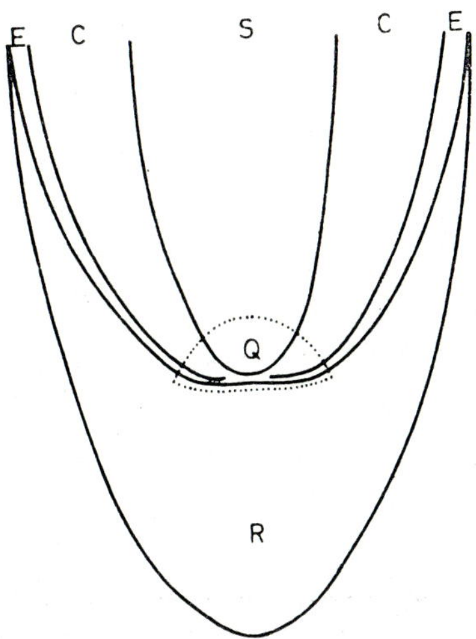

FIG. 9.1. Diagrammatic median section of the root apex of *Zea mays* (maize). The sites of the initial cells of the promeristem are indicated by dots. E, piliferous layer; C, cortex; S, stele; Q, quiescent centre; R, root cap. (Drawn by Dr. L. Clowes, from H. E. Street, as Fig. 7.3.)

cell division on the one hand and the rates and directions of cell expansion on the other. The orientation of division walls seems to be determined by the shape of the spindle arising during the nuclear division (mitosis) which precedes cell division (cytokinesis). Meristem size is either stable or changing with time according to the balance established between division rate and the rate at which the daughter cells mature to the point where they differentiate directly into the root tissue cells.

The discovery of *mitosis* around 1880, mainly as the result of the elegant studies of the botanist Eduard Strasburger and the zoologist Walther Flemming, was a landmark in the history of biology, and combined with the discovery some fifteen years later of meiosis (reduction division) laid the basis from which modern knowledge of cytogenetics was to develop. The recognition that multicellular plants have localised growth centres or meristems where cells continue to divide and that these are distinct from regions where division ceases and cell behaviour is to be described in terms of expansion and differentiation leads us on to other very important aspects of nuclear behaviour and function. It poses such problems as what changes are involved in the preparation of nuclei to undergo mitosis and what naturally "triggers" this process to commence or inhibits it from occurring. To answer such questions it is necessary to consider what we know of the metabolism of the nucleus both during interphase and during mitosis.

The reduplication of the chromosomes, including reduplication of their genetic material (their DNA), takes place during *interphase* (in the "resting" nucleus) so that when the nucleus enters upon prophase each chromosome normally contains twice the material it contained at telophase. This also implies that synthesis of new chromosome material does not take place during mitosis. The chromosomes of the interphase nucleus are in the "extended" condition (in contrast to the highly spiralised "condensed" form of the chromosomes during mitosis) and it is while in this state that both their DNA and other constituents including the characteristic basic protein of the chromosomes (the histone) are doubled (as far as we can judge from experimental studies *exactly doubled*). It is at this stage also that the chromosomes exert their controlling influence on cell metabolism. The description "resting nucleus" for the interphase nucleus therefore only has meaning in that it indicates that the nucleus is not involved in division.

The mechanism of the *replication of the DNA molecules* of the chromosomes is at present entirely a matter for speculation. It has been suggested that this could involve an unwinding of the double helix (Fig. 9.2) and the synthesis from the appropriate precursors and under the influence of the appropriate enzyme(s) of a complementary spiral to each exposed single spiral to give finally two complete double helices. This duplication of chromosomal material, as indicated by the doubling of the DNA content of the nucleus, must take place if division is to occur. A failure of this aspect of nuclear metabolism can be shown to occur in some cells which fail to divide. However, doubling of the chromosomal material when completed does not, of itself, cause the nucleus to enter the prophase of mitosis. DNA replication is completed before prophase and there are many

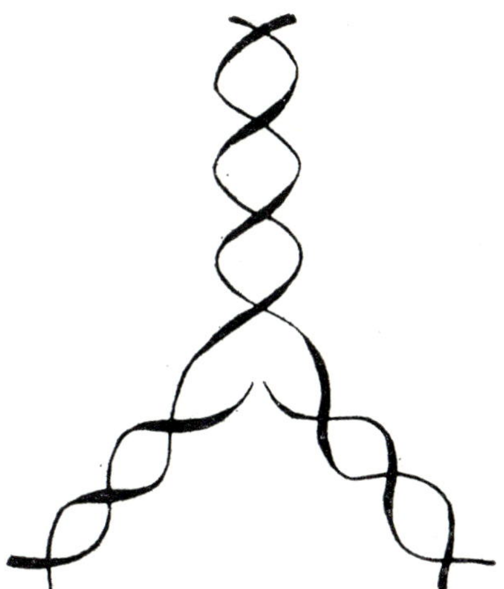

FIG. 9.2. Diagram of the replication of the DNA molecule.

instances known where DNA replication goes on to give $4\times$ or even a higher multiple of the telophase content without prophase being initiated. For instance, by appropriate illumination and nutrition, cells of the alga, *Chlorella*, can be caused to grow large and their nuclei accumulate abnormally large amounts of DNA. Then on transfer to darkness these cells divide rapidly to give 16 smaller cells with the normal "telophase" DNA content. It is now recognised that many of the differentiated tissue cells of higher plants are polyploid although the cells in their meristems are uniformly diploid. In such polyploid cells DNA replication has continued without mitosis intervening to restore the normal DNA content. The substance *kinetin* (6-furfurylaminopurine), discovered as a decomposition product of nucleic acid by Professor F. Skoog of the University of Wisconsin and subsequently shown to stimulate the division of certain plant cells, can, in certain cases, induce divisions in such polyploid plant tissue cells.

There is evidence that the centres (poles) from which the spindle develops are cytoplasmic structures and that they divide in the cytoplasm either before mitosis is complete or very early in interphase. Further, it is during interphase that the proteins which will go to form the spindle and its fibres are synthesised. This must mean that a very considerable part of protein synthesis in meristematic cells must be concerned with the *synthesis of the spindle proteins.*

One very interesting aspect of the metabolism of dividing cells is their very low rate of respiration during mitosis and the insensitivity of mitosis, once initiated, to inhibition by inhibitors of respiration. This means that the massive intracellular movements involved in mitosis can and normally do proceed without the cell simultaneously generating energy by respiration. This has led to the view that an "*energy reservoir*" is filled during interphase, as a result of respiration in the cytoplasm, and that this when filled is capable of powering the process of mitosis. The chemical nature of this energy reservoir is still uncertain. It does not seem to be the building up of a sufficient store of ATP although nuclei contain an ATP-ase (an enzyme which can split the terminal phosphate bond of ATP) and, therefore, ATP could be the immediate donor of energy for mitosis. Two hypotheses have been put forward as to the nature of the reservoir, one that it is some unknown high-energy compound (possibly a compound with high energy S-bonds such as occur in co-enzyme A) whose breakdown could be linked to the synthesis of ATP from ADP; the other that the energy required is locked up in the protein molecules of the chromosomes and the spindle and which, therefore, have the ability to combine and contract when the signal is given (when the "trigger" is pulled to initiate mitosis).

Clearly, the changes during interphase which prepare the cell for mitosis involve the metabolic activities of the cytoplasm. If the interphase nucleus is the control tower for the metabolism of the cytoplasm, the latter is in turn the source of energy and of precursor molecules for the essential metabolic events which must proceed both within the nucleus and outside it prior to division. The further elucidation of this two-way inter-relationship between the nucleus and the cytoplasm is a major objective of current research in cell physiology.

During the prophase the two most prominent events are the coiling and associated condensation of the chromosomes so that they become cytologically defined and the disappearance from the *nucleolus* (or nucleoli) of the RNA which is a prominent constituent of this structure during interphase. Often the nucleoli completely disappear as they become separated from the "nucleolar organiser" regions of the chromosomes during prophase condensation of the chromosomes. Then towards the end of prophase the nuclear envelope breaks down.

Certain changes which occur in the cytoplasm at this time may be related to the loss of nucleoli (particularly if these are important in the flow of RNA from nucleus to cytoplasm) and to the breakdown of the nuclear envelope. The nuclear envelope is, perhaps, the most constant feature of the membrane system which we call the endoplasmic reticulum (ER) (see Chapter 2) and associated with the breakdown of the nuclear envelope there is usually some disintegration of the ER which may become reduced to a discontinuous system of vesicles confined to the periphery of the cytoplasm. These changes can be correlated with independent evidence that the synthetic activities of the cytoplasm are at their lowest ebb during mitosis. It has also been observed that there is a marked decrease in cytoplasmic viscosity, particularly at the time of the differentiation of the spindle, and the build-up of —SH compounds in the cytoplasm which occurs during interphase is reversed during mitosis and again particularly during organisation of the spindle.

The *mitochondria* do not become enmeshed in the spindle but usually either aggregate around the poles or along the surface of the spindle and are fairly accurately divided between the daughter cells. It seems that in many plant cells there is a significant *decrease* in the number of mitochondria prior to cytokinesis (division of the cell into two daughter cells).

The disorganisation of cytoplasmic structure and function referred to above in describing mitosis is probably even more marked at the meiosis (reduction division) which leads to the origin of haploid pollen grains and ova. This would ensure the primacy of the nucleus in heredity. In those cases where there is evidence of maternal inheritance (cytoplasmic inheritance) this may reflect the incompleteness of cytoplasmic simplification so that some replicating cytoplasmic particles or *plasmagenes* (the identity of these with recognised cytoplasmic structures is uncertain) are carried over and function in the fertilised egg with the messages received from the diploid nuclei of the female parent.

At the conclusion of mitosis (*telophase*) the nuclear membrane is reconstituted from the persistent vesicles of the ER and this seems to be catalysed in some way by the extension of the chromosomes during this period. If any chromosomes become detached from the main group they also become associated with ER vesicles and can form micronuclei. Again, as the chromosomes extend nucleoli become apparent at the "nucleolar organizer" sites of the chromosomes apparently by the aggregation there of newly synthesised nucleolar substance.

The cell as it exists at telophase and before cytokinesis has double the functional potentialities of the parent cell and is about to reconstruct in two separate cells a new cytoplasmic organisation. There has been a doubling of physiological potential (*a physiological reproduction*). In this connection, Professor D. Mazia of the University of California has pointed to the probable importance of the duplication of the nucleoli. It is the nucleolus which during mitosis has gone through a cycle from "*oneness*" to "*nothingness*" to "*twoness*" and it is with the reappearance of the nucleoli, and in time very closely correlated with this, that there is a reactivation of cellular metabolism and the initiation of a reconstruction of the ER. Probably the nucleolus is an active "middleman" between the genes and their expression in the cytoplasm.

The beginning of *cytokinesis* is nicely adjusted to telophase and the plane of division is normally exactly along the equator of the spindle. If the spindle is not centrally located, unequal daughter cells are produced as has been clearly shown in studies of the growth of

Spirogyra filaments. The first evidence that cytokinesis is to occur is the appearance of a zone (the phragmoplast) across the equator of the spindle. This zone is rich in RNA and contains ER structures which have invaded the spindle and clustered in its equator. It is thought that it is from these ER structures that the plasmalemmae of the daughter cells are formed and that it is between these membranes that there is secreted the *cell plate* which becomes the middle lamella. At this stage the cytoplasm is clearly synthesising pectins and very quickly cellulose begins to be deposited by the daughter cells on either side of the cell plate to give two daughter cells, each with its own cellulose wall.

It not only accords with observation but would appear to be essential to the very persistence of the cells that normally cell growth must intervene between successive mitoses. Further, the evidence strongly supports the view that a nucleus is only capable of controlling a certain mass of cytoplasm and that when by growth this mass is attained then physiological reproduction becomes essential. This may also be expressed by saying that there is a critical *nuclear/cytoplasmic ratio* which is not exceeded in cells which continue to divide. It is in line with this concept that the dividing cells of polyploids are often larger than those of the normal diploids. Attention may also, in this connection, be directed to the fact that cell expansion is a finite process both in extent and time even in cells which enlarge beyond the point where they can divide; also that tissue elements (such as vessel units, tracheids, fibres) which are developed by very marked and predominantly unidirectional cell expansion are dead units when fully differentiated (their growth leads to a breakdown of cellular organisation).

There is considerable evidence that the normal development of cells is for them to undergo *limited* growth and then divide and that cells which do not continue this life history are cells in which mitosis has become *blocked* in some way. This hypothesis is supported by the ability of many such non-dividing cells to resume mitosis if appropriately stimulated. Such stimulation is illustrated by the response of certain tissues to mechanical wounding or insect attack (some of the uninjured cells dividing to heal the wound or, in the case of insect attack, to form a gall) and by the plant tumours which result from the

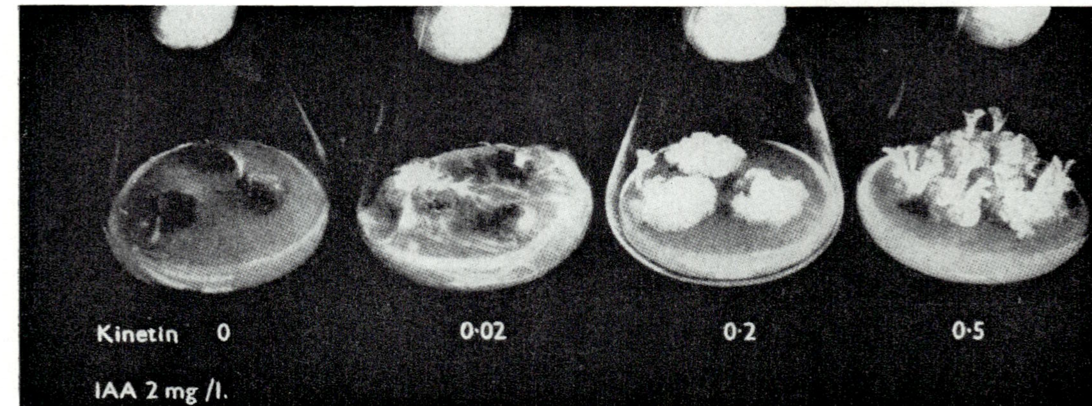

Fig. 9.3. The effect of kinetin (0–0·5 mg/l.) in the control of growth and organ initiation in tobacco callus cultured on a nutrient agar containing 2·0 mg/l. IAA. Extensive root formation at low kinetin concentration (0–0·02 mg/l.). Active callus growth at 0·2 mg/l. Shoot formation at 0·5 mg/l. (From F. Skoog and C. O. Miller, *The Biological Action of Growth Substances*, p. 118 *et seq*. Symposia of the Society for Experimental Biology, XI, Cambridge University Press, 1957.)

invasion of plant tissues by the Crown Gall bacterium. Explanted tissues can often also be induced to divide by transference to an appropriate nutrient medium supplemented with essential *growth factors*. One source of such growth factors which has proved particularly valuable in the initiation and maintenance of actively growing tissue cultures from such explants is the liquid endosperm of the coconut (*coconut milk*) whose normal metabolic function is to support the growth of the immature coconut embryo. A classical example of the chemical stimulation of cell division in mature expanded cells is the induction of an actively growing tissue culture from an explant of tobacco stem pith cells by the use of a culture medium supplying not only sugar and essential mineral ions but supplemented with appropriate concentrations of both kinetin (p. 268) and the plant growth hormone, indol-3yl-acetic acid (IAA) (p. 287). Further, in the tissue mass that develops there can develop organised meristems giving rise to roots and shoots (Fig. 9.3). Examples such as this show that differentiation is a reversible process in that starting from mature tissue cells we can, by appropriate stimulation, obtain again cells whose structure and behaviour are identical with the meristematic cells from which they were originally differentiated. Our consideration of the metabolism of mitosis has led us on to the problem of the physiological inter-relationship between the nucleus and the cytoplasm which is determinative not only in mitosis but in cell growth and differentiation.

CELL GROWTH AND DIFFERENTIATION

Behind the promeristem of the root tip is the region where cell enlargement and differentiation are the dominant processes. For instance, in the seedling maize root very few cell divisions are observed beyond 1·5 mm from the root cap and it is in the 2nd and 3rd mm from the root cap that the most active linear extension by cell expansion takes place. By 5 mm from the root cap most of the roots cells are fully expanded and already their differentiation into the root tissues is well advanced. As earlier described the meristematic cells are isodiametric, small, have a high nucleus to cell volume ratio and are filled with cytoplasm. During cell expansion, not only is the cell volume increased dramatically but the cells change in shape and develop vacuoles. The mature cortical cells of the root may have 20 or more times the volume of the promeristem cells. The process of cell enlargement is associated with the uptake of water leading first to the appearance of a number of small vacuoles and ultimately of a single central vacuole. This uptake of water, which is the consequence rather than the cause of the cell expansion, is accompanied by large increases in cell dry weight, protein content and cell wall material. The expansion process involves real growth and some degree of differentiation (change towards specialised function). The time course of this growth follows a characteristic pattern (Fig. 9.4); at first

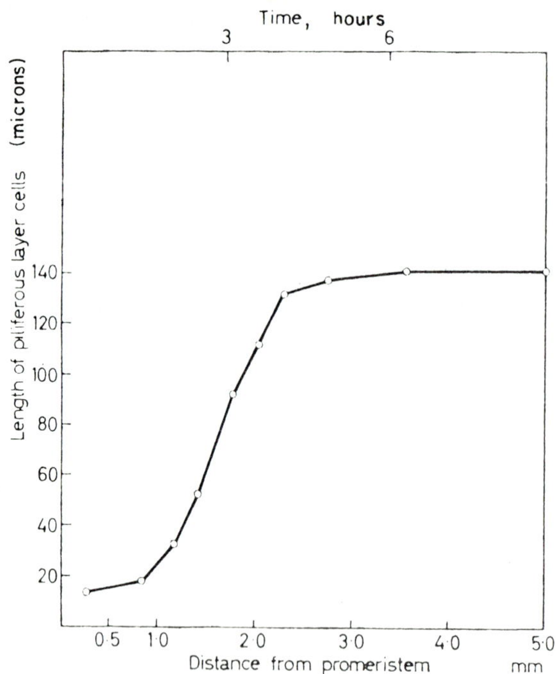

FIG. 9.4. Curve showing the course of elongation of the piliferous layer cells of excised tomato roots growing in a culture medium containing 1·5% sucrose. (From H. E. Street, as Fig. 7.3.)

growth is slow, then it accelerates to reach and maintain a high velocity until again as the growth process reaches its completion, the rate quickly declines to zero. Studies on root hair development emphasise rather nicely the principle already enunciated that cells have a limited expansion potential. The longest cells of the piliferous layer are either without hairs or produce the shortest hairs, while the shortest piliferous layer cells produce the longest hairs. Each piliferous layer cell appears to have a certain capacity for growth which may be expressed in either a longitudinal or horizontal direction.

The cells as they expand and differentiate become part of the continuing pattern of arrangement of the primary root tissues. It could be that the primary tissues impose their pattern upon the differentiating cells by some apically directed influence. When, however, the embryo root is initiated or when root meristems arise in tissue culture masses it is clear that the root meristem precedes the existence of mature root tissues. Further, it has been shown that if root tips are cut off, re-orientated and then replaced on the root stump, the new tissues as they appear are out of line with those of the stump. It is in the apical millimetre or so of the root, in the region of active cell division, that the tissue pattern is determined. The destiny of cells in the meristem is determined by their micro-environments in that region.

The cells as they differentiate into the functional tissue cells, diverge from the cells of the meristem and from one another in size, shape, structure and physiology. Cells initially "identical" in genetics and physiology embark upon separate pathways of differentiation. Various workers, using thin serial sections of seedling roots, have revealed a changing pattern of enzyme activity during the process of cell expansion. Not only are there significant changes in the relative activities of certain enzymes per cell but also per unit of cellular protein (Fig. 9.5). Further, it has been shown that although during the process of expansion the synthesis of new protein and of cytoplasmic RNA run parallel, the composition of the RNA changes. The proportion of purine to pyrimidine bases in the nucleic acid rises as the cells mature. There are, also, changes in the respiratory activity both per cell and per unit of protein and changes in the sensitivity of the respiration to inhibitors.

These observations on the changing enzyme activities of cells during differentiation prompted the distinguished American bacteriologist Sol Spiegelmann to characterise differentiation as the controlled production of unique enzymatic patterns. This implies that studies on the induction and suppression of enzyme synthesis (see Chapter 8) could hold the key to understanding differentiation. They cannot, however, at this stage in our knowledge go very far to explain

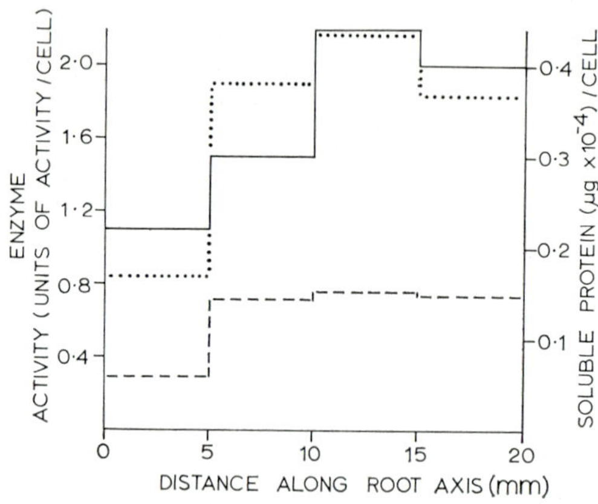

FIG. 9.5. Changes in enzyme activity (phosphofructokinase ——, malic enzyme ----) and soluble protein content (····) during the growth and differentiation of cells of seedling pea roots. (Data from M. W. Fowler, Carbohydrate oxidation and biosynthesis in pea roots, Ph.D. Thesis, University of Cambridge, 1969.)

differentiation in higher plant cells. We do not know whether such systems are involved in the control of the synthesis of *all* enzymes, we have not elucidated the *mechanisms* of induction and repression and we have no experimental evidence that these phenomena are the cause of the "switch-overs" which occur during differentiation from one pattern of enzymes to another equally functional, and perhaps more stabilised pattern, associated with a different metabolic status. This highlights a major gap in our understanding of differentiation. Differences in the "environment" at different points in a meristem such as differences in carbon dioxide and oxygen tensions might well be expected to alter the relative concentrations of metabolites and this, in turn, could modify enzyme synthesis and the activities of enzyme molecules. What we cannot do is to describe how such changes can lead to differentiation in the right sense and to the right extent and over the requisite cell masses to give the functional tissue pattern of a plant organ. Further, and still accepting the reversibility of differentiation, we have also to explain the quite considerable stability of the characteristic metabolic patterns of the different tissues of which a multicellular plant is built up and which can persist through successive cell divisions.

Initially all the cells of a multicellular organism may be presumed to be genetically identical, being derived from the mitosis of the fusion nucleus of the ovum. Tissue and organ differentiation also proceed with a uniformity and precision which seems to rule out the involvement of genetic change (mutation) in differentiation. The ability of mature tissue explants to regenerate root and shoot meristems of normal size and function also argues for the genetic stability of the body cells. It is, however, known that two kinds of nuclear change can occur during cell differentiation. The first of these is the occurrence of *polyploidy*. While meristems and potentially meristematic tissue (for instance, the pericycle of the root from which lateral root meristems arise) remain uniformly diploid, nevertheless, polyploidy is a widespread phenomenon amongst the cells of specialised tissues (root cortical and endodermal cells are frequently tetraploid, vessel units of the metaxylem often tetra- or octaploid). When explants, involving differentiated cells, are used to initiate tissue cultures and these, in turn, to initiate organised meristems then it is a general finding that despite the high degree of polyploidy in the explant, nevertheless, the meristems are built up of diploid cells. These observations suggest that the polyploidy which develops during differentiation is either a barrier to the division of such cells or modifies their behaviour so that they do not participate in meristem initiation. It may also be suggested that polyploidy has the biological significance, in large tissue cells, of extending the range of control of the nucleus. There is, however, *no* evidence that this tissue polyploidy is a cause rather than a consequence of differentiation or that it has any role in stabilising particular patterns of metabolic activity.

The second kind of nuclear change is usually designated *nuclear* (or *chromosomal*) *differentiation* and our knowledge of this comes mainly from three lines of work with animal cells. First we have nuclear transplantation experiments such as those brilliantly executed by Drs. T. J. King and R. Briggs at the Lankenau Hospital, Philadelphia. These workers transplanted nuclei from frog embryos at various stages of differentiation into enucleated frog eggs and found that nuclei from the early stages of development permitted normal animals to develop. However, nuclei transplanted from later in embryo development resulted in the development of abnormal embryos and these abnormalities were intensified as nuclei from older and older embryos were transplanted. Further, nuclei taken from the abnormal embryos continued to cause abnormal development when transferred to enucleated eggs. The conclusion was drawn from these findings that nuclei from differentiated embryo cells had altered capacities as compared with egg nuclei and that these differences were persistent despite repeated nuclear division, indicating strongly that the chromosomes themselves were "differentiated". Secondly, there is evidence of chromosomal changes during differentiation from detailed microscopic studies of the giant chromosomes of the larval tissues of certain insects. The banding pattern of these chromosomes is modified during development by the appearance of enlarged areas called *Balbiani rings*; the occurrence and distribution of these Balbiani rings is characteristic of the tissue and its stage of development. The Balbiani rings are often associated with fine loops which extend out laterally from the chromosomes and these loops can be shown to synthesise droplets of RNA which are probably released into the cytoplasm. The extent of these loops can change and they can form and regress, suggesting that the formation of the loops is related to the "activity" of the chromosome sites where they occur. Thirdly, we have the evidence that the tissues within certain animals have characteristic and different levels of RNA in their nuclei, and that the nuclear RNA content changes as the tissues differentiate.

These observations strongly suggest that the total gene complement may rarely, if ever, be fully and simultaneously committed in the control of metabolism; that the metabolism of the cell is controlled by those particular genes which are "activated" in its nucleus. One can then speculate that the interphase chromosomes may be at certain points along their length fully extended and actively involved in the transmission of messages to the cytoplasm and that other regions which are condensed or coiled may be inactive. Nuclear differentiation could then be visualised to involve relatively stable changes in the configuration and chemical composition of the chromosome threads corresponding to different gene activation patterns.

James Bonner and his colleagues at the California Institute of Technology have developed the view that inactive chromosome segments are ensheathed by the basic proteins (the histones) of the chromosomes and that the pattern of histone distribution determines the pattern of gene activation and repression. Their work is based on the isolation of plant nuclei and the preparation from such nuclei of *chromatin*, which like the chromosomes from which it comes can promote RNA synthesis, a property which is rapidly lost if its DNA is degraded with DNA-ase. To synthesise RNA, the chromatin must be provided with the riboside triphosphates of the four RNA bases (guanine, adenine, cytosine and uracil). The rate of this RNA synthesis is greatly enhanced by adding RNA-polymerase of bacterial origin. If now additionally we add 18 amino acids (one of which is ^{14}C labelled), ribosome and transfer-RNA fractions prepared from *E. coli*, activating ions and appropriate buffer, then the chromatin promotes DNA-dependent protein synthesis. The capacity for protein synthesis of the chromatin is increased, if it is freed from the histones with which it is associated; a greater proportion of the DNA is then involved in m-RNA synthesis.

The activity of chromatin in protein synthesis is characteristic of the cells from which it is isolated. Thus chromatin from pea cotyledon cells synthesises pea seed globulin (identifiable by the techniques of immunochemistry) whereas chromatin isolated from pea bud cells does not. However, deproteinised pea bud cell chromatin will synthesise pea seed globulin; in the chromatin of the pea bud cells the gene(s) controlling synthesis of the globulin is masked by histone—remove the histone and the gene(s) functions. An objection has been raised to the view that histones are selectively removed when groups of genes are activated (derepressed) and added when groups of genes are repressed, on the grounds that there appear to be only a very small number of different nuclear histones. However, if the basic histone is combined with RNA the specificity of histone binding to the chromosomes could be a property of the RNA and not of the histone component.

In considering the above hypothesis it should perhaps be borne in mind that there are a limited number of different kinds of specialised tissue cells in the plant body; an extremely small number compared with the vast number of active genes in any living cell. Each specialised cell may have its unique regiment of genes but many of the genes in this regiment may be common to all cells. Again the "essential" genes may be duplicated many times in the genome. The discovery of isoenzymes possibly points to such multiplication. In specialised cells the active regions of the chromosomes may include these essential enzymes together with the "specialised" genes. It is possible therefore that there is need only for as many histone masks as there are types of tissue cell.

PROGRAMMING OF DIFFERENTIATION

A sequence of metabolic changes precedes the establishment of the "steady state" metabolism of a mature tissue cell. A more complex programme of organised cell division, cell expansion and cell differentiation is involved in the initiation and development of a plant organ.

If the changes occurring during the growth and differentiation of cells reflect changes in their content of skeletal and enzymic proteins then programmes of differentiation are at base programmes of protein synthesis and turn-over and of enzyme activation and inhibition. The basis of any discussion of the biochemistry of differentiation is then our knowledge of the mechanisms of protein synthesis (Chapter 5) and of metabolic regulation (Chapter 8). This knowledge enables us to visualise the points where controls could operate.

The synthesis of proteins is dependent upon the *transcription* of the genetic information by DNA-dependent synthesis of the messenger, ribosomal and transfer RNAs involved. This proceeds in the chromosomes and plastids and probably also in nucleoli and mitochondria. For activity the genes must be "switched on" and accessible to and supplied with the precursor molecules needed for RNA synthesis. Controls in transcription could operate by exposing (histone removal?) or masking the genes (histone synthesis?), by repressing or derepressing the genes via the activity of regulator and operator genes (see below), and through the synthesis and transport to the genes of precursors. The synthesis of proteins is, however, also dependent upon the functioning of the RNA species to promote polypeptide synthesis (*translation*). Controls could therefore also operate via the assembly of ribosomal sub-units (function of the nucleolus?), presence or absence of factors blocking association of ribosomes with messenger RNA, transport of the RNAs to the sites of protein synthesis, availability of essential enzymes (amino-acid activating enzymes) and inorganic ions, availability of the 20 essential amino acids and availability of ATP. The functioning of the skeletal proteins and enzymes will in turn depend upon their protection from or exposure to degradative enzymes. The activity of the enzymes will also be subject to the "fine" control mechanisms discussed in Chapter 8 (p. 252).

The relative importance, in the initiation and progress of differentiation, of these many possible points at which control of metabolic activity could operate cannot at present be assessed. However, the postulation of "model systems" in which feasible types of control are employed to switch cells from one pattern of metabolism to another can lead to the planning of experimental work on the "molecular" aspects of differentiation.

The phenomenon of end-product inhibition of enzyme activity (Chapter 8, p. 253) where the final product of a reaction chain inhibits the activity of an earlier enzyme in the chain can be elaborated to achieve interaction between two metabolic pathways. Consider (Fig. 9.6) two independent metabolic pathways giving rise to metabolites a, b, c, d and α, β, γ, δ. Then it can be postulated that the enzymes (E_1 and E'_1) catalysing the first reaction in each pathway could be inhibited by the final product (d and δ) of the *other* pathway. This is a "cross feed-back" where one of the two pathways, provided it once has a head-start or a temporary metabolic advantage, will permanently inhibit the other. *Switching* of one pathway to the other could be accomplished by a variety of methods; for instance, by inhibiting temporarily any one of the enzymes of the active pathway. Here we have considered only two interacting pathways. Is it feasible to consider a "chain reaction" in which activation of one pathway (by enzyme activation or synthesis) would suppress a whole alternative pattern of metabolism? This could profoundly affect the

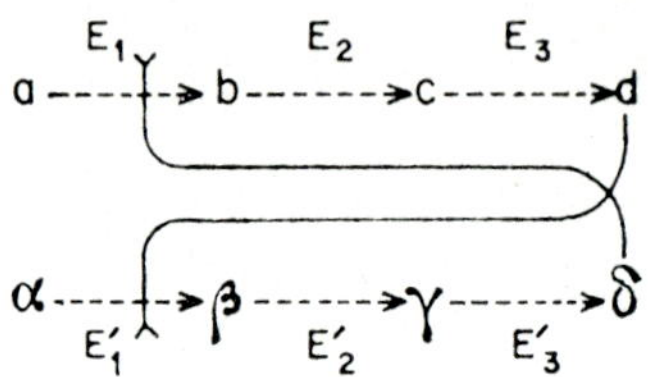

FIG. 9.6. Model I. The reactions along the two pathways $a \rightarrow b \rightarrow c \rightarrow d$ and $\alpha \rightarrow \beta \rightarrow \gamma \rightarrow \delta$ are catalysed by the enzymes E_1, E_2, E_3 and E'_1, E'_2, E'_3. Enzyme E_1 is inhibited by δ, the product of the other pathway. Conversely, enzyme E'_1 is inhibited by metabolite d, product of the first pathway. (From J. Monod and F. Jacob, in Cellular regulatory mechanisms, *Cold Spring Harbor Symposia in Quantitative Biology*, vol. 26, Cold Spring Harbor, New York, 1961.)

availability of precursors for RNA and protein synthesis and thus not only modify enzyme activity but the pattern of protein synthesis.

A second model (Fig. 9.7) can be based upon the concept of the role of "structural" and "regulator" genes in the induction and repression of enzyme synthesis (Chapter 8, p. 251) and involves two reaction sequences interconnected in that each produces an inducer *of the other*. The two systems, not necessarily otherwise closely related in metabolism (they could be involved in widely different metabolic systems), are mutually dependent. One system automatically induces the other by inactivating the appropriate repressor substance. If a metabolic pattern was the outcome of a number of reaction sequences interconnected in this way, a "point" activation of one system would lead to the growth of the mutually dependent systems. Similarly, if we consider the opposite case where a reaction product of one reaction sequence acts as an activator of the repressor substance then the systems become mutually inhibitory. The "live" system inhibits the second system and the only way the second system can come into operation is by eliminating temporarily a substrate of the "live" system, and once activated it represses the first system.

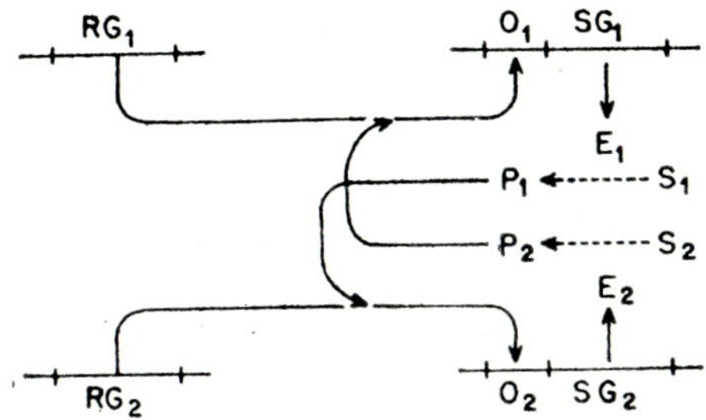

FIG. 9.7. Model II. Synthesis of the enzyme E_1 depends upon activity of the structural gene SG_1. The activity of this gene is suppressed by regulator gene RG_1 via a repressor molecule acting on the operator gene O_1. The synthesis of enzyme E_2 is similarly controlled by the genes, O_2, SG_2 and RG_2; P_1 product of action of E_1 on substrate (S_1) suppresses the functioning of RG_2 and hence its presence induces enzyme E_2. P_2 can similarly induce formation of the enzyme E_1. (From J. Monod and F. Jacob, as Fig. 9.6.)

Clearly models of this kind, although showing how a metabolic effect may be the first step in a consequential programme of metabolic change, do not identify the initiating stimuli nor indicate how the programme is monitored to reach a steady state.

The precision of the morphology of plant organs and of their tissue patterns has led to the recognition that cells become what they are because of where they are within the organism. Further the changes which cells undergo during differentiation, provided they do not result in cell death (as occurs in the differentiation of xylem conducting elements and fibres), are probably reversible. This has been convincingly demonstrated by inducing cultured somatic cells isolated from different plant organs to undergo segmentations recapitulating embryogenesis and giving rise thereby to new plants. What then are the critical factors in the environment of the cells of multicellular plants which control their growth and development ? Clearly amongst the extra-cellular factors the plant hormones are of special importance.

HORMONES AND REGULATION OF GROWTH AND DIFFERENTIATION

We have so far concerned ourselves with considering cellular mechanisms which may operate in the control of cell growth and differentiation. However, in multicellular plants many adjacent cells often follow the same course of growth and differentiation to give rise to tissues, and further, these tissues differentiate as part of the development of organs like the shoot and root. It is in this relationship of the cell to its tissue environment and the tissue to its organ environment that we must look for the factors which initiate and determine the growth and direction of differentiation of the individual cells. A cell at a certain site in an organ is not only in a particular environment in regard to light, temperature, water, radiation and such factors but it is part of a protoplasmic continuum (a symplast) and hence imports and exports nutrients and metabolic products from and into this system.

Within a photosynthetic unicellular alga function mechanisms which control the integrated activities of its nucleus and cytoplasm and within its cytoplasm of the chloroplasts, mitochondria, microsomes, ER membranes and hyaloplasm. Within the multicellular organism *extra-cellular controls* determine the distribution and activity of centres of organised cell division (meristems), of organised cell expansion and of the specialised tissues concerned with photosynthesis, the bio-synthesis of particular metabolic products, solute and water absorption, translocation and so on. Thus, superimposed upon the nutritive inter-relationships of the specialised structures *within* the cells we now have nutritive inter-relationships *between* cells fulfilling different and partial functions in the economy of the whole organism.

The multicellular organism presents us with a different kind of complexity to that presented by the algal unicell. The higher plant cell has a genetic equipment which differs in a very interesting way from that of the unicell. The fertilised ovum has the ability to develop into the whole multicellular autotrophic plant (it has, we say, *totipotency*). However, neither the ovum nor the body cells can give an autotrophic unicell. The meristematic cell of a flowering plant is a highly heterotrophic cell only capable of maintaining its meristematic activity in a very special environment. Removed from this special environment it dies, moving out of it within the framework of the organism it differentiates in an exclusive way. Totipotency is expressed in the potentiality of the cell to differentiate in many different ways, but any one pathway of differentiation is to the exclusion of an alternative pathway. The emergence of the multicellular organism involves the determination and balanced development of cell lines along these many and complementary pathways of differentiation.

It is very easy in a general way to demonstrate nutritive inter-relationships between the separate organs and tissues of a plant. This was done long ago by ringing experiments and by defoliation which interrupted the flow of nutrient ions and metabolites. More recently such inter-relationships have been studied by such techniques as the growth in culture of isolated root tips and leaf primordia. For instance, the growth in culture of excised root tips has demonstrated the dependence of the root upon the shoot, not only for a supply of carbohydrate but for certain vitamins, particularly for vitamins of the B group which are known to function as co-enzymes and prosthetic groups of essential enzymes. It is clear, however, that many root systems must have additional and, as yet, unknown nutrient requirements, for a supply of essential nutrient ions, sugar and B vitamins only permits the roots of a very limited number of species to grow in culture. There are many aspects of the nutritive inter-relationships between plant organs which have yet to be elucidated. This is even more so when we come to the tissue level. We are very far from understanding qualitatively, let alone quantitatively, the extent to which the different tissue cells are dependent upon metabolites received from other parts of the organism or of understanding the totality of the contributions they separately make to the organism as a whole.

Clearly, this interdependence of cells for metabolites means that tissues must interact during differentiation and one aspect of this interaction is that each line of differentiation has its own permissive nutrient and metabolite requirements. The tasks before plant physiologists in this field are not only to be able to describe in *quantitative* terms the input and output of metabolites from all the different tissue cells, not only to understand what determines the rates of movements of these substances between cells and tissues within the organism but also by culturing isolated cells, to find out how far these substances permit of *or* determine (programme) differentiation itself. It is the great interest of these questions which motivates the intensive work now proceeding in a number of botanical laboratories to obtain growing cultures of separated higher plant cells (free-cell cultures) and to induce single isolated plant cells to divide, grow and differentiate.

The word, metabolite, is as all-embracing as its mother word, metabolism. Some metabolites can clearly be described as essential in that if the cell has lost the capacity for their biosynthesis its continuing life depends upon a supply from other cells. Such metabolites, since they enter into the metabolism of all cells might for that very reason not be the direct determinants in differentiation, although differentiation may profoundly affect the cells' ability to carry out their synthesis. Again, because there is strong evidence that *patterns of differentiation are determined in meristems* rather than under the influence of specialised tissues (p. 275), it would seem to be in meristematic cells that the final steps take place in the synthesis of the *chemical determinants* of tissue differentiation. This, of course, is not to say that special precursor molecules are not synthesised outside of the limits of the meristems. For instance, leaves undoubtedly supply substances to the shoot meristems which are essential for the transformation of a vegetative into a flowering apex. Nor should this hypothesis be regarded as at variance with the strong experimental evidence that the maintenance of an active state of cell division is dependent not only upon essential metabolites but also upon determinant molecules synthesised in and then transported from mature tissue cells to the meristems.

Our knowledge of the role of the meristem in the chemical determination of growth and differentiation has its beginnings in the classical studies of tropisms and particularly of phototropism (the growth curvature of plant organs in response to, and determined in direction by, unilateral illumination). It was the experimental study of the phototropism of the coleoptiles of grass seedlings, initiated in 1881 by Charles Darwin, which led on to the discovery of the natural occurrence of the plant growth hormones now called *auxins*. This discovery is usually placed at 1928 when F. W. Went working in the Department of Botany of Utrecht University described how a chemical agent essential to the expansion growth of coleoptile cells could be collected from the coleoptile tips by diffusion into agar jelly. This work confirmed hypotheses which had gradually developed during the previous 15 years and which are exemplified by the following quotation from a research paper by A. Paál of the University of Budapest in 1919: "the tip is the seat of a growth-regulating centre. In it a substance (or mixture) is formed and internally secreted, and this substance, equally distributed on all sides, moves downwards through the living tissue. If the movement of this correlation carrier is disturbed on one side, a growth decrease on that side results, giving rise to curvature of the organ."

Subsequent work showed that the substance synthesised in the coleoptile tip and moving from there to initiate expansion in the cells below was indol-3yl-acetic acid (IAA) (Fig. 9.8), that the cells of the coleoptile could not embark upon cell expansion in the absence of an external supply of this substance and that it was effective in minute (catalytic) amount. IAA, by its activity at great dilution and by expressing a physiological effect at a site spacially removed from its site of synthesis clearly called to mind the animal hormones and hence was described as a plant growth hormone.

The new growth hormone clearly controlled the linear growth of the grass coleoptile and its tropic curvatures were the outcome of an unequal distribution of the hormone induced by various unilateral stimuli. Quickly, evidence accumulated that auxin was universally distributed in higher plants. Secreted by the apical meristems of both shoots and roots it controlled in these organs the expansion of the tissue cells. Further, its physiological activities were not confined to the control of cell expansion. It was implicated in apical dominance

AUXINS

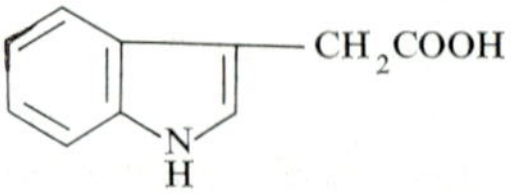

Indol-3yl-acetic acid (IAA); also indolylacetonitrile, indole aldehyde and other indole derivatives, some of which have still to be chemically characterised.

CYTOKININS

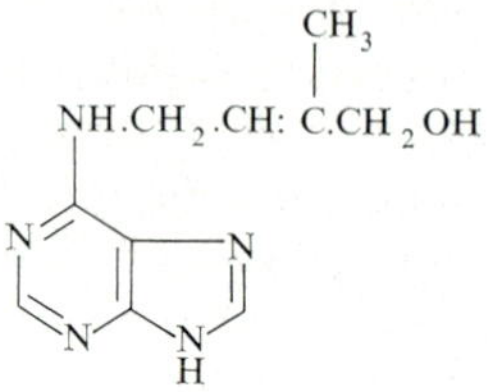

Zeatin [6-(4-hydroxy-3-methyl-2-butenyl) aminopurine] which occurs also as zeatin nucleoside and zeatin nucleotide; also other 6-substituted aminopurines and related compounds yet to be identified. The synthetic substance, kinetin, is 6-furfurylaminopurine.

GIBBERELLINS

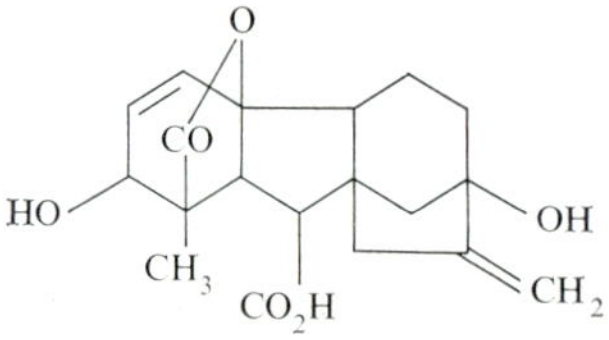

Gibberellic acid (gibberellin A_3), found in seeds of barley and *Echinocystis* and in *Festuca pratensis.*

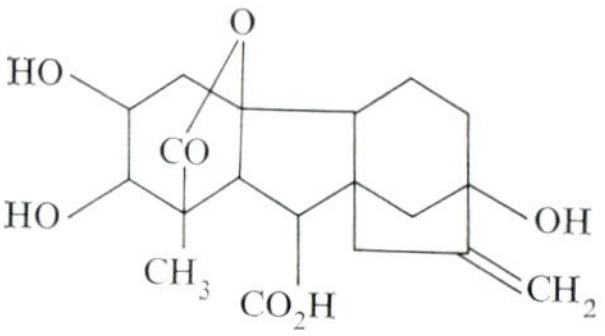

Gibberellin A_8, found in immature seeds of *Phaseolus multifloris* and *Phaseolus vulgaris.*

Gibberellins A_1, A_4, A_5, A_6, A_7, A_8, A_{17}, A_{19}, A_{20} also occur in higher plants, together with other unidentified gibberellins.

ABSCISINS

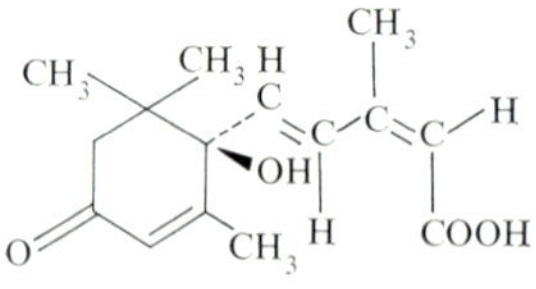

Abscisic acid (abscisin II, dormin), isolated from cotton fruits and sycamore leaves.

Fig. 9.8. Chemical structure and notes on the occurrence of some natural plant growth-regulating hormones.

(the inhibition of lateral bud and lateral root development by the active apical meristem), in the initiation of adventitious roots at the base of stem cuttings and probably also in lateral root initiation, in the seasonal activity of the cambium in trees, in the retention or falling of leaves and flower buds, in flower development and in the initiation and continuance of fruit development. Auxin could inhibit or promote not only cell expansion but cell division. The reaction it produced varied both with the concentration of the auxin and the sensitivity of the tissue and this sensitivity or responsiveness of tissues to auxin clearly depended upon their kind and stage of development and their position within the organism. More recent experimental work has emphasised the role of auxin in organ initiation and, particularly in work with tissue cultures, shown that it is concerned in tissue differentiation. For instance, vascularisation (the development of strands of xylem elements) in growing tissue culture masses seems to take place only along lines where a sufficiently steep gradient of auxin concentration becomes established.

For a long time IAA remained the only hormonal factor known to exert a controlling influence in plant growth and development although even during this period there was obtained evidence that some of its physiological activities were dependent upon the simultaneous presence of catalytic amounts of other growth factors, particularly of certain vitamins and purine bases. More recently, we have witnessed the isolation of certain metabolic products (*gibberellins*) from cultures of the fungus *Gibberella fujikuroi* (the causative agent of the *bakanae* or "foolish seedling" disease of rice) and the demonstration that these substances were responsible for inducing in diseased rice plants the excessive elongation growth. These substances, starting with gibberellic acid (gibberellin A_3) but later involving other higher plant gibberellins (Fig. 9.8), were shown to markedly enhance internode elongation in other species, being active, for instance, against dwarf but not against tall varieties of the garden pea. The interest in these gibberellins was markedly increased when evidence was obtained that such compounds are natural constituents of flowering plants and that they are probably essential to cell expansion. This permitted the hypothesis to be advanced that when gibberellins promote cell expansion, gibberellin is and auxin is not a "limiting factor" and, conversely, where auxin is essential for expansion then the cells are deficient in auxin but not in gibberellin. The gibberellins, like auxin, also have effects on cell division. For instance, there is evidence that gibberellin is involved in the initiation of cell division in the cambium and that the main effect of auxin in cambial activity may be to promote the differentiation of the products of cambial activity into xylem elements rather than to initiate the divisions. Following the discovery of the gibberellins came the description of the activity of kinetin as a cell division factor by Professor Skoog (see p. 268). Not only kinetin but a number of other related purines have this activity and Skoog has, therefore, coined the term *cytokinins* (from the older word, cytokinesis, used to describe the division of the cell into two daughter cells) for these compounds. One natural cytokinin, zeatin (Fig. 9.8), has been isolated and there is strong evidence for the occurrence of further natural cytokinins and that they are also substituted aminopurines. Skoog, using isolated plugs of tobacco pith showed not only that both kinetin and auxin were essential for the formation of a growing tissue culture but that the relative concentrations of these two substances had determinative effects on the cultures, particularly in regard to their capacity to initiate root and shoot meristems (Fig. 9.3). In other systems interactions between auxin, kinetin and gibberellic acid on growth by division and expansion and on cell differentiation have been exposed.

Is the list of interacting plant growth hormones complete? Almost certainly it is not. Only very recently *abscisic acid* (Fig. 9.8), which is involved in leaf and fruit abscission and in bud dormancy in woody plants, was discovered. There is considerable evidence for the involvement of a flowering hormone (*florigen*) in the induction of flower initiation and for the involvement of additional hormones in the functioning of vascular cambia.

Clearly, the plant hormones are candidates for consideration as determinants in growth and differentiation in higher plants. If we are to elucidate their role in the organism, what further information do we need and how may it be obtained? Well, first we need to know where and in what amounts they are synthesised, what concentrations of each are established in the cells whose growth and differentiation

is reflected in plant development, and how far, at different sites within the organism, can we correlate the relative and absolute concentrations of these substances with the patterns of growth and differentiation which occur. It must immediately be stressed here that there are very great technical difficulties in the way of obtaining this information. For the moment, therefore, it may well be that advances in this direction will come from the study of such problems in the apparently simpler systems represented by organ, tissue and free-cell cultures.

The problem of the chemical identification of the natural regulators of growth and differentiation also faces formidable technical obstacles. Firstly, those regulators so far detected are present in extremely low concentration in plant cells. This often precludes their recognition or estimation in plant extracts by chemical means. Hence their presence and their attempted purification have to be followed by sufficiently sensitive biological tests (bioassays). The specificity of these bioassays can, however, only be assessed against known compounds such as are available to the plant physiologist in pure form. One such bioassay for auxin activity depends upon the measurement of the longitudinal extension, in a suitable sugar–salt medium, of coleoptile segments (short segments cut from that region of the etiolated grass coleoptile which is capable of most marked elongation). These segments show only very limited extension in the basic medium, but addition of IAA (or of other "synthetic auxins") at appropriate concentration markedly enhances this extension. If such a test is used in a search for natural growth regulators it is those which similarly stimulate coleoptile segment extension which will be detected. It does not follow that the detected substances would resemble IAA in its other physiological properties and "fractions" of the plant extract which fail to promote or may inhibit the growth of the coleoptile segments may do so because they contain impurities masking the activity of any "auxins" they contain.

Consequent upon the development of paper partition chromatography (see Chapter 5, p. 145) and the demonstration that, with appropriate solvents, this technique can effectively separate known indole compounds, many workers have submitted plant extracts to this procedure in an attempt to separate out natural growth regulators.

However, when such chromatograms have been cut into sequential segments and the segments separately washed (eluted) with water or alcohol it has been shown, by bioassay of such eluates, that there are along the chromatograms a number of discrete regions showing growth-regulating activity, some at least of which give no chemical reactions indicative of their indole nature. The chemical constituents occurring in these active regions of the chromatograms have hardly ever been positively identified, although frequently the evidence has strongly indicated their lack of identity with known indoles and the plant gibberellins. Such studies have, therefore, supported the view that plant cells contain a number of unidentified substances which can give, apparently at very low concentration, positive responses in the bioassays used to detect and estimate auxins and gibberellins. The significance of these substances, however, just cannot be assessed until they are isolated in pure form to permit of their wider biological testing and determination of their chemical structure. Clarification of this apparently very complex situation urgently needs the extraction and fractionation of plant extracts on a much larger scale than hitherto usually attempted in biological laboratories.

THE MECHANISM OF ACTION OF PLANT HORMONES

If plant hormones are the critical determinants in growth and differentiation great interest attaches to their mechanism of action, to how they control metabolism. Do they exert a direct chemical action on cell structures (such as the cell wall or cytoplasmic membranes) altering their physical properties, do they promote or inhibit enzyme activities, do they intervene in the sequence of chemical events by which the genes control the synthesis of enzymes?

Very formidable technical obstacles face attempts to identify the primary actions of hormones in both plant and animal cells. These are in part due to the limited resolving power of existing biochemical techniques but also due to the special problems associated with obtaining satisfactory experimental systems. Diffusion patterns establishing complex concentration gradients of hormones are probably built up in developing plant tissues—these may be destroyed

in attempting to devise experimental systems. The hormones are active in very low concentrations and as emphasised in the discussion above interact with one another. A system is sought which is highly responsive to the added hormone, in which it can perhaps be assumed that the hormone is a limiting factor to growth and differentiation. However, in all such cases there is the problem that the action of the added hormone will be modified by its interaction with the endogenous and unknown levels of other hormones. Again there is the problem that each of the known plant hormones appears to have a multiplicity of effects; the nature of the response being determined by the "target" tissue. The action of gibberellin on the expanding cells of a stem internode is quite different to its effect on the aleurone layer of cereal grains. Does this diversity arise from a common "master" reaction or can hormones intervene at many different points in metabolism in unrelated ways? The fact that the nature of the tissue determines its responses to applied hormones has prompted the suggestion that plant hormones amplify (or "damp down") the activity of patterns of metabolism but do not play a primary role in determining pathways of differentiation (Wareing and Phillips, 1970).

Although it is now possible to obtain pure IAA and "synthetic" auxins, several pure cytokinins and rather more pure gibberellins, nevertheless, isotopically labelled hormones, particularly where only particular C or N atoms are labelled, are as yet obtainable only with great difficulty. Again the hormones which have been isolated are probably the forms in which they are transported within the plant and may not be their active forms. This being so, the many negative results of experiments on the effects of the hormones on isolated enzymes and on *in vitro* multi-enzyme synthesising systems (e.g. *in vitro* protein synthesising systems) may be of no significance. Recently it has been claimed that auxin is a positive allosteric effector (see Chapter 8, p. 254) of the enzyme *citrate synthase*, but this claim needs more critical biochemical examination and an assessment of the probability that such an effect could be of physiological significance *in vivo*.

The responses to hormones may occur very rapidly. Auxin can increase protoplasmic steaming in 2 min, an increase in the rate of cell expansion may be recorded in less than 15 min and an increase in the rate of respiration in less than 30 min. Changes which can only be detected after longer periods of time may merely reflect the enhanced growth and metabolism of the treated cells and be only distantly related to the primary action of the hormone. This must be borne very much in mind when reviewing the evidence that plant hormones increase RNA synthesis; the enhanced RNA synthesis may result from, rather than initiate, the enhanced growth or change in metabolic activities.

Action of Auxin

IAA has been available to plant physiologists for 40 years and during this period many experiments have been undertaken to try to identify its primary action on cells. As early as 1930 the German botanist A. N. J. Heyn showed that auxin enhanced the plastic (irreversible) extensibility of plant cell walls. He measured the bending of horizontal plasmolysed oat coleoptiles under weights and their recovery on removing the weights. The degree of recovery was a measure of elastic extension, the incompleteness of recovery a measure of the plastic extension. He then demonstrated that immediate pretreatment of the coleoptile with auxin increased particularly the plastic component of extension. The auxin "softened" the cell walls probably by destroying certain structural chemical bonds. One way in which this could happen would be by esterifying (methylating) carboxyl groups of the wall pectins and it is therefore particularly relevant that auxin has subsequently been claimed to promote the incorporation of methyl groups from the amino acid methionine into cell walls and to enhance the activity of esterase enzymes and particularly of *pectin methylesterase*. Some workers have contested these findings, and those who have favoured the view that auxin acts on cell walls by its effect on this enzyme have variously considered this to be due to activation of the enzyme, increased synthesis of the enzyme or to promotion of attachment of the enzyme to the cell wall. In rejecting the hypothesis that auxin acts on pectin methylesterase, it has been suggested that auxin affects the cell wall by its action on cytoplasmic membranes increasing the availability at the site of wall synthesis of the essential precursors (themselves synthesised within the cytoplasm).

The promotion of cell expansion is blocked by actinomycin D (which blocks DNA-dependent RNA synthesis), by puromycin (which prevents normal growth of the polypeptide chains at the ribosomes) and by 8-azaguanine (which becomes incorporated into m-RNA and renders it useless for directing the synthesis of the native proteins of the cell) (see Chapter 5, p. 175). Such observations, while indicating that active RNA metabolism and protein synthesis are essential to cell expansion, do *not* indicate that auxin is directly implicated in these processes. However, in studies on the enhanced *cellulase* activity of stem apices of bean and maize seedlings treated with IAA, evidence was obtained that only polyribosomes isolated from the IAA treated apices were capable of cellulase synthesis *in vitro*. Studies with internode segments of sugar-cane stalks have pointed to auxin as an inhibitor of their *β-fructofuranosidase* and *peroxidase* activity by preventing the synthesis or functioning of the m-RNAs involved in the renewed synthesis of these enzymes which normally follows excision of the segments. Auxin applied to cultured tobacco stem pith cells is reputed to inhibit the formation of two of the isoenzymes of peroxidase but to induce the formation of a third isoenzyme. Here auxin does seem to be intervening in enzyme synthesis but such observations face the perennial difficulty of assessing how indicative they are of the primary action of auxin, particularly as other and more rapid effects of the hormone have not been sought in the test plant material used.

Action of Gibberellins

The gibberellins are particularly interesting as plant hormones because the list of different higher plant gibberellins continues to grow. Of course, all these may be precursors or degradation products of a single active gibberellin, but alternatively they could each have a distinctive action and hence each could control the activity of a different gene or group of genes. Such specificity, if it could be demonstrated, would enable us more readily to conceive of the gibberellins playing a key role in the determination of pathways of growth and differentiation.

For our present purpose we shall confine ourselves to a consideration of a particular effect of gibberellic acid (GA), which has been rather thoroughly studied. This is the role of GA in inducing α-amylase activity in the aleurone layer of barley grains.

During the early stages of germination of the barley grain, a gibberellin is released by the embryo and transported to the aleurone layers of the endosperm, where it causes the appearance of *α-amylase* activity and enhancement of the activity of a number of other hydrolytic enzymes (proteases, ribonucleases, β-glucanases and pentosanases). The same result can be obtained by treating the moist isolated aleurone layers with GA within the concentration range 10^{-7} to 10^{-11} M. The appearance of α-amylase activity is due to *de novo* synthesis of the enzyme; this synthesis can be inhibited by inhibitors of RNA and protein synthesis. The α-amylase activity can be detected 7–8 hr after the GA application and GA must remain present if synthesis and release of the enzyme is to be maintained. However, it is only during the initial lag phase of 7–8 hr that actinomycin D prevents α-amylase synthesis and it can be suggested that it is during this lag period that the messenger-RNA coding for α-amylase is formed and that once formed, it is fairly stable. If this is so, the need for GA after this initial period is due to another effect of the hormone on enzyme synthesis (an effect other than on the DNA-dependent RNA synthesis). The activity of GA in increasing the activity of the other hydrolases released by the aleurone layers may be of this second kind since these enzymes are (unlike α-amylase) present initially so that GA may not be needed to activate the genes producing their m-RNAs.

Very recently it has been shown that although the barley embryo produces gibberellin, nevertheless its growth is enhanced by 10^{-4} M GA. When barley embryos are moistened they begin to synthesise protein within 30 min; by contrast, no new RNA is synthesised for some 12 hr, so that presumably masked m-RNA is present in the dry embryo and is activated during the initial imbibition of water by the embryo. The GA addition which enhances growth enhances the initial rate of protein synthesis in the embryo, but apparently *not* by promoting precocious RNA synthesis or the effectiveness of the ribosomes. This may be a similar phenomena to that of the GA enhancement of the synthesis and release of hydrolytic enzymes by the aleurone layers after the m-RNA for α-amylase has been synthesised.

One very interesting observation made during this study of the enhanced protein synthesis in barley embryos resulting from added GA was that the activity of an *in vitro* enzyme-synthesising system prepared from the embryos was not directly enhanced by adding GA but was more active if prepared from embryos pretreated with GA. A similar observation was made in a study of the RNA-synthesising activity of nuclei isolated from stem segments of dwarf pea plants. GA did not directly enhance RNA synthesis by the nuclei, but nuclei prepared from stem segments which had previously received GA were more active in RNA synthesis and produced RNA with a different base composition. Both these studies seem to show that the activity of GA is dependent upon some cofactor *in the cytoplasm*; that the active hormone could be a gibberellic acid–cofactor compound.

The insect hormone ecdysone, which promotes the moulting of insects, is considered to act by initiating the synthesis of specific messenger-RNAs at sites in the chromosomes which are observed to develop loops associated with Balbiani rings in response to the presence of the hormone (see p. 278). Ecdysone has low but significant activity in the dwarf pea bioassay for gibberellins; gibberellic acid has low but significant activity in the locust moulting assay. This would suggest that GA, like ecdysone, can act as a derepressor of genes.

Activity of Cytokinins

Active cytokinins are N^6-adenine derivatives, either the free base or riboside or ribonucleotide. In addition to zeatin (known also to occur as the riboside or ribonucleotide) there is evidence for the natural occurrence of N^6-methylaminopurine, 6-(Δ^2-isopentenyl amino)-purine (IPA) and other and as yet unidentified cytokinins.

A number of workers have reported rapid changes in the content of cellular RNA following cytokinin treatment; kinetin application can double the RNA level of onion root tip cells within 30 min. Associated with these changes, the cytokinins increase or alter the pattern of protein synthesis. The ability of kinetin to preserve protein in the detached leaves of a number of species (to arrest their senescence) is associated with enhanced RNA and protein synthesis in the treated leaves; it may also involve a kinetin activation of amino acid retention by and transport to the responding cells. There is some evidence, from the use of inhibitors such as actinomycin D, that cytokinins promote messenger-RNA synthesis and from electron microscopy that kinetin can preserve polyribosomes and maintain ribosome numbers in conditions otherwise conducive to their breakdown. As might be expected from such observations, kinetin has also been shown to enhance the synthesis of enzymes (for example the enzymes, *tyramine methylpherase* in barley roots) and to suppress the activity of other enzymes (for example the nucleic acid degrading enzymes, *ribonuclease* and *deoxyribonuclease*).

Recently a new consideration has arisen following the demonstration that transfer RNAs (for a brief description of the structure of t-RNAs see Chapter 5, p. 180, and Fig. 9.9) from yeasts, bacteria, mammals and higher plants when degraded by enzymes or acid hydrolysis yield active cytokinins. This was followed by the demonstration that t-RNAs contain unusual bases and that certain individual t-RNAs contain a single cytokinin riboside and that this occurs immediately adjacent to the anti-codon triplet of the RNA. Thus the t-RNAs coding for serine and tyrosine in yeast contain the riboside of IPA (isopentenyladenosine) and the t-RNA serine of a number of higher plants the *cis*-isomer of zeatin riboside. These findings led to the hypothesis that cytokinins acted by being incorporated into particular t-RNAs and that the cytokinin unit by virtue of its position in the molecule influenced the activity of the t-RNA in the interaction of its anticodon with the complementary codon of m-RNA. Further study has, however, shown that the isopentenyl side chain of IPA is synthesised in cultured tobacco pith tissue from mevalonic acid and that there is no direct incorporation of IPA riboside into the t-RNA being synthesised. This has led to the conclusion that there is no direct connection between the physiological activity of IPA and its presence in t-RNA although opening up the possibility that one pathway of IPA biosynthesis is via t-RNA synthesis followed by release of the free active compound by t-RNA degradation. Thus, despite the presence of cytokinins in t-RNAs, it seems that the growth-regulating activity of cytokinins is not to be explained by their controlling the synthesis of these molecules essential to protein synthesis.

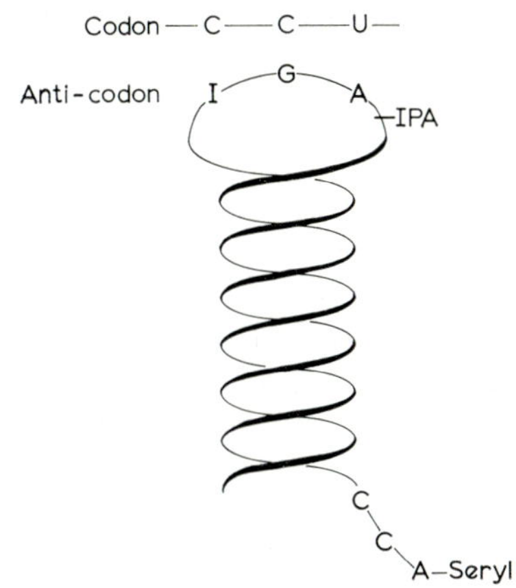

FIG. 9.9. Simplified possible structure of a t-RNA (serine specific) showing the nucleotide chain where the amino acid is attached, the anti-codon triplet and the position of the IPA riboside.
Key: A, adenine; C, cytosine; G, guanine; U, uracil; I, inosine; IPA, isopentenylaminopurine.

Looking over the experimental data relevant to the question of the primary modes of action of plant hormones, it must be admitted that these hormones cannot be identified with any certainty as the factors controlling the patterns of gene activation and repression presumably involved in programming growth and differentiation. There is still no convincing evidence that they intervene directly in either the transcription or translation of the genetic information, although the evidence in relation to the gibberellins probably points more strongly in this direction than in the case of the other known plant hormones. The present intensive studies of the nucleic acid metabolism of plants will clearly advance our knowledge of the metabolic events involved in growth and differentiation, even though they may exclude hypotheses which imply that the primary action of the plant hormones is to be found in this area of metabolism.

It may be that other factors such as gradients of O_2 and CO_2 tensions and of metabolites (such as sucrose and amino acids) and of vitamins and physical factors such as temperature, light and pressure (compressions and tensions) are of importance in programming the behaviour of cells and that the hormones activate or inhibit the programmes thereby preselected. However, at this stage in our knowledge the operation of as yet unknown determinants cannot be excluded. The discovery that the volatile compound ethylene is a product of many, if not all, cells and that it can, at extremely low concentration (fractions of a part per million), exert powerful effects upon growth should, perhaps, prompt a more intensive study of the chemical nature and physiological activity of the many volatile compounds which are obviously produced by plant tissues. Is it not possible to identify many plants from the smell of their vegetative organs?

FURTHER READING

L. J. Audus. *Plant Growth Substances.* Leonard Hill, London, 1959.

E. W. Sinnott. *Plant Morphogenesis.* McGraw-Hill, New York, 1960.

H. E. Street and H. Öpik. *The Physiology of Flowering Plants: Their Growth and Development.* Edward Arnold, Ltd., London, 1970.

P. F. Wareing and I. D. J. Phillips. *The Control of Growth and Differentiation in Plants.* Pergamon Press, Oxford, 1970.

MORE ADVANCED READING

F. A. L. Clowes. *Apical Meristems.* Botanical Monographs, vol. 2. Blackwell, Oxford, 1961.

D. Mazia. Mitosis and the physiology of cell division, in *The Cell*, vol. 3, pp. 77–412, edited by J. Brachet and A. E. Mirsky, 1961.

J. Monod and F. Jacob. General conclusions: Teleonomic mechanisms in cellular metabolism, growth and differentiation, in *Cold Spring Harbor Symposia on Quantitative Biology*, 26. *Cellular Regulating Mechanisms.* The Biological Laboratory, Cold Spring Harbor, New York, 1961.

G. E. Fogg (Ed.). *Cell Differentiation.* Symposia of the Society of Experimental Biology, No. 17. Cambridge University Press, 1963.

H. E. Street. Excised root culture. *Biol. Rev.*, **32:** 117–155, 1957.

H. E. Street. Special problems raised by organ and tissue culture. Correlation between organs of higher plants as a consequence of specific metabolic requirements. *Encyclopaedia of Plant Physiology*, **11**: 153–178. Edited by W. Ruhland, Springer-Verlag, Berlin, 1959.

H. E. Street. *The Repertoire of Plant Cells*. Leicester University Press, 1969.

D. N. Butcher and H. E. Street. Excised root culture. *Botanical Reviews*, **30**: 513–586, 1964.

R. M. Klein. *Plant Growth Regulation*. Iowa State University Press, Ames, Iowa, 1961.

A. W. Galston and W. K. Purves. The mechanism of action of auxin. *Annual Review of Plant Physiology*, **11**: 239–276, 1960.

C. O. Miller. Kinetin and related compounds in plant growth. *Annual Review of Plant Physiology*, **12**: 395–408, 1961.

M. B. Wilkins (Ed.). *Physiology of Plant Growth and Development*. McGraw-Hill, London, 1969.

F. C. Steward (Ed.). *Plant Physiology—a Treatise*, vol. 5B. Academic Press, New York, 1969.

J. Heslop-Harrison. Differentiation. *Annual Review of Plant Physiology*, **18**: 325, 1967.

J. Bonner. *The Molecular Biology of Development*. Clarendon Press, Oxford, 1965.

F. Wightman and G. Setterfield (Eds.). *Biochemistry and Physiology of Plant Growth Substances*. Runge Press, Ltd., Ottawa, 1968.

Index

Main references to subjects in bold numerals (f. indicates "and following pages"). References to text figures in italicised numerals. Authors' names in capitals and small capitals. Plant names in italics

AAA-enzymes 179, 281
α-helix *65*, **65**, 67
Abbreviations ix
Abscisic acid 257, *288*
Abscisins *288*
Absorption of solutes **212** f.
Absorption spectra 115, 119, *142*
Accumulation of ions 13, 213
  sites of, within cell **224**
Acetaldehyde 73, 85, 93, 97
Acetate 198
Acetoacetate 197, 198
Acetyl co-enzyme A 100, 105, 109, 113, 165, 187, 197, 202, 253
ACP *see* Acyl carrier protein
ACP-SH *see* Reduced acyl carrier protein
Actinomycin-D 295, 296, 298
Action spectrum 89, *139*
Activated state 138
Active acetate 98, 100
Active movement 9, 16, 213, 219, 245
Active sulphate 173
Active transport 9, 16, 34, 37, 213
Acyl carrier protein (ACP) 165
Adaptation 250
Adenine (A) 176, *177*, *182*, 279
Adenoid 37, 214
Adenosine diphosphate (ADP) 98, 106, 129, 230, 257
Adenosine diphosphate-glucose 162
Adenosine diphosphoglucose pyrophosphorylase 163, 164
Adenosine monophosphate (AMP) 121
Adenosine-5′-phosphosulphate (APS) 173
3′-adenosine-5′-phosphosulphate (PAPS) 173
Adenosine polyphosphate 70, 122
Adenosine triphosphate 14, *96*, 101, 106, 120, 128, 129, 148, 149, 152, 156, 173, 175, 197, 230, 243, 252, 254, 256, 257, 268
Adiantum pedatum 200
ADP *see* Adenosine diphosphate
ADP : O ratio 123
Aerobic respiration 85, 158
Aglycones 202
Alanine 114, 146, 170
β-alanine 199
Albizziine 199
Alcohol dehydrogenase 73, 80, *82*, 97
Aldolase 95
Aleurone grain 25, 296
Alkaloids 82, **189** f.

Allosteric 75, 254, 293
Amides 172
Amino acid activating enzymes *see* AAA-enzymes
Amino acid decarboxylases 69
Amino acids 8, 24, 49, **61** f.
activation of 179
non-protein **199** f.
4-amino-butanal 193, *194*
Amino-group 62, 67
Aminopurines 290
Amino-transferases *see* Transaminases
Ammonia 61, 114, **167**
Amphoteric carriers 227, **231**
α-amylase 7, 70, 107, 296
β-amylase 7, 107, 162, *163*
Amylopectin 7, 107, 163, *163*
Amyloplast 25, *36*, *45*, 46, 161
Amylose 7, 107, 162, *163*
Anabolism 10, 14, **131** f.
Anaerobic respiration 85
Anaplerotic reactions 101
Ancient symbiont theory 178
ANDERSON, D. B. 18
ANDERSON, J. M. 155
Anion respiration 228
Ankistrodesmus 143
Anthocyanidins 202
Anticodon 179, 298
Anti-metabolite 252
Aphids 236
Apical dominance 264, 287
Apigenin 202
Apparent photosynthesis 141
Apples 90, 102
Apposition 54
APS *see* Adenosine-5′-phosphosulphate
APS-kinase 173
Arabinose 160
Arecoline 191
Arginine 66, 67
ARNOLD, W. 134
ARNON, D. I. *152*, 153
Arum 117
Ascorbic acid oxidase 69, 74
Asparagine 8, 172, 201
Asparate carbamyltransferase 250, 254
Aspartate kinase 68
Aspartic acid 8, 66, 67, 77, 146, 169, 201, 253, 254
Assimilation starch 47
Asymmetric synthesis 61
Atmungsferment 116
ATP *see* Adenosine triphosphate
ATPase 127, 128, 268
ATP-sulphurylase 173
Atropa belladonna 191, 193
Atropine 190
AUDUS, L. J. 300
Autotrophs 132
Auxin 34, 54, 287, 291
mechanism of action of **294**
Avena sativa *6*, *7*
AVRON, M. 185
Azetidine-2-carboxylic acid 199, 201
8-azoguanine 295

Bacillus lactis aerogenes 251
Bakanae disease 289
BAKER, J. W. *30*
BAKKE, A. C. 6
Balbiani rings 278, 297
BALL, N. G. 237
BANDURSKI, R. S. 184
Bark 236
Barley *5*, 296, 297, 298
Base composition 49
BASSHAM, J. A. *144*
BAYLEY, S. T. *53*
Beet *4*, *24*, *30*, 106, 220, 231, 233
BEEVERS, H. 111, 129, 130
BEEVERS, L. 185
Belladonna 191, 192, 193

BENNET-CLARK, T. A. *30*, *32*, 58, 232
BENSON, A. A. 147
BERTRAND, S. 95, 116
β-form of polypeptide chain *65*
Beta vulgaris 3
Betel nut 191
Bioassays 291
Biochemistry 2
Biological clock 259
Biotin 165, 174
BLACKMANN, F. F. 130, 134, 137
Bladderwort 34
Bleeding of phloem sap 239
Blue-green algae 178
BOARDMAN, N. K. 155
BOHNING, R. H. 18
BONNER, J. 18, 130, 279, 301
BONNER, W. D. *87*, *89*, 130
Boron 12, 213
BOULTER, D. 185
BOURQUELOT, E. 116
Boyle's Law 28
BRACHET, J. *39*, 58
BRACKETT, F. S. *137*
Branching enzyme 162
Brei 26
BRIGGS, G. E. 86, 130, 245
BRIGGS, R. 278
Broad bean *21*, 295
BROWN, A. J. 5
BROWN, A. S. 209
BROWN, R. 58
BROWN, ROBERT (1831) 19
Bryophyllum fedtschenkoi 260
BUCHNER, E. 59, 93
BUCHNER, H. 59, 93
BUTCHER, D. N. 301
BUTLER, J. A. V. 58
BUTT, V. S. 130
BUVAT, R. 58

^{14}C *see* Radioactive carbon
Cabbage 174
Caesium 222
Caffeic acid 205, *208*
Caladium *30*
Calcium 12, 213, 222
Callose 236
Callus *272*, 289
Calvin cycle **146** f., *147*, *150*, 255
CALVIN, M. *144*
Cambium 87, 289
Canavanine 199, 201
Cane-type plants 149
CANNY, M. J. 245
Carbohydrate 6, 91, 187
Carbon dioxide 11, 60, 85, 90
  assimilation paths of *152*
  dark fixation of 259
Carbon dioxide assimilation, rate of *137*, 255
Carbon monoxide 73, 116, 228
Carbonium ion 77
Carbonyl group 63, 67
Carbonylcyanide-m-chlorophenyl hydrazone (CCCP) 233
Carboxydismutase *see* RuDP carboxylase
Carboxyl group 62, 95, 146, 294
Carboxyl phosphate bond 121
β-carotene *140*, *142*, 195
Carotenoids 47, 89, 139, *140*, 187, 196, 197
Carrier-ion complexes 222, **225** f.
Carriers 117, *127*, 222, 225, 248, 257
Casparian strip 235
Castor bean 106, 112
Catabolism 10, **84** f.
Catechol oxidase 116
Cauliflower 102
CAVENTOU, J. B. 133
CCCP *see* Carbonylcyanide-m-chlorophenyl hydrazone
C_4-dicarboxylic acid pathway *see* Hatch and Slack pathway
Cell differentiation **273** f.

Cell division 10
Cell expansion **273,** *274*, 287, 289
Cell plate 271
Cell sap 24
Cell structure **19** f.
  and control of metabolism **258**
Cell theory 19
Cell wall *21*, 23, **50,** 119, 206, 271, 294
Cell walls
  primary 52
  secondary 207
Cellulase 295
Cellulose 23, 50, *52*, 160, **164**
Cereals 106
Chaetomorpha melagonium *56*
Chain reaction 282
Chara 217
Charge separation 229
Chemical determinants of differentiation 286, 290
Chemical potential 28
Chemiosmotic hypothesis **126** f., *127*, *128*, 230
Chemotaxonomy 189
Chloramphenicol 175, 231, 233
Chlorella pyrenoidosa 144
Chlorella vulgaris 252
Chlorine 12, 70, 213
p-chloromercuribenzoic acid 74
Chlorophyll 12, 14, 47
Chlorophyll *a* 47, 139, *140*, *142*, 155
Chlorophyll *b* 47, 139, *142*, 156
Chlorophyll complex 138
Chlorophyll P690 154, 156
Chlorophyll P700 *154*, 156, 157
Chloroplast dimorphism 143
Chloroplasts 14, *24*, 25, *45*, *46*, **47,** 126, 146, 152, 169, 197, 258
Choline 167, *232*
Choline acetylase 233
Choline esterase *232*
Chromatin *21*, 49, 278
Chromatium 157
Chromoplast *45*
Chromosomes 178, 266, 279
Cinchona ledgeriana 191
Cinnamic acid 187, 193, 204, *208*
Circadian rhythms *259* f.
Cistron 178
Citrate synthase 293
Citric acid 98, 253
Citric acid cycle *see* Tricarboxylic acid cycle
Citrulline 199
Citrus fruits 102
Claviceps purpurea 191
CLOWES, F. A. L. *41*, *42*, *45*, *51*, 58, *265*, 300
Coarse control of metabolism **249** f.
CoA-SH *see* Co-enzyme A
Cocaine 193
Coconut milk 273
Codeine alkaloids 191
Codon 178
Co-enzyme I *see* Nicotinamide adenine dinucleotide
Co-enzyme II *see* Nicotinamide adenine dinucleotide phosphate
Co-enzyme A (CoA—SH) 70, *96*, 99, 174, 199, 268
Co-enzyme Q 118
Co-enzymes 14, **69** f., 95
Co-factor 70, 253
COHEN, G. N. 262
Colchicine 190
Coleoptile *4*, 287, 291
COLLANDER, R. 212
Colligative 29
Colloid 25, 95
Companion cells 243, *244*
Compartmentalisation 102
Competition in ion uptake 221
Competitive inhibition *75*, 104
  in ion uptake 223
Compression 300
Coniferin 210
Conifers 207

Coniferyl alcohol 207, *208*
Coniine 191
Conium maculatum 191
Constitutive enzymes 249
Convallaria majalis 201
Copper 12, 69, 73, 116, 213
Corn *87*
Cotton *53, 288*
Cotyledon *4*
p-coumaric acid 204, 205, *208*
Coumarin 187, 204, **205**
p-coumaryl alcohol 205, *208*
Coupling 123
Co-valent bonds 63
Co-zymase 70
CRAFTS, A. S. *27*
Crassulacean plants 102, 259
Cristae *43*, 258
CROFTS, A. R. 185
Cross feed-back 281
Crossover point 120, 249
Crown gall 273
Cultured cells 283, 286
Cyanide 73, 93, 116, 117, 134, 201, 228
Cyanide-resistant respiration 117
β-cyanoalanine 200
Cyclic photophosphorylation 157
Cysteine 66, 74, 78, 174
Cystine 174
Cytidine triphosphate 254, 255
Cytochrome *a* 117, *118*
Cytochrome a_3 *see* Cytochrome oxidase
Cytochrome *b* 117, *118*, *154*, 155, 228
Cytochrome *c* 116, 117, *118*
Cytochrome *f* *154*, 155, 156
Cytochrome–anion complex 228
Cytochrome oxidase 73, 116, 117, *118*
Cytochromes 69, 115, 228
Cytokinesis 265, 270, 290
Cytokinins *288*, 290
  mechanism of action of **297** f.
Cytoplasm 23
Cytoplasmic viscosity 269
Cytosine (C) 176, *177*, *182*, 279

DAINTY, J. *218*, 245
Dark reactions 15, 134, 146
DARWIN, C. 263, 287
Datura 193
DAVIS, B. D. 59, 247
Day length 261
DCMU *see* Dichlorophenyldimethylurea
Deadly nightshade 192, 193
Decarboxylation 92, 98, 103, 113, 193
Dehydrogenase 61, 95, 97, 98
Denaturation 69, 70, 89
Deoxyribonucleic acid (DNA) 44, 49, 176, *177*, 178
  replication of **266,** *267*
Deoxyribose phosphate 176
DE SAUSSURE, N. T. 85, 133
Deuterium 81
DE VRIES, H. 29
Dewar flask 11
Dextrin 162
Dialysis 70
α,β-diaminopropionic acid 199
Dicarboxylic acids 150
Dichlorophenyldimethylurea (DCMU) 233
Dicoumarol 205, 206
Dictyosomes **41**
Diethyldithiocarbamate 73
Difference spectroscopy 155
Differential centrifugation 119, 175
Differentiation 10, 25, 82, 105, 247, **263** f.
  programming of **280** f.
Diffusion 214, 234
Diffusion pressure deficit 28, *32*
Digitonin 155

Dihydroxyacetone phosphate 95, 108, 196
Dimethylallyl pyrophosphate 196
Dinitrophenol 123, 221, 233, 257
Diphosphoglyceric acid 81, 97
Diphosphopyridine nucleotide *see* Nicotinamide adenine dinucleotide
Disulphide linkage 66, 174
Diterpene 197
DIXON, H. H. 237
DIXON, M. 83, 262
DNA *see* Deoxyribonucleic acid
DNA-ase 50, 279, 298
DNA-dependent protein synthesis 279, 280
DNA-dependent RNA synthesis 296
DNA polymerase 177
Donnan equilibrium 220
Dormancy 236, 257
Dormin *see* Abscisic acid
DOTY, P. *65*
Double helix 176, *177*
DPN *see* Nicotinamide adenine dinucleotide
DUNHILL, P. M. 211
DUTROCHET, R. J. H. 133

Ecballium elaterium 200
Ecdysone 297
Echinocystis *288*
Ectoplast 31
Effectors 75, 254, 293
Efflux 213, 215, 235
Elaioplast *45*
Electrical conductivity 214
Electrical potential gradient 215
Electrochemical potential 215
Electron microprobe analysis 233
Electron microscope 22, 50, 225, 237
Electron transport 45, 109, **115** f., *118*, 168, 257
Electron transport chain **117** f., 154, 228
Electrons 63, 85, 119
Electro-osmosis 243
Elodea canadensis 231
Embden–Meyerhof–Parnas (EMP) pathway **98** f., *99*, 146
Embryo 3, 5
EMERSON, R. 134
EMP pathway *see* Embden–Meyerhof–Parnas pathway
Endergonic 174
Endodermis 235, 277
Endoplasmic reticulum *36*, **37**, *39*, *40*, 80, 225, 234, 237, 269
Endosperm *4*, 7, 273, 296
End-product inhibition 254, 281
Energy
  conservation of **120** f.
  liberation in respiration **120** f.
  of activation 76
Energy reservoir in mitosis 268
Energy-rich bonds 121, 221
Energy-rich phosphorus compounds 11, **120** f., 221
ENGLEMANN, T. W. 133
Enucleate eggs 278
Enzyme
  changes in activity during differentiation *276*
  factors controlling activity **71** f.
  induction and repression of synthesis **249** f., 282
  reaction kinetics **75**
Enzyme–inhibitor complex 74
Enzymes 6, 58, **59** f.
  activation of **252** f.
  active sites of 67, 76
  adaptive 250
  constitutive 249
  inhibition of 74, **252** f.
  nature of **61**
  sub-units of 68, 255

Enzyme–substrate complex 72, 222, 254
Eosin 236
EPSTEIN, E. *223*
Equilibrium constant (*K*) 73
ER *see* Endoplasmic reticulum
Ergot alkaloids 191
Erythrose-4-phosphate 105, 172, 186, 187
Erythroxylaceae 193
ESAU, K. *24*
Escherichia coli 250, 279
Essential elements 213
Ethanol 60, 73, 85, 93, 97
Ethanolamine 201
Ethylamine 200
Ethylasparagine 200
Ethylene 300
Eukaryotic cells 178, 184
Excited state 153, 154
Extinction point (EP) 90
Extra-cellular controls 284

FAD *see* Flavin adenine dinucleotide
Families of amino acids 170
Far red light 261
Farnesyl pyrophosphate 197
Fats 6, 8, 23, 92, 106, 108
  respiration of **113**
Fatty acid spiral *109*, *112*
Fatty acids 108, *109*
  synthesis of **165** f.
Fatty acyl-CoA 110
Fatty acyl-CoA dehydrogenase 110
Fermentation 85, 90, 129, 157, 254
Ferredoxin 156, 168
Ferredoxin-NADP-reductase 156
Fertilised egg 270
Ferulic acid *208*
Festuca pratensis *288*
Ficin 78
Fine control of metabolism **253** f.
First singlet state 153
Flavin adenine dinucleotide (FAD) *109*, 110, 113, 118, 167, 174
Flavin adenine mononucleotide (FMN) 168
Flavones 202
Flavonoid pigments 89, 187, **201** f.
  biological significance of **203**
Flavonols 202
Flavoproteins 69, 117, *118*, 228
Flavylium ion 202
FLEMMING, W. 266
Florigen 290
Fluoracetate 101
Fluorescein 236
Fluorescence 153
Fluxes of ions **215** f.
FMN *see* Flavin adenine mononucleotide
FOGG, G. E. 300
Formylmethionine 183
FOWDEN, L. 185, 186, 211
FOWLER, M. W. *99*, *276*
Fraxinus 87
Free energy, change in 85, **120** f., 161
Free space 224
Freeze-etching 22, 37, *41*, 46
Freezing point, depression of **29**
FREI, E. *56*
FREUDENBERG, K. 207
Frog's eggs 278
β-fructofuranosidase 106, 161, 295
Fructose 78, 94, 106, 160
Fructose-1-6-diphosphate 94, 107, 121, 147, 196, 254
Fructose-6-phosphate 106, 254
Fucoxanthol 139
Fumarase 113
Fumaria officinalis 102
Fumaric acid *100*, 102
Fumitory 102
6-furfurylaminopurine *see* Kinetin

GA *see* Gibberellic acid
Galactokinase 252
Galactose 160, 202, 252
Galactose-6-phosphate 252
β-galactosidase 251
GALSTON, A. W. 18, 211, 301
Garlic 174
Genes 82, 248, 281
  control of enzyme synthesis by 82
  operator 178, 281
  regulator 178, 251, 281, 282
  structural 178, 251, 282
Genetic code 178
Gentiobiose 202
Geranyl pyrophosphate 196
Germination **3** f., 256
Gibberella fujikuroi 289
Gibberellic acid 68, *288*, 289
  mechanism of action of **295** f.
Gibberellic acid-cofactor 297
Gibberellin 7, *288*, 289
  mechanism of action of **295** f.
Gibberellin A_3 *see* Gibberellic acid
Gibberellin A_8 *288*
Gibbs free energy 85, 120, 121, 122
Globular protein 67
Globulin, of pea seed 279
β-glucanases 296
Glucolipid 164
Glucopyranose 162
Glucose 7, *52*, 85, 92, 94, *99*, 106, 161, 202, 252
Glucose-1-phosphate 78, 108, 159, 256
Glucose-6-phosphate 103, 105, 106, 108, 122, 159, 256
Glucose-6-phosphate dehydrogenase 253
Glucose-6-phosphate isomerase 159
Glucosidic linkage 107
Glucosylglucan 162
Glutamate dehydrogenase 61, 114, 169
Glutamic acid 8, 61, 66, 67, 77, 114, 169
Glutamine 8, 172, 200, 257
Glutamine synthetase 172, 199, 200, 257
γ-glutamyl-β-cyanoalanine 200
Glutathione 200, 256
Glyceraldehyde-3-phosphate 81, 95, 97, 121, 146, 196
Glyceraldehyde-3-phosphate dehydrogenase *see* Triose phosphate dehydrogenase
Glycerol 35, 108
α-glycerolphosphate 108
Glycine 62
Glycollic acid 143
Glycosidic bond 77, 107, 202
Glyoxylate cycle 109, *110*, 111, *112*
Glyoxylic acid 110, 111
Glyoxysome 110, 111, *112*, 113
GODDARD, D. R. *87*, 130
GOLDACRE, R. J. 230
Golgi apparatus *36*, *40*, **41**, *42*, 57
GOVINDGEE 184
Grana *46*, *48*, *50*, 258
GRANICK, S. *50*
GREENWOOD, A. D. *30*, *49*
GRIFFITHS, D. E. 130
Growth
  limitation of 271
  process of 10, **263** f.
Growth correlation 264
Growth hormones 9, 273
  and regulation of growth **284** f
  mechanism of action of **292**
Guaiacum 206
Guaiaretic acid 206
Guanine (G) 176, *177*, *182*, 279
Guanosine phosphates 183
Guanosinediphosphate-glucose 164
Guard cells 233
Guayule 188, 198
Guttation 16

Haem 73
Haemoglobin 116
HAGEMAN, R. H. 185
HAGEN, C. E. *223*
Half-life 153, 183
HARDEN, A. 70, 94
HARTIG, T. H. 236
HARTREE, E. F. 116, 130
HARVEY-GIBSON, R. J. 3
Hatch and Slack pathway **150**
HAYWARD, H. E. 4
Heartwood 189
Heat inactivation 7, 60
HEATH, O. V. S. 184
Heavy hydrogen 81
Heavy nitrogen (^{15}N) 169, 233
Heavy oxygen (O^{18}) 136, 143
Helianthus annuus 87, *244*
Hemicellulose 23, 50, 160
Hemlock 191
Henbane 193
Heptose 147
Heptulose 103
Heroin 191
HESLOP-HARRISON, J. 51, 301
Heterocyclic compounds 190
Heterotrophs 132
Hevea brasiliensis 188
Hexokinase 106, 122
Hexose gain, in photosynthesis *147*, 148
Hexose phosphates 147, 152
HEYN, A. N. J. 294
HIGUCHI, T. 209
HILL, R. 136, 151
Hilum 161
Histidine 67
Histones 178, 266, 279, 281
HOAGLAND, D. R. 212, 245
Homoarginine 199
Homocysteine 199
Homoserine 68, 199, 253
Homoserine kinase 253
HOOKE, R. 19
HOOVER, W. H. *137*, 139
HOPE, A. B. 245
HOPKINS, F. G. 1
Hordeum vulgare *5*
Hormones 7, 9, 34, 54
  and regulation of growth **284** f.
  mechanism of action **292** f.
HOUWINK, A. L. *53*
hν *see* Quantum
Hyaloplasm 23, 25
Hydathodes 34
Hydrodictyon 217, 226
Hydrogen 158
Hydrogen bond 51, **63,** 64
Hydrogen electrode 117
Hydrogenase 158
Hydrolysis 61, 107, 161
Hydrosulphate 116
cis-o-hydroxycinnamic acid 205
Hydroxyethylasparagine 200
Hydroxylamine 168
γ-hydroxy-methyleneglutamic acid 200
β-hydroxy-β-methyl glutarate 198
Hygrine 193, *194*
Hyoscine 195
Hyoscyamine 190, 195
Hyoscyamus niger 193
Hypocotyl *4*
Hyponitrite 168

IAA *see* Indol-3yl-acetic acid
IAA oxidase 203
Imbibition 3, 6, 7
Imide group 63
Immunochemistry 279
Incipient plasmolysis *33*
Indole alkaloids *187*, 190, 191
Indol-3yl-acetic acid 68, 203, *272*, 273, 287, *288*, 291
Induced absorption 219
Inducer 250

Induction of flowering 261
Influx 212, 215
INGEN-HOUSZ, J. 131, 133
Inhibition of enzymes 75
Inosine 181
Interference 20
Interphase 266
Intussusception 54
Inulin 82
Invertase *see* β-fructofuranosidase
Iodoacetate 74
Ion absorption
  factors controlling **219** f.
  linkage of metabolism with *225* f.
Ion clusters 234
Ion pumps 226, 233
Ion-complexes 222
Ion-rich vesicles 234
Ions, rates of uptake, *223*
Iron 12, 69, 73, 115, 136, 156, 213
Islands of synthesis 54
Isoamylase 107
Isocitrate lyase 110, 111
Isocitric acid **100,** 110, 111
Iso-electric point 62
Isoenzymes **68,** 295
Isoleucine 253
Isopentenyl pyrophosphate *187*, 196, 197
Isopentenyladenosine 298
Isopentenylaminopurine *180*, 181, 297, 298, *299*
Isoprene unit 196, *198*
Isoprenoid compounds **195** f., 198
Isotonic 30

JACOB, F. *282*, *283*, 300
JAGENDORF, A. J. 126, 130
JAMES, A. T. 184
JAMES, W. O. *90*, 129, 130
Janus Green B 46
JOHNSON, R. P. C. 241, 245
JOHNSTON, E. S. *137*
JONES, R. L. 244
JUNIPER, B. E. *41*, *42*, *45*, *51*, 58

K_m *see* Michaelis constant
Kaempferol 202
KANDLER, O. 184
KEILIN, D. 115, 116
KENNEDY, J. S. 236
α-keratin 66
β-keratin 63
Keto acids 114, 170, 193, 200
α-ketobutyric acid 253
Key enzymes 249
KIDD, F. 86, 130
Kinases 79, 97
Kinetin 268, *272*, 273, 290
KING, T. J. 278
KLASON 207
KLEIN, R. M. 301
KNOP, W. 213
KORNBERG, H. L. 262
KOSTYCHEW, S. 93
KOSUGE, T. 185
KRAMER, P. J. 18
Krebs cycle *see* Tricarboxylic acid cycle
KREBS, H. A. 98, 100, 101, 253

Laccase 209
Lactate dehydrogenase 67, 68, 81
Lactic acid 85
Lactonisation 205
Lamellae, of chloroplast *46*, *48*, *50*, *51*, 158
Lamium album 151
Latex 191, 198
Lathyrus odoratus 200
LEACH, W. 84
Leaves 92
Lecithin 26, *232*
Lecithinase *232*

LEECH, J. H. *36*, *38*, *40*, 58
LEETE, E. 211
LEHNINGER, A. L. *44*
Lethals 248
Leucine 170
Light energy, conversion to chemical energy in photosynthesis **151** f.
Light flashes 134, 158
Light gathering pigments *154*
Light saturation 137
Lignans *187*, 204, **206**
Lignin *187*, 204, **206** f.
Lignin monomers *208*, 209
Liliaceous plants 201
Limiting factors **137**
Limiting plasmolysis *30*
α-linkage between sugars 162
β-linkage between sugars 164
Lipase 8, 108
Lipids 10, 26, 47, 166
LIPMANN, F. 100
Lipo-protein membranes 26, 35, 225, *232*
Lithium ions *223*
Lolium 261
LUNDEGÄRDH, H. 227
Lupin 106
Lutein *140*
Lycopersicon esculentum *4*
Lycopodium 191
Lysine 66, 67, 68
Lysozyme **67**, 77

MCCULLY, M. E. *21*, *49*, 58
Macrofibrils *52*
Magnesium 12, 70, 172, 213, 222
Maidenhair fern 200
Maize *4*, 143, 169, *265*, 295
Malate synthase 110, 111
Malic acid 92, *100*, 146, 152
Malic dehydrogenase 113, 253
Malic enzyme 101, 152, *276*
Malonic acid 74, 104, 165
Malonyl-CoA 165, 187, 202
Maltase 107
Maltose 7, 107, 162, *163*
Mandragora officinalis 191, 193
Mandrake 191, 193
Manganese 12, 136, 165, 213
MARTIN, A. J. 145
Mass flow 9
Mass-flow hypothesis **238** f.
Master reaction 293
Maternal inheritance 270
Matrix materials 57, 211
Matrix potential 3
Maximum velocity (*V*) 72
MAYER, D. M. 18
MAZIA, D. 270, 300
Meiosis 266, 270
Membrane potential **215** f.
Membranes 26, 31, 34, 225
Meristematic cells 23, *36*, *38*, *40*, *41*, 264, 273
Meristems **264** f., 284
Messenger RNA 176, *182*, 248, 252, 279, 295, 297, 298
Metabolism 1
Methionine 68, 174, 184, 199, 252, 294
β-methyl crotonic acid 198
Methylation 193, 294
γ-methyleneglutamic acid 200
Methyl-β-thiogalactoside 250
Mevalonic acid 186, *187*, *197*
MEYER, B. S. 18
Micelle 51
Michaelis constant 72
Microelectrodes 217
Microfibril 51, *52*, *55*, *56*
Micronutrients 12, 213
Microscopes 20
Microsome 38, *39*, 80, 234
Microspectrophotometer 22
Microtubules 54, 56, 165
Middle lamella 27, 271
Miller, C. O. *272*, 301

Mimosine 199, 201
Mimosoideae 199
MIRSKY, A. E. *39*, 58
Mitchell hypothesis 126
MITCHELL, P. 126, 130, 230
Mitochondria 23, *24*, 25, *36*, *38*, *39*, *40*, **43**, *44*, 80, *112*, 119, 225, 229, 257, 258, 269
Mitosis 88, **265** f.
MITTLER, T. E. 236
Model systems 281
MOLLENHAUER, H. H. *36*, *38*, *40*, 58
Molybdenum 12, 167, 213
Monochromatic light 138
MONOD, J. *282*, *283*, 300
MOORE, S. 83
Morphine alkaloids 191
Morphogenesis 203
MUHLETHALER, K. *52*
Multi-enzyme units 80, 247, 293
Multinet structure *53*
MÜNCH, E. 238
Mung bean 102, 201
Mutation 177, 188, 248, 277
Myelin process 26
Myosin 230

^{15}N *see* Heavy nitrogen
N-acetylmuramic acid 77
NAD *see* Nicotinamide adenine dinucleotide
NADP *see* Nicotinamide adenine dinucleotide phosphate
NADP reductase 168
β-naphthylamine 93
1-naphthyl-2-sulphonic acid 94
Native proteins 69
Negative feed-back 253
NEISH, A. C. 211
Nernst potential 215
Neurospora 248
Neutral genes 188
Nicotinamide adenine dinucleotide (NAD) 70, 80, 95, *96*, 113, 167, 253
Nicotinamide adenine dinucleotide phosphate (NADP) *96*, 103, 147, 152, 166, 167, 174, 253
Nicotine 191
Nitella translucens *33*, 218, 221, 224
Nitrate, assimilation of **167** f.
Nitrate reductase *167*, 251
Nitrilase 200
Nitrite 167
Nitrite reductase 168
Nitrogen 12, 213
  assimilation of **167** f.
N^6-methylaminopurine 297
NOECKER, N. L. 6
NOLLET, M. 27
NOMURA, M. 185
Non-competitive inhibition 75
Non-competitive interaction between ions 222
Non-cyclic photophosphorylation 157
Non-free space 224
Non-polar 67
Non-protein amino acids **199** f.
NORTHCOTE, D. H. *41*
Nuclear differentiation 278
Nuclear envelope *21*, 26, *36*, 37, *40*, 49
  pores in *40*
Nuclear histones **279** f.
Nuclear transplantation 278
Nuclear/cytoplasmic ratio 271
Nucleolar organiser 269, 270
Nucleolar vacuole *21*
Nucleolus *21*, 49, 269, 281
Nucleoplasm 49
Nucleoproteins 12
Nucleotide sugars **159** f.
Nucleotides 176, 177
Nucleus *21*, 23, *24*, 25, *36*, *38*, **49**, 119, 178
Nutrients 11

O^{18} *see* Heavy oxygen
Oat *6*
O'Brien, T. P. *21*, *49*, 58
Ochoa, S. 122, 152
O'Connor, C. M. 262
Oleyl 166
Oligosaccharide 159, 236
Onion 27, *53*, 174, 297
Operon 251
Öpik, H. 18, 300
Opium poppy 191
Optical isomers 63
Organ initiation 289
Organelles, isolation of **119**
Organic acids 92, 102
Ornithine *187*, 199
Osmiophilic globules *46*, *49*, *50*
Osmosis 5, 13, **27** f.
Osmotic model of translocation *239*
Osmotic potential (π) **28** f., 119
Osmotic pressure 28
Overton, C. E. 35, 37, 214
Oxaloacetic acid 98, *100*, 169, 253
β-oxidation 108, *109*, 166
Oxidation–reduction potential 117, *154*, 229
Oxidative phosphorylation *118*, 125, 254
2-oxoglutarate dehydrogenase 70
2-oxoglutaric acid 61, 71, *100*, 169
Oxygen absorption 11, 85, *89*, 90, 221
Oxygen electrode 117
Oxysomes 45

P690 *see* Chlorophyll P690
P700 *see* Chlorophyll P700
Paál, A. 287
Pacemaker reactions 249
Palladin, V. I. 93
Palmitic acid 110
Palmitoyl-S-CoA 113
Papain 74, 78
Paper chromatograms *144*, *145*, 291
PAPS *see* 3′-adenosine-5′-phosphosulphate
Parenchyma *24*
Parthenium argentatum 188, 198
Passive movement 213, 219, 235
Pasteur effect 129, **254**
Pasteur, L. 85, 93, 129, 254
Pauling, L. 59
Pea 276, 279, 289, 297
Peanut 167
Pectin 23, 50, 294
Pectin methylesterase 294
Pelletier, J. 133
Pentosanases 296
Pentose 103, 147
Pentose phosphate pathway *99*, **103**, 172, 187
Peonidin 202
PEP *see* Phosphoenolpyruvic acid
PEP carboxylase 101, 149, 260
Peptide bond 62, 78
Peptides 8
Pericycle 277
Period of rhythms 259
Perisperm *4*, 6
Permeability 34, 225
Permease 248
Peroxidase 209, 295
Pfeffer, W. 28, 84, 93, 133, 214
PGA *see* Phosphoglyceric acid
pH 62, 73, 126, 229
Pharbitis 261
Phase-contrast 20, *21*
Phaseolus aureus 43
Phaseolus multifloris *288*
Phaseolus vulgaris *288*
Phenylalanine 170, *171*, 172, 187, 204, *208*
Phenylalanine ammonia lyase (PAL) 204, *208*
Phenylpropanoids 201, **204** f.
Phenyl-β-D-thiogalactoside 251
Phillips, D. C. 83

PHILLIPS, I. D. J. 293, 300
Phloem 8, 15, 236
respiration of 242
structure of **237** f.
Phosphatidic acid 167, *232*
Phosphoenolpyruvate carboxylase *see* PEP carboxylase
Phosphoenolpyruvic acid (PEP) 97, 101, 112, 121, 146, 149, 172, *187*, 260
Phosphofructokinase 106, 254, *276*
Phosphoglucomutase 108, 256
Phosphogluconate dehydrogenase 103, 253
Phosphogluconic acid 103
Phosphoglyceraldehyde *see* Glyceraldehyde-3-phosphate
Phosphoglyceric acid (PGA) 97, 102, 121, 145, 146, 164, 255
Phosphohexoseisomerase 107
Phospholipid 26, 42, 43, 47, 167, *232*
Phosphorus 213
Phosphorus compounds 12, 26, 94, 129
Phosphorylase 107
Phosphorylation, efficiency of 123
Phosphorylation sites *118*, 125
5-phosphoshikimic acid 172
Phosphotransferases 97
Photochemical reactions 14, 47, 134
Photon 135, 153
Photoperiodic induction 261
Photoperiodism, timing reaction in **261**
Photophosphorylation 125, 157
Photorespiration 143, 149
Photosynthesis 12, 14, 47, 89, **132** f.
discovery of **132** f.
influence of temperature on 141
influence on ion uptake 221
path of carbon in **144** f.
rate of **136** f.
Photosynthetic apparatus, location of **158** f.
Photosynthetic electron transport *154*
Photosynthetic unit 158
Photosystem I *154*, 155, 156
Photosystem II *154*, 155, 156
Phragmoplast 271
Phycobilins 139
Physiological reproduction 270
Phytoalexin 203
Phytochrome **261**
Picea excelsa 205
Pinocytosis 234
Pipecolic acid 199
Pisatin 203
Planck's constant 138
Plasmagenes 270
Plasmalemma 26, 31, **42**, 165, *218*, 225, 229, 237, 271
Plasmatic filaments *240*, 241
Plasmodesmata 10, *27*, *38*, 54, 234
Plasmolysis 26, *30*
Plastic extensibility 34, 294
Plastids 119
interrelationships of *45*
Plastocyanin 154, 156
Plastoquinone *154*, 156
Pleated-sheet structure 63, 67
Ploidy 267, 271, 277
Plumule 3, *4*
P : O ratio 123
Point activation 282
Polar 67
Polarised flow of ions 243
POLJAKOFF-MAYBER, A. 18
Polyglucan 162
Polygonatum multiflorum 201
Polypeptide 63, 68, 183
Polyploidy 277
Polyribosomes 175, 183, 298
Polysaccharides 6, 10, 23, 41, 77 94, *159*
Polysome *see* Polyribosome
Polyuronic acids 10
Pool 102

Porphyrins 69, 101, 115
PORTER, K. R. *39*
Positive feed-back 255
Potassium 12, 213, 222
  uptake sites for 222, 235
Potassium ions
  in stomatal regulation 233
  polarised flow in sieve tubes 243
Potato *27*, 91, 106, 220, 233, 237
PREISS, J. 185
Prephenic acid 172
Pressure gradients in sieve tubes 241
PRESTON, R. D. *55*, *56*
PRICE, C. A. 185
PRIESTLEY, J. 132
Primary absorption 219
Primary pit field *21*
Primary structure of proteins 63
Primer 108, 162, 256
Prokaryotic cells 178
Proline 66, *187*, 199
Promeristem 264, *265*
Proplastid 23, 25, *38*
Prosthetic group **69** f.
Protein turn-over 234
Proteins 7, 12, 23, 25, 35, 42, 43, 47, **61** f., 92, 106
  contractile 230, *231*
  respiration of **114**
  synthesis of **175** f., 281
Proteolytic enzymes 8, 114, 256, 296
Proton 63, 85, 126
Proton gradient *127*, *128*
Proton transport **126**
Protoplasm 20
Protoplasmic streaming 230
  action of auxin on 293
Protoplast 23, 29
Pseudomonas saccharophila 78
Pseudouridylic acid 181
Pteridine 156
Punctuation of genetic code 178, 184
Purine 173, 275
Purine/pyrimidine ratio 275
PURVES, W. K. 301
Pyridine-related alkaloids 191
Pyridoxal phosphate 69, *170*
Pyridoxamine phosphate 170
Pyrimidine 176, 275
Pyrophosphatase 173
Pyrophosphate 179
Pyrophosphate bond 151
Pyrrolidine alkaloids *187*
Pyrrolidine-piperidine alkaloids 191
Pyrus malus 102
Pyruvic acid 85, 93, 98, 100, 152, 170

Q_{10} *see* Temperature coefficient
Quantasomes 49, 159
Quantum (*hν*) 135, 138, 153
Quantum efficiency 138, 155
Quaternary structure of proteins 67
Quiescent centre 264, *265*
Quinine 191
Quinone 118,136

RABINOWITCH, E. I. *137*, *145*, 184
Radicle 3, *4*, 11
Radioactive carbon (^{14}C) 102, 104, 136, 144, 145, 175, 198, 200, 209, 279
Radioactive phosphorus (^{32}P) 79
Raffinose 236
Rauwolfia 191
Red drop in photosynthesis 155
Red–far red interaction 261
Reduced acyl carrier protein (ACP-SH) 165
Reducing activity 136, 151, 169
Regulation of metabolism **247** f.
Repressor 250, 251
Reserpine 191
Resin canals 189

Respiration 11, 34, 85
  aerobic 85
  anaerobic 85
  chemical pathways of **93** f.
  control of **128** f.
  effect of light on 90, 143
  evolution of heat in 11, 86, 258
  rate of 3, 5, 6, **86** f., 257
Respiratory quotient (RQ) **92,** 113
Rhamnose 202
Rhoeo 30
Rhoeo discolor 35
Rhythms **259** f.
Riboflavin 69
Ribonuclease 66, 175, 296, 298
Ribonucleic acid (RNA) 43, 49, 175, 234
Ribonucleoprotein 38, 175
Ribose-5-phosphate 105
Riboside triphosphates 279
Ribosomes **38,** *39*, 44, **175** f., 176, **181,** 248
  subunits of 181, 281
Ribulose-1-5-diphosphate (RuDP) *147*, 148, 255
Ribulose-5-phosphate 103, 152
Rice 108, 289
Ringing experiments 236, 285
RNA *see* Ribonucleic acid
RNA polymerase 179, 279
m-RNA *see* Messenger RNA
s-RNA *see* Transfer RNA
t-RNA *see* Transfer RNA
ROBERTSON, R. N. 229, 245
ROELOFSEN, P. A. *53*, 58
Root 13, *53*, *87*, 89, 91, 221, 228, 234, *265*, *272*, *274*, 275
  meristem of *265*, 275
Root cap *265*
Root cultures 285
Root hair development 275
Root pressure 16
Rooting of cuttings 289
RQ *see* Respiratory quotient
Rubber *187*, 188, 195, **198**
RUBEN, S. 136
Rubidium ions *223*
RuDP *see* Ribulose-1-5-diphosphate
RuDP carboxylase 148, 151
RUHLAND, W. 35, 211
Rye 191

SACHS, J. 133, 213
Salt linkages 66
Salt pumps 13
Salt uptake 13
Salts, translocation of **234** f.
Scenedesmus obliquus *144*, *145*
SCHLEIDEN, M. J. 19
SCHWANN, T. 19
Second singlet state 153
Secondary metabolism 186
Secondary structure of proteins *65*
Secondary valency 63
Secretion 9, 34
Seed coat 3, *4*, 11
Seeds 3, *4*
Selective uptake 13, 221, 235
Selenomethionine 252
Semi-permeable 26, 28, 31, 35, 238
SENEBIER, J. 133
Senescence, arrest of, by kinetin 298
Serine 170, 174, 199, 200, 256, 298
Serine sulphydrase 174
Sesquiterpene 197
SETTERFIELD, G. *53*, 301
SHANNON, L. M. 83
Shikimic acid 105, *171*, 172, 187
Shikimic acid pathway *171*, 172, 204
Sieve plate 237, *240*, *244*
Sieve tube 9, 236
  models of structure of *240*
  structure of **237** f.
Sinapic acid 205, *208*
Sinapyl alcohol 208

Sink 8, 243
SINNOTT, E. W. 263, 300
SKOOG, F. 268, *272*, 290
Slime 241
SMITH, I. K. 211
Sodium 222
Sodium–potassium pump 226, 233
Soil solution 235
Solanaceae 191
Solanum tuberosum *27*
Soluble enzymes 80, 81
SOMERS, G. F. *60*
Sorbose 78
Source 8, 243
SPANNER, D. C. 243, *244*, 245
Sparing action 129
SPIEGELMANN, S. 276
Spinacea oleracea *49*
Spinach 152, 153
Spindle 178, 268
Spirogyra 271
Spruce 210
Squaline 197
Stachyose 236
STADTMAN, E. R. 262
Standard chemical potential 215
Starch 6, *24*, 47, 94, 106, **107** f., 152, 160, **161** f.
Starch grain 161
Steady state 247, 280
Stearoyl 166
STEIN, W. H. 83
Stellaria media 151
Steroids *187*, 197
STEWARD, F. C. *32*, *87*, 219, 234, 301
STILES, W. 84
STOLL, A. 134
Stomata 15, 233
STRASBURGER, E. 20, 266
STREET, H. E. 18, *226*, *231*, 245, *265*, *274*, 300, 301
Stroma 47, *50*, 158, 258
Strontium 222
Strychnine 190, 191
Strychnos nux-vomica 191
Substrate 5, 11, 70, 72, 74, 89, 91, **105** f., 256
Substrate phosphorylation 124
Succinate dehydrogenase 74, 113, 118, 253
Succinic acid 74, *100*, 104
Succinyl–CoA 71, *100*, 124
Sucrose 5, 8, 15, 35, 78, *91*, 106, 112, **159** f., 203, 236
Sucrose glucosyltransferase 78
Sucrose phosphate 161
Sucrose phosphate synthetase 161
Sucrose synthetase 160
Suction force 31, *32*
Suction pull 16
Sugar cane 106, 143
Sugars, translocation of **236** f.
Sulphate 173
Sulphate reductase 174
Sulphide 173
Sulphite 174
Sulphite reductase 174
Sulphur 12, 213
  assimilation of **173** f.
Sulphuretted hydrogen 135, 158
Sulphydryl group 74, 78, 256, 269
SUMNER, J. B. *60*, 61
Sunflower 87
SUTCLIFFE, J. F. *232*, 234, 245
Svedberg unit (S) 181
SWAN, G. A. 211
SWANSON, C. P. 58, 245
Sweet clover 206
Swiss chard 152
Sycamore 257, *288*
Symplast 235, 284
SYNGE, R. L. M. 145
Synthetic auxins 291
SZENT-GYÖRGYI, A. 116, 131

TAMUYA, H. *33*

Tannin cells 189
Tannins *187*
Target tissue 293
Telophase 267, 270
Temperature coefficient (Q_{10}) 73, **88,** 134, 220, 260
Tension 16, 300
Terminal oxidation **115** f., 257
Terpenes *187*, 195
Tertiary structure of proteins 66
Tetrahydrophytoene 197
Tetrose 103, 105, 147
THAINE, R. 242, 246
Thermochemical reactions 14, 134, 220
Thiamin 174
Thiosulphate 158
Thornapples 193
Threonine 68, 199, 253
Threonine kinase 253
Thylakoid system 47, *51*, 126
Thymine (T) 176, *177*, *182*
Tissue culture 289, 290, 295
Tissue patterns 276
Tobacco *24*, 143, 233, 290, 295
Tomato *4*, *91*, 192, *274*
Tonoplast 26, 31, 42, *218*, 229
Totipotency 284
TPN *see* Nicotinamide adenine dinucleotide phosphate
Training 251
Transaldolase 103
Transamidation 78
Transaminases 69, 114, **170**
Transamination 69, 114, **170,** 199
Transcarboxylase 150
Transcellular strands *240*, 241, 242
Transcription 176, 280
Transfer cells 245
Transfer-RNA 176, 179, *180*, *182*, 251, 279, 298, *299*
Transfer-RNA (serine specific) *299*
Transketolase 103
Translation 176, *182*, 281
Translocation 8, 13, 213
of organic compounds **236** f.
of salts **234** f.
Transphosphorylases 70
Transpiration 15, 235
Transpiration stream 15, 236, 239
Tricarboxylic acid cycle **98** f., *100*, 109, 129, 187, 253
Tri-N-acetylglucosamine 67, 77
Triose phosphate dehydrogenase 70, 81, 95, 97, 121
Triplet of bases 178
Triterpene 197
Triticum *41*, *137*
Tropane alkaloids 191, 192, **193** f.
Tropane skeleton, synthesis of *194*
Tropic acid 193, 195
Tropine 193, *194*, 195
Tropinone 193, *194*
Tropisms 287
Tryptophane 172, 187, 190
Turgor potential 31, *33*
Turgor pressure 31, 233
Tyramine methylpherase 298
Tyrosine 172, 204, *208*, 298
Tyrosine ammonia lyase (TAL) 204

UDPG *see* Uridine diphosphate glucose
Ultramicrotomes 22
Ultrasonic disintegration 119
Ultra-violet microscope 21
UMBARGER, H. E. 262
Uncoupling agent 123, 221
Units ix
Unsaturated fatty acids 166
Uracil (U) 179, *182*, 255, 279
Uranium acetate 94
Urea 35, 61, 70, 255
Urease *60*, 61, 70
URIBE, E. 130
Uridine diphosphate glucose 112, 160

Uridine triphosphate (UTP) 160
Uronic acids 50, 160
Ussing–Teorell equation 216
UTP *see* Uridine triphosphate
Utricularia 34

Vacuolar non-free space 224
Vacuole 23, *24*, *38*, 189, 225
Valentine bean *89*
Valine 170
Valonia ventricosa *55*
VAN NIEL, C. B. 135, 136
Vapour pressure 29
VARNER, J. E. 130
Velocity constant 72
VENNESLAND, B. 83
Vicia faba *21*, *46*, 89, 200
Vicia sativa 200
VISHNIAC, W. 152
Vitamins 9, 69, 170, 174, 285
Volatile metabolites 300
VON MOHL, H. 20

WALKER, D. A. 184, 185
Wall pressure 31
WALSH, E. O'F. 262
WARBURG, O. 116, 134
WAREING, P. F. 293, 300
Warfarin 206
Water
  as electron donor 157
  content of 3, *6*
  culture 213
  molecule *64*
  oxidation of 154
  uptake of *5*, 11, 23, **27** f., 274
Water potential (Ψ) **29** f., *33*
  of protoplast 31, *32*, 33
Wavelength of light (λ) 138
Waxes 167
WEATHERLEY, P. E. 241, 245
WEBB, E. C. 83
WEHRMEYER, W. 287
WENT, F. W. 287
WERKMAN, C. H. 136
WESLEY, J. 83
WEST, C. 86, 130
WHALEY, H. G. *36*, *38*, *40*, 58
Wheat *41*, *137*, *139*, 209
WIGHTMAN, F. 301
WILKINS, M. B. 262, 301
Willardiine 199
WILLSTATTER, R. 134
WOLSTENHOLME, G. E. W. 262
WOOD, H. G. 136
WORLEY, F. P. 5

Xanthium 261
X-ray diffraction 50, 67
Xylem 12, 15, 199, 207, 234, 235, 236, 283, 289
Xylose 160, 202

Yeast 60, 80, 93, 100, 103, 115, 116, 173, 298
Yeast press juice 60, 94
Yellow enzyme 117
Young, W. J. 70, 94

Zea mays *4*, *87*, *265*
Zeatin *288*, 290
Zeatin riboside 297, 298
Zinc 12, 213
Zwitterions 62
Zygnema 35
Zymase 95